SO-AEY-788

AN INTRODUCTION TO WIRELESS TECHNOLOGY

Gary S. Rogers

John Edwards

Macon State College

Prentice-Hall PTR
Upper Saddle River, New Jersey 07458

Library of Congress Cataloging-in-Publication Data

Rogers, Gary S. (Gary Strawder), 1954-
 An introduction to wireless technology / Gary S. Rogers, John Edwards.
  p. cm.
 Includes bibliographical references and index.
 ISBN 0-13-094986-8 (casebound)
  1. Wireless communication systems. I. Edwards, John Solomon, 1939- II. Title.

TK5103.2 .R64 2003
621.382—dc21   2002029821

Editor in Chief: Stephen Helba
Assistant Vice President and Publisher: Charles E. Stewart, Jr.
Production Editor: Alexandrina B. Wolf
Design Coordinator: Diane Ernsberger
Production Coordination: Carlisle Publishers Services
Cover Designer: Bryan Huber
Cover art: Digital Image
Production Manager: Matthew Ottenweller
Marketing Manager: Ben Leonard

This book was set in Palatino by Carlisle Communications, Ltd. It was printed and bound by R. R. Donnelley & Sons Company. The cover was printed by Phoenix Color Corp.

Pearson Education Ltd.
Pearson Education Australia Pty. Limited
Pearson Education Singapore Pte. Ltd.
Pearson Education North Asia Ltd.
Pearson Education Canada, Ltd.
Pearson Educación de Mexico, S.A. de C.V.
Pearson Education—Japan
Pearson Education Malaysia Pte. Ltd.
Pearson Education, *Upper Saddle River, New Jersey*

Copyright © 2003 by Pearson Education, Inc., Upper Saddle River, New Jersey 07458. All rights reserved. Printed in the United States of America. This publication is protected by Copyright and permission should be obtained from the publisher prior to any prohibited reproduction, storage in a retrieval system, or transmission in any form or by any means, electronic, mechanical, photocopying, recording, or likewise. For information regarding permission(s), write to: Rights and Permissions Department.

10 9 8 7 6 5 4 3 2 1
ISBN 0-13-094986-8

Janice—Thanks!
Kim and Cheryl—Your cherished love
and tireless support are unequalled!

PREFACE

We believe this textbook covers all the primary areas in the wireless technology arena. It is the result of many years of research and work experience as well as years teaching college students all about the wonders of wireless technology.

We begin with a brief overview of networking. This not only sets the stage for later discussions on wireless, but also provides an opportunity for those students who do not already have a background in networking. The chapters that follow address the chief wireless technologies such as wireless application protocol, Bluetooth, cellular telephony, emergency services, wireless local area networks, satellite communications, global position system, and paging systems.

This text is well suited to Computer Information System (CIS) and Information Technology (IT) programs that emphasize a practical approach to this field. We have deliberately developed this book so that readers do not require any mathematical foundation. The concepts are presented thoroughly and clearly.

Being in the education field ourselves, we realize that instructor supplements are important. We never seem to have enough time, it seems! An extensive set of supplemental materials, including lecture notes, PowerPoint presentations, many review and test questions, hands-on exercises, and team exercises is found in the Instructor's Manual that accompanies this text. In addition, we have gathered many wireless-oriented third-party software tools and placed them on the accompanying CD. It is truly amazing how many third-party vendors are available and more than willing to share their expertise.

ACKNOWLEDGMENTS

We would like to thank all those who have contributed to this book. Denise Edwards-Wilson, Rene Roberts, Karyn Morgan, David Williams and Terry Zimmerman worked tirelessly at creating vibrant and useful instructor supplements and figures. The many students in our wireless classes at Macon State College have contributed their sweat and expertise while we "tweaked" these wireless topics. The administration of the college, especially the President, David Bell, and our Division Chair, Bill Elieson, has been extremely supportive of our efforts. Alex Wolf and Kelly Mulligan, editors extraordinaire, have pulled out a lot of hair dealing with first-time textbook authors such as ourselves.

Charles Stewart, vice president of Prentice Hall, certainly deserves our undying gratitude and respect. He trusted in two first-time authors and has displayed the highest levels of integrity, professionalism, and even friendship through this time of innumerable questions and interruptions.

We would also like to thank the following reviewers for their valuable feedback: Phillip Davis, Del Mar College; Lee Rosenthal, Farleigh Dickinson University; and Costas Vassiliadis, Ohio University.

Finally, our families have paid a toll in missed opportunities to go out together while we were laboring on this book.

Gary S. Rogers
John Edwards

ABOUT THE AUTHORS ▮

John Edwards has a Ed.D. degree from the University of Georgia and holds the rank of Associate Professor of Information Technology at Macon State College. He has 35 years experience installing and managing local and wide area networks in education and business, and has published twenty-seven technical articles in the field of computing. He is a principal in a firm which provides e-commerce consulting, network configuration, installation and management, and website development.

Gary S. Rogers has a Ph.D. degree from Walden University and holds the rank of Assistant Professor of Information Technology at Macon State College. He specializes in telecommunications, possessing more than 20 years experience in the computer networking field and many years designing and implementing wired and wireless networks. Dr. Rogers has designed and implemented computer networking systems in both the private and public sectors. He has authored many articles and technical presentations in the area of networking, network security, and telecommunications standards.

CONTENTS

AN INTRODUCTION TO WIRELESS TECHNOLOGY

1 Introduction to Networking

OBJECTIVES

After reading this chapter and completing the exercises, you will be able to:

- Understand the evolution, role, and current state of the Internet
- Understand the purpose, operation, and role of reference models in networking
- Identify essential network types, media, topologies, and connectivity devices
- Understand the increasing role of wireless networking

This chapter provides you with a solid foundation in networking basics, which will enable you to have a deeper appreciation and understanding of the more complex wireless discussions that follow in subsequent chapters.

SEEKING ALTERNATIVES TO SNAIL MAIL

AS THE ONGOING THREAT of anthrax injects fear and danger into the circulation of letters and packages, online delivery alternatives are getting a second look from businesses and government agencies seeking reliable and safe ways to communicate with customers and partners.

Some of the benefits of e-mail and secure online delivery services include cost savings and convenience, according to Dana Gardner, research director at Boston's Aberdeen Group.

"Now is the time to re-evaluate the benefits of securely delivered e-mail," Gardner says. "It is unfortunate that [the anthrax threat] needs to be a reason for people to reconsider or re-evaluate e-mail; it is a good [idea] on its own without these unfortunate circumstances."

And using e-mail for customer communications can help forge a digital bond with customers that will prove beneficial in the long run, Gardner adds.

"The benefit goes beyond cost savings and convenience," Gardner explains. "The fact that [e-mail] is a two-way link and provides a directory entry that includes e-mail will benefit organizations in many ways down the line, particularly as Web services and context and collaboration become more prevalent."

The anthrax incidents, coupled with technology improvements in the past year and passage of the e-signature law, are expected to help speed the adoption of online-delivery services in enterprises, according to Sue Barsamian, senior vice president of marketing at Critical Path, a San Francisco-based messaging infrastructure provider.

"This is going to be a very strong catalyst for hastening the migration to secure online delivery service. The migration is inevitable because it is more cost effective than the physical delivery alternative," Barsamian says.

Critical Path's recently released Registered Mail Server, which enables businesses to deliver and receive documents via a secure Internet connection, aims to address the issue of integrating secure-message delivery with existing messaging systems, says Michael Serbinis, CTO and vice president of engineering at Critical Path.

"The stumbling block of most secure messaging is that it has not been integrated with e-mail and directory technology [for providing] security credentials," Serbinis says. "Registered Mail Server is an integrated platform, so secure messaging is just another type of messaging now available."

Another vendor, Digital Signature Trust, offers a secure document delivery system allowing users to send any size electronic document and track the delivery and receipt, says Greg Worch, senior vice president of Salt Lake City-based Digital Signature Trust.

Dubbed CertainSend, the system operates as simply as e-mail but allows businesses to guarantee the delivery of large documents that could otherwise overwhelm an e-mail connection, Worch adds, noting that the system uses TrustID digital certificates to guarantee identity and security.

"We can make any delivery legally binding," Worch explains. "[CertainSend] can deliver documents in business situations, whether it is a contract or sensitive data."

Online bill payment is also gathering momentum. According to Gardner, recent mail safety concerns are encouraging both billing companies and consumers to pursue electronic billing and payment channels more aggressively.

The Stamford, Connecticut-based research firm says its clients have reported a 20 percent increase in electronic bill presentment enrollment since the outset of the anthrax scare. Gardner also predicts that the number of American consumers viewing bills electronically will jump from 32 million at the end of 2001 to 64 million by year-end 2003.

Toronto-based messaging software vendor Metamail this month rolled out XML-based e-mail-client software designed to enhance Web-based electronic billing systems. The Metamail technology works with messaging systems, including Microsoft Outlook, Lotus Notes, Netscape, Eudora, and Novell Groupwise, to enable the secure online delivery of realistic-looking financial statements. To retrieve bills, users enter a password which allows them to view, pay, and print from within the messaging in-box, according to officials at Metamail.

With the dozens of online delivery systems nipping at its heels, the U.S. Postal Service has been steadily ramping up online services during the past five years. The recent safety concerns may provide the impetus to develop a national electronic postal service, says Bill Robertson, senior vice president of Boulder, Colorado-based NETdelivery, which provides the technology behind Canada's electronic postal service.

The anthrax threat "is having an untold impact on businesses and governments across the United States because of the interruptions of the mail distribution system, not to mention the terrible human toll," Robertson says.

As a result, Robertson says, "businesses and governments will be looking at alternative methods that might reduce the risks and provide a more uninterrupted flow of communication."

With the two-year-old success of the Canadian ePost as a model, and messaging technology capable of supporting billions of transactions per year, "There is no reason why a large scale national electronic post implementation could not be installed in the United States," Robertson adds.

Source: Moore, Cathleen. Retrieved November 21, 2001, staging.infoworld.com/articles.

TECHNOLOGY AND CHANGE

As you are undoubtedly aware, we live in the midst of an information revolution. This time can be challenging or even frightening. This change is evident in every sector of society, including the entertainment industry. In a 1998 science fiction series, *Babylon 5*, a notable character sums up our situation nicely: "The past tempts us, the present confuses us, and the future frightens us."(United Paramount Network. 1998. Babylon 5 TV series). Alternatively, as Alvin Toffler indicates in his *Third Wave*, this advent of technology and its associated information can be thought of as a third wave of information and is the dawn of a new civilization (Toffler 1991).

It is an event as profound as that first wave of change unleashed thousands of years ago by the invention of agriculture, or the earthshaking second wave of change touched off by the industrial revolution. Information technology forms the basis of this new age of civilization—the information age. Just as the industrial age caused mass movements of people from farm to factory, the information age has produced dramatic changes in the way we work and live (Toffler 1991). Certainly, major technological advances await us in this millennium, such as those in the biotechnology, cloning, and nanotechnology fields.

With this technology, however, comes something else—something perhaps as tangible as the technology itself—namely *change*. We live in a time of technological change, from the workplace to the home. On the work scene, this change even affects a tradition-ruled organization like the U.S. Congress. Members of Congress are so inundated with e-mail messages from constituents and special interest groups—80 million last year alone—that lawmakers routinely ignore most of them, according to a new study (Tech crisis 2001).

At home, how many would have seriously envisioned receiving 500+ TV channels from an 18-inch satellite dish located in the backyard? DishTV is a major player in this arena, using six satellites to transmit the TV channels and associated services such as Internet connectivity (Our Satellites 2001). It is clear that people, organizations, and society as a whole must learn to adapt to this change and learn from it.

THE INTERNET

Technological change has increased the rate at which computer networks and their accompanying applications are becoming not only accepted and prevalent but, in many cases, taken for granted. Many people in the United States are so accustomed to an electronic mail service that they would have difficulty living without it. Recent surveys indicate that electronic mail may soon become as important as the telephone.

Evolutionary Growth of the Internet

Perhaps the network experiencing the most change is the **Internet.** Contrary to public belief, the Internet is not one huge cable that travels to every major city in the country. Instead, it is a network of networks. In fact, the use of the Internet has become both intriguing and explosive in terms of popularity. The Internet links millions of computers via an array of network equipment, routing devices, and communications methods. It offers information from governments, universities, companies, and individuals. In fact, the estimates on the exact number of computers comprising the Internet vary widely. A recent study of unique information on the World Wide Web estimates over 1 billion pages (Tech report 2000). Moreover, once connected to the Internet, most of the information can be accessed free of charge. For less than $20 per month, access to this bonanza of networks is readily available. In fact, several Internet service providers (ISPs) have been offering free Internet access. Perhaps the most notable of these are NetZero (www.netzero.com) and Juno.com. In June 2001, these two companies merged. The two survivors now form a single company called United Online, which will continue to offer free services under the NetZero brand. Together, the two will reach 7 million active subscribers—people who have logged on to the service at least once in the last month—giving them a larger base than any other ISP except America Online. Only about 1 million of those people represent paid subscribers (Borland 2001).

At some moment in the past 2 years (2000–2001), the United States passed an important threshold: A majority of households now have e-mail (E-mail catches up 2001). Free or inexpensive Internet service, however, is not limited to the United States. It is common in the United Kingdom and Brazil as well (Free Internet access 2000). Internet access in areas of high competitiveness is becoming less expensive, but this may be changing. There is a retrenching of free ISPs because advertising dollars, intended to subsidize the free ISP service to subscribers, is not proving to be the financial panacea originally promised. It remains to be seen what will transpire in this arena.

In some rural areas even in the United States, where access is limited to one ISP, the cost may exceed $25 per month. In other areas, residents must also pay long-distance charges in addition to the ISP fees because they are too far from a local access point (the local number established by the ISP for you, a customer, to dial into in order to use their service) associated with any ISP.

The growth of the Internet has been phenomenal as shown in Figure 1–1. In Brazil, Internet use is increasing at over 80 percent per year. Specifically, the Internet is a network of over 30,000 networks. It had 1.7 million computers interconnected in the summer of 1993, with an estimated 30 million users worldwide. One person could send a message to all 1.7 million computers, to one of the 11,000 networks, or address any one of the computers, or any single user of any such computer. As of 1998, the number of computers (not users) increased to 30 million, and then to 90 million by 2000. The number of users using these computers varies widely, ranging from 75 million to 500 mil-

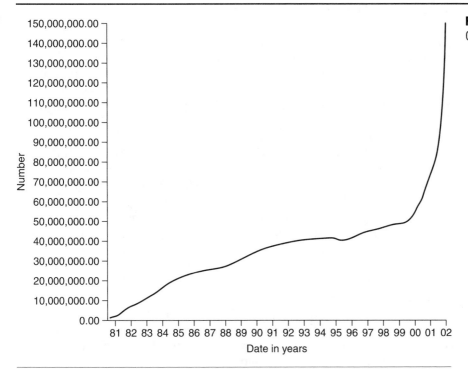

Figure 1–1
Growth of Internet Hosts

lion. The rate at which new computers are being added to the Internet now exceeds one per second. Some interesting data exist on this phenomenon. As of 1995, Norway, for example, had almost 5 Internet hosts per every 1,000 persons. The United States was second with 3.8 per 1,000. China only trails the United States in the actual number of people with Internet access at home, according to a recent study (China in 2nd). According to this study, the number of Chinese with Internet access at home is up to 56 million. A higher percentage of people in many developed nations can access the Internet at home, but China's sizable population placed it above Japan, Germany, and Great Britain in terms of overall at-home Internet access. The United States leads the world with more than 166 million Americans able to access the Internet from home. Extrapolating from previous studies, such as Naisbitt's Global Paradox book in 1994, would mean that there will be around 2 billion people accessing the Internet by 2005.

Many areas of the world depend on access to information via the Internet. In Japan, for example, more than 2 million people utilize their wireless handset to send e-mails and to access the Internet. In Finland, where 90 percent of the teenagers have wireless phones, a wireless user can purchase everything from soft drinks to car washes to bubble gum. As you can see, this market penetration is much more extensive than in the United States, where wireless access to the Internet is still a fledgling industry. Moreover, this phenomenon is not unique to Finland. There are over 300 million wireless handsets in use throughout the world, and this number is expected to grow to 1 billion by 2005 (Naisbitt 1994). The Gartner Group (2001) says more mobile phones were shipped in 1999 than the number of cars and personal computers combined.

Advanced Research Projects Agency (ARPA)

ARPANET

The early beginnings of the Internet are rooted in the formation of the network for the Department of Defense's (DOD's) **Advanced Research Projects Agency** (**ARPA**), titled **ARPANET.** The department's mission was to advance technology to serve the U.S. military. The dilemma, however, was that ARPA had only a small staff of scientists. To solve this problem, ARPA issued millions of dollars in contracts and grants to eligible universities and firms that could provide innovative approaches to creating a wide range of technology from enhanced military weaponry to high-speed satellite communications. For example, the DOD awarded several sizable grants to universities to investigate a new theory on message communications called packet switching. After initial research proved the idea sound, the DOD released for bid a contract for firms to build this network. ARPA awarded the contract to Bolt, Baranak and Newman (BBN), a technology firm in Massachusetts, in December 1968. Work began immediately on this effort. The evolution of the Internet is shown in Figure 1–2.

protocol

Transmission Control Protocol/Internet Protocol (TCP/IP)

Once developed and deployed by BBN in the summer of 1969, ARPANET grew rapidly. The subsequent development of a standardized communications **protocol,** or **Transmission Control Protocol/Internet Protocol (TCP/IP),** expanded the growth of this network as it allowed the interconnection of vastly dissimilar networks. A protocol is a special set of rules used by software programs in a telecommunications connection when they communicate. Both end points must recognize and observe a protocol. Protocols are often described in an industry or international standard by organizations such as the following:

- American National Standards Institute sets standards for all types of products, including computer networking equipment.
- Institute of Electrical and Electronics Engineers, an international organization, sets communications standards.
- Consultative Committee on International Telegraph and Telephone (CCITT) sets standards for modems, e-mail, and digital telephone systems.
- International Standards Organization sets standards in scientific, technological, and economic areas.

military network (MILNET)

Growth of ARPANET continued into the 1980s, when in 1983, it was renamed the Internet. This network consisted of hundreds of host computers. At that time, the strictly military portion was sectioned off to form the **military network (MILNET).** Also (Tanenbaum 1996) during the decade of the 1980s, other networks joined the Internet. For example, in 1984, the National Science Foundation (NSF) began to build a network titled NSF Network (NSFNET), because of the immense success of the ARPANET. To get onto the ARPANET, a university was required to have a research contract

Year	Event
1957	ARPA created
1960	Research activities on packet switching
1968	BBN wins ARPANET contract
1968–1969	BBN develops and deploys ARPANET
1970	5 ARPANET sites active
1971	12 ARPANET sites active
1972	E-mail created and becomes most used application on ARPANET
1974	TCP/IP developed
1975	ARPANET transferred to Defense Comm. Agency
1982	TCP/IP adopted as official protocols on APARNET
1983	MILNET created, Internet created
1984	Over 1,000 nodes on Internet
1984	NSFNET created
1985–1990	Commercial networks proliferate
1986	Over 5,000 nodes on Internet
1988	Over 60,000 nodes on Internet
1990	ANSNET created
1990	ARPANET officially retired
1991	Over 600,000 nodes on Internet
1992	Over 1M nodes on Internet
1993	Over 2M nodes on Internet
1995	NAPS initiated
1996	Over 9M nodes on Internet
1997	Over 16M nodes on Internet
2000	Over 35M nodes on Internet

Figure 1–2
Evolution of the Internet

with the DOD. Because many universities had no such contracts, the NSF provided an alternative solution for them to share their research information. Originally, the following six universities cooperated by connecting their own supercomputer centers.

1. Boulder, Colorado—University of Colorado
2. Champaign, Illinois—University of Illinois
3. Ithaca, New York—Cornell University
4. Pittsburgh, Pennsylvania—Carnegie Mellon University

5. Princeton, New Jersey—Princeton University
6. San Diego, California—University of California

Each supercomputer was front ended by a specialized microcomputer and all were connected via 56 kilobytes per second (Kbps) leased lines. In this setup, a specialized microcomputer performs a particular function and then interacts with a larger computer. Specifically, the microcomputer performs all telecommunications functions, which leaves the main computer free to perform all other functions. NSF funded several other regional networks that also connected to NSFNET to allow access to other universities, research laboratories, libraries, and selected museums. The participants that were allowed legitimate access to ARPANET were provided a gateway so they could access that network as well.

Advanced Networks and Services (ANS)

ANSNET

Soon after NSFNET became operational, it also became overburdened. Because of funding concerns in its attempt to keep up with the demand, NSF encouraged a nonprofit corporation, the **Advanced Networks and Services (ANS),** to manage the network. In 1990, it assumed control of the NSFNET and immediately upgraded its backbone (the major cable lines from major places, or nodes, on a network) and renamed it **ANSNET.** In 1995, ANSNET was sold to America Online.

In addition to the NSFNET activity, commercial vendors recognized the importance of a regional or national network and subsequently established their own TCP/IP-based networks. This action of establishing networks proved to be an issue in that interoperability was increasingly becoming important. After all, why should anyone have all these networks and not be able to connect from one to another? To solve this dilemma, the NSF awarded contracts to several network operators (Bell South, Ameritech, MFS, and Sprint) to connect all the networks via gateways or special access links called network access points (NAPs). Finally, various government and commercial NAPs were also established, providing access to even more information. It was during this time that these networks collectively became known as the Internet.

The Internet2

In the mid-1990s, the tremendous growth of the Internet caused serious network performance problems. Because of this and political factors—such as both Congress and the scientific community realizing the need for a faster network—a project was established. A consortium of over 180 universities began working in partnership with industry and government to develop and deploy advanced network applications and technologies. This new so-called Internet2 is initially for use only by the research community. The primary purpose is to build a high-speed digital-video network for the research community and higher education. The current Internet simply is not capable of achieving this purpose because of bandwidth constraints. Ironically, this same research community established the original Internet. It is hoped by many in the educational and scientific communities that Internet2 will relieve some of the performance problems currently experienced on the original Internet.

REFERENCE MODELS, LAYERING, AND PROTOCOLS

During the time the Internet was taking shape, experts from the **International Standards Organization (ISO)** developed a model of protocol standards called the open systems interconnect (OSI) reference model (Stallings 1987). The International Standards Organization, established in 1947, is headquartered in Geneva, Switzerland, and represents over 100 countries. The purpose of this nongovernmental organization is to establish communications and networking standards.

International Standards Organization (ISO)

The committee specified a set of standards for the exchange of information among systems that are "open" to one another because of their mutual use of the appropriate standards. The ISO established a committee for this purpose in 1977, and in 1984, the **OSI reference model** was born. The OSI reference model is shown in Figure 1–3.

OSI reference model

The OSI reference model incorporates existing standards including IEEE 802.3 (Ethernet) or IEEE 802.5 (Token Ring). The OSI model provides a conceptual and functional blueprint, and is not a set of implementation rules. The model employs the concept of layering. The OSI committee determined that communications hardware and software on a host must perform the totality of communications-related activities. Therefore, this functionality or

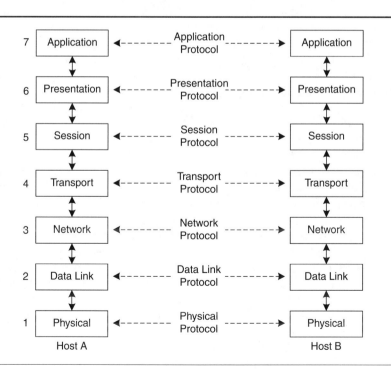

Figure 1–3
OSI Reference Model

set of activities is divided into seven vertical **layers,** essentially grouping similar functions in a single layer. For example, all functions or activities related to message routing are set in layer 3 of the OSI reference model, the network layer.

The principles used by ISO in developing the model are as follows:

- Keep the process of describing the layers as easy as possible.
- When creating these layers, make sure the edges are "clean," meaning the interface between the layers should be as efficient as possible.
- Combine similar functions into a layer.
- Create the layers to better utilize newer technologies.
- Create a boundary in which at some point in time the corresponding interface may be standardized.
- Create a layer in which there is a need for a different level of abstraction in handling data (e.g., syntax, semantics).
- Allow function or protocol changes without other layers being affected.
- Ensure that each layer has only two boundaries.

The ISO subcommittee had the task of defining a set of layers and the services performed by each layer. The number of layers was an important issue. There should be enough layers to make each one small for manageability purposes, but not so many layers that the processing overhead imposed by these layers would be too demanding on network resources. The subcommittee's goal was admirable, but unfortunately, most protocol suites created by software developers adhering to the model ran slow because of the model's massive overhead. Even though the ISO committee had established seven layers, not all are used. Specifically, layers 5 and 6 are not generally used.

The OSI model describes the flow of data from a sending network node to a receiving node. First, the application programs (e.g., Microsoft Word) interface to the communications stack (all communications software) via the operating system. Then, the appropriate layer 7 program (e.g., e-mail or file transfer) is called and performs the assigned processing. For example, the network layer determines how to route messages over a network. The software

header
then attaches a **header** (control information) to the data. Then layer 7 calls the layer 6 program. This program performs its assigned function, attaches its header, and then calls the layer 5 program, and so on until the data, with all associated headers, are transmitted over the communications media to the destination computer. Thus, each layer attaches a header and incorporates

peer process
into its header the information necessary for its **peer process** on the destination computer. A peer process is a process on another computer that performs an identical function to the process on the first computer. As an example, the OSI layer 3 software program (typically IP or IPX on a Novell network) adds its own source Internet address and destination address. In most networks, this layer 3 software program is the Internet Protocol (IP). Figure 1–4 illustrates this process. A brief description of the roles of each layer follows.

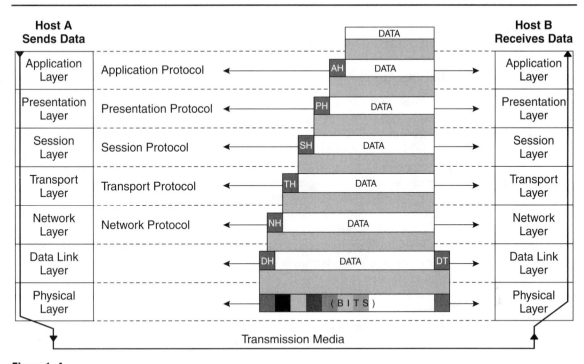

Figure 1-4
OSI Model with Headers

Physical Layer

The **physical layer** deals with mechanical and electrical properties, and the physical transmission medium. The physical layer simply transmits each bit of data over the attached media, whether this media is coaxial, twisted pair, fiber optics, or others such as microwave systems transmitting in the atmosphere. It also describes specifically how many volts should be used to represent either a 1 or a 0, the number of pins present in the network connector, and the function of each pin.

Data Link Layer

The **data link layer** performs error correction and formatting on the bits transmitted by the physical layer. The data link layer performs this function by breaking the data into relatively small data frames, transmitting the frames in a sequential fashion, and processing the **acknowledgment frames** sent back by the receiver. An acknowledgment frame is a frame that is sent to the sender of data to acknowledge receipt of a data frame. The data link layer uses special characters to indicate the beginning and end of frames. These

acknowledgment frames

special characters usually would not occur in the data, otherwise confusion would result. Finally, the data link layer tracks frame delivery and may retransmit frames if necessary because of non-delivery.

Network Layer

network layer

The **network layer** determines the route that packets of data follow from source to destination. This can be a difficult endeavor in that routing can be either static or dynamic. Static routing is when network administrators program routing into routing tables of routing devices on a network such as the Internet. That device then routes data over the network based on these algorithms. One such example would be to send all data from one routing device to another based solely on whichever path is the closest. Dynamic routing results from installing more sophisticated series of routing algorithms that will react to a changing network. These routing algorithms can involve criteria such as shortest path from sender to receiver, traffic load on the network, error rates, and administrative aspects such as cost of transmitting data along a transmission line.

The network layer must also control congestion. If too many packets are present in the network at the same time, they will get in each other's way, forming bottlenecks.

Transport Layer

transport layer

The **transport layer** accepts data from the session layer, divides this data into smaller fragments if needed, and passes the fragments (via whatever internal software mechanism is being employed) to the network layer. In addition, this layer ensures that all fragments arrive correctly at the destination. This error checking is called end-to-end reliability. Furthermore, this task must be done efficiently and in a way that isolates the upper layers (5 through 7) from any changes in the hardware technology.

The transport layer is the first layer to perform end-to-end functions along the network. In other words, a program on the source computer holds a conversation with a similar program on the destination computer using the message headers and control messages. In the previous layers (1 through 3), the communications occur between each machine and its immediate neighbors, not the ultimate source and destination machines, which may be separated by many routers.

Session Layer

The **session layer** permits users on distinct machines to set up sessions, or conversations, on the network. A session allows ordinary data transfer, as does the transport layer. It also enables enhanced services, which are quite useful in some applications such as logging in to a remote timesharing system or transferring a file between two computers.

This layer also manages conversations or dialogues. Sessions can allow data traffic to go in **full duplex** or **simplex mode.** Duplex mode means that transmissions occur simultaneously in both directions along a transmission line. Simplex mode means transmission can occur only in one direction at a time.

full duplex

simplex mode

Presentation Layer

The **presentation layer** translates the syntax and semantics of the data transmitted into codes which the network understands. Different computers have different codes for representing character strings (e.g., ASCII = American National Standard Code for Information Interchange, and EBCDIC = Extended Binary Coded Decimal Interchange Code) or integers (e.g., 1's complement and 2's complement). The presentation layer then converts from the representation, or code, used inside the computer to the network standard representation, or code, and vice versa.

presentation layer

Application Layer

The **application layer** contains a variety of protocols, such as **Simple Mail Transport Protocol (SMTP)** and **File Transport Protocol (FTP),** that the underlying layers support and that are important to users. Protocols at this layer involve electronic mail, network management, and remote login capability. The software at this level is most noticeable to users because it involves that which they can see, such as the electronic mail program Qualcomm's Eudora or Microsoft's Outlook Express. Software at other layers exists, but works behind the scenes and so is invisible to the user.

application layer

Simple Mail Transport Protocol (SMTP)

File Transport Protocol (FTP)

OSI AND OTHER MODELS

Theoretically, all protocol suites must accomplish the same totality of functions in the transmission of data from the communications media (cabling, and so on) to an application program. As shown in Figure 1–5, the TCP/IP reference model does not utilize session or presentation layer functions, such as session control or data conversion, because the TCP/IP world only recognizes the ASCII character set. This eliminates any need for data conversion at layer 6. Figure 1–6 shows an example of this model-to-model protocol mapping.

An Example of Layering and the OSI Reference Model

Assume a user is using an e-mail application such as Microsoft's Outlook Express, or a similar product such as Qualcomm's Eudora. Figure 1–4 shows this process graphically.

Figure 1–5
TCP/IP Reference Model

Layers	Protocols	Notes
Application	Telnet, SMTP, FTP, SNMP	Applications such as e-mail, etc.
Transport	UDP, TCP	Provides end-to-end reliability TCP = connection oriented UDP = connectionless
Internet	IP	Provides message routing
Network	802.3, 802.5	Existing LAN/MAN standards
Physical		

Figure 1–6
Comparison of the OSI and the TCP/IP Reference Models

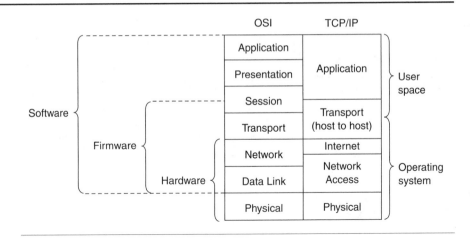

The user types the message, "Hi Susie!" The application layer program, the e-mail program, passes the data ("Hi Susie!") to the presentation layer[1] (layer 6). The presentation layer translates and encrypts the data. Through substitution, the process of **encryption** creates unrecognizable data that are transmitted to the recipient. The recipient must decrypt the data. **Decryption** is the process of deciphering the original encrypted data. Let's say you want your message to remain confidential, so you encrypt it. You may select to change every A to a Z, B to a Y, and so on. Then you transmit the data and the recipient uses that identical formula, reading Z as A, and so on. The presentation layer passes this data, in this case "Hi Susie!" (and the header that the

encryption

decryption

---

[1] Even though the presentation and session layers are not generally used, this functionality is incorporated in layer 7 application programs.

presentation layer attaches to the data), to the session layer where the dialogue is set for full-duplex communication (since chats function in this fashion). Next, the session layer calls the transport layer that packages the data as segments. The recipient's name is matched to its corresponding IP address by accessing a table linking these two items. The transport layer also adds error checking because, as discussed, this layer is responsible for end-to-end reliability. The transport layer then calls the network layer (layer 3), which has responsibility for message routing.

The network layer packages the data as datagrams (fragments or "chunks" of the message). If, after an analysis of the IP address, the destination is on a network that is not local, then the network layer determines the IP address for the next intervening device and adds it to the header as the next destination. This process is crucial in that this layer (the network layer) is concerned with the next destination on the route that is necessary to send the datagram on its way to the final destination. The data link layer then packages (or chops up) the datagram into frames. The physical address of the device is resolved via a table, linking physical PCs with their corresponding data link layer addresses. On an Ethernet LAN, for example, this is the media access control (MAC) address assigned by the manufacturer. The data link layer passes the data to the physical layer, which packages it as bits (0's and 1's), and the network adapter (usually a network interface card) transmits it across the transmission media. This transmission media may be copper or fiber optic cable. Fiber optics are specially designed and manufactured glass strands that can transmit light, as a media, at speeds exceeding 20 gigabits per second (Gbps) (Hecht 2000).

At the other end of the transmission media, the same process is repeated, with each layer analyzing its header and passing the data up to the next layer (layer 2 to layer 3, and so on). Once the data travels from layer 1 to layer 3, it is then sent to the transport layer. The transport layer calls the session layer, which acknowledges that the data have been received. Next, the presentation layer translates and decrypts the data, and then the application layer finally calls the e-mail application, for example Microsoft's Outlook Express, and the message "Hi Susie!" appears when the user activates the e-mail software. This description demonstrates the amount and complexity of the communications software needed just to send a simple message.

NETWORK TYPES

The three generally recognized network types are local area networks (LANs), metropolitan area networks (MANs), and wide area networks (WANs). Each is described in more detail in the following sections.

Local Area Networks

Local area networks (LANs) generally have limited geographical scope. LANs may encompass one building, a small campus of buildings, or a single room within a building, as shown in Figure 1–7. The IEEE 802.3 or 802.5

Figure 1–7
Typical LAN

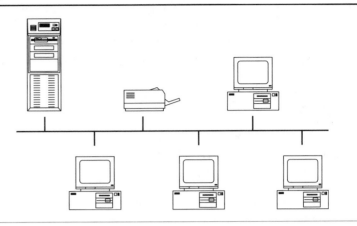

standards govern LANs. You may go to www.ieee.org for more information on these standards.

LANs connect workstations in offices or laboratories as in a college or at a research and development facility. They are also typically distinguished from the other two types by four characteristics.

1. Transmission technology (e.g., Ethernet, Token Ring, or ARCNET)
2. Limited size
3. Relatively small error counts
4. Unique topology (A **topology** is the shape or the physical layout of the network elements, which include the computers, network devices, and cabling.)

topology

Metropolitan Area Networks

metropolitan area
networks (MANs)

Metropolitan area networks (MANs) are essentially a larger version of the local area network, but they are not as limited geographically. A typical MAN could be found on a large industrial campus where several LANs in each building are connected via a backbone network, as depicted in Figure 1–8. Another example could involve multiple offices that are occupied by one firm within the same city. MANs are usually slower in speed than a LAN, but faster than a WAN. The IEEE 802.6 standard governs MANs. You may find more information on this standard at www.ieee.org.

Wide Area Networks

wide area networks
(WANs)

Wide area networks (WANs) span large geographical areas such as among states or worldwide. For example, worldwide networks of automated teller machines (ATMs) service the banking system. ATMs are relevant in that half of adult Americans use them on a regular basis (Arizona Vending 2001). A

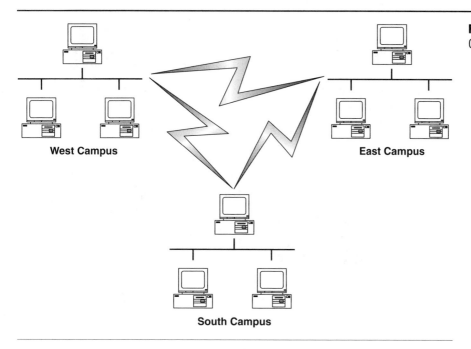

Figure 1–8
Campus MAN

West Campus

East Campus

South Campus

multinational firm may use a WAN to share data among offices in the United States, Paris, Japan, Singapore, and Cairo. WANs generally have higher error rates than the other two types of networks (because of more devices and longer runs of cabling) and are slower than the other two. This speed difference is typically because of cost considerations. Could you design and operate a WAN at the highest LAN speeds? Absolutely! It is done in some instances today. However, the return on investment would be difficult to justify because of the immense costs involved in equipment, cabling, and purchasing the right-of-way.

Certain terms are used strictly in the WAN world. For example, *end systems* are the end points in the network, whereas the intervening network is called the *subnet*. The end systems can be any one of various computer systems. Figure 1–9 depicts a typical WAN. Numerous standards govern WANs, including CCITT's X.25, or the ATM Forum. You may go to www.atmforum.org for more information on ATM.

NETWORK MEDIA

Several types of network media are available in the marketplace today. The most prevalent are coaxial cable, twisted-pair cable, and fiber optic cable.

Figure 1–9
Typical WAN

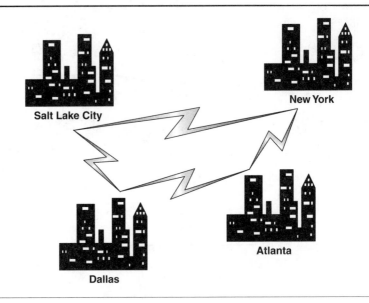

Coaxial Cable

coaxial

cable

If you watch television you may be familiar with a type of cabling called **coaxial,** or coax **cable.** If you look behind your television, you will see a series of cables, most likely coax. Coax consists of two conductors that are parallel (on the same axis) to each other. The center conductor in the coax cable is almost always copper, although aluminum can be used. Aluminum is an excellent conductor of electricity and was used, particularly during a span in the 1970s, for various electrical projects, including home wiring and communications cabling. It was essentially discontinued for this use, however, because it had an unfortunate tendency to create a catalytic condition when in contact with other metals that could cause heat. In fact, some homes during that period apparently caught fire from the wiring because of this effect. The copper can be either a solid wire or a stranded material. Outside this central conductor is a nonconductive material. It is usually a white, plastic-like material used to separate the inner conductor from the outer conductor. The outer conductor is a fine mesh made from copper, which is used to shield the coax cable from electromagnetic interference (EMI). Outside the copper mesh is the final protective cover, usually a plastic or fabric-like substance, as shown in Figure 1–10.

Twisted-Pair Cabling

twisted-pair wiring

megabits per second
(Mbps)

The most popular network cabling used today is **twisted-pair wiring.** It is lightweight, easy to install, inexpensive, and supports many different types of networks. It can also support speeds easily exceeding 100 **megabits per**

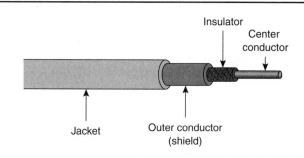

Figure 1–10
Coaxial Cabling

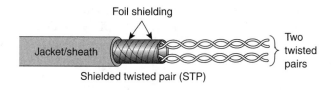

Figure 1–11
UTP and STP Cabling

second (Mbps). Also, the recently developed category 6 cabling supports speeds up to 1 Gbps. This is important in that networks continue to transmit data in ever-increasing speeds due to increasing user demands.

Twisted-pair cabling consists of pairs of copper wire, twisted around each other. The twists alleviate EMI and **crosstalk.** Crosstalk results when signals from one wire leap over and mix with signals in an adjoining wire. The number of pairs of wire in the cable depends on the type of cabling purchased. The user should purchase the specific type of twisted wiring needed. There are two types of twisted-pair cabling: unshielded twisted pair (UTP) and shielded twisted pair (STP). Figure 1–11 depicts both types of twisted-pair cabling.

crosstalk

Unshielded Twisted Pair

UTP is the more prevalent and much less expensive of the two types. It can be used for either voice or data communications. UTP is readily available anywhere from heavy-duty industrial supply shops to the local Radio Shack at reasonable rates. For example, 1,000 feet of this cabling can be purchased for as little as $120.[2]

---

[2] http://catalog.blackbox.com/BlackBox/Templates/blackbox/mainscreen.asp.

There are six levels of data grade cabling based on their performance capabilities.

- Category 1 is used in older telephone lines and low-speed data cable.
- Category 2 is for lower speed networks of up to 4 Mbps.
- Category 3 is popular and is used for older Ethernet network and home wiring.
- Category 4 is capable of higher speeds and of handling up to 20 Mbps. Because of timing and cost issues, it is rarely used and has been superceded by category 5.
- Category 5 is used for networking applications that require a higher speed than the previous categories 1 through 4 cabling. It is enhanced to allow for Gigabit Ethernet speeds can support up to 100 Mbps.
- Category 6, an emerging standard, is enhanced for speeds above Gigabit Ethernet. This cabling can withstand higher data rates and has a higher signal-to-noise ratio.

Shielded Twisted Pair

STP is mainly used in Token Ring Networks. Token Ring Networks adhere to the lesser-used network standard (IEEE 802.5). STP is similar to UTP, but contains a mesh shielding that protects it from EMI. This protection is valuable for use in a factory environment, for example, in locations where electrical motors and devices can emit significant levels of EMI. Such interference produces static that can seriously degrade network performance.

Fiber Optic Cabling

Fiber optic cabling consists of numerous specially designed glass strands. The number of glass strands depends on the amount purchased from a fiber optics vendor. The center of the fiber cable consists of a glass core. The light propagates through this core. In multimode fibers, the core measures 50 microns in diameter, about the thickness of a human hair. In single-mode fibers, the core measures 8 to 10 microns. Figure 1–12 depicts a single fiber.

Fibers are typically grouped in bundles that are protected by an outer sheath. A glass cladding surrounds the glass core with a lower index of refraction than the core to keep all light inside the core. Surrounding the

Figure 1–12
Fiber Optic Cabling

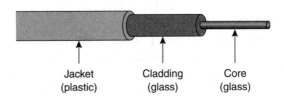

Jacket (plastic)   Cladding (glass)   Core (glass)

cladding, then, is a thin plastic jacket. The core usually consists of twelve or twenty-four strands. It is common to install either twelve or twenty-four strand cabling and light only the number of strands currently required. The remaining strands can be used as bandwidth needs increase.

In a fiber cable, light moves only in one direction. Two-way communication requires two strands of cable. Each strand is responsible for one direction of communication. A laser at one device sends pulses of light through this cable to the other device. These pulses are translated into l's and zeros at the receiving network node.

The two types of fiber optic cabling are multimode and single mode. The differences between the two are the quality of manufacturing and the application. Multimode is the lesser quality of the two, but is less expensive and is primarily used for shorter distance networking. The transmitter for multimode fiber is typically light-emitting diode (LED) boards. Single-mode fiber is more expensive than multimode and is typically seen in long-distance networking applications using lasers. For example, most transoceanic fiber optic cabling installations use single-mode fiber with lasers. In this way, fewer repeaters are needed, because of the powerful beam of light produced by the laser combined with the superior light-carrying characteristics of single-mode fiber.

Two kinds of light sources are typically utilized to do the signaling—LEDs and semiconductor lasers. Lasers are more powerful; thus, the light generated lasts longer before needing amplification by a repeater. Lasers also are more expensive than LEDs. A summary of lasers versus LEDs follows in Figure 1–13.

Item	LED	Semiconductor laser
Data rate	Low	High
Mode	Multimode	Multimode or single mode
Distance	Short	Long
Lifetime	Long life	Short life
Temperature sensitivity	Minor	Substantial
Cost	Low cost	Expensive

Figure 1–13
Lasers versus LEDs

Comparison of Copper and Fiber Optics

It is important to compare fiber to copper, because the decision on which media to use affects other technical and business decisions. Fiber has many advantages. For example, fiber can handle much higher bandwidths than copper. Most research on increasing bandwidth is assuming a cable media of fiber optics, not copper, and is rapidly increasing the rate of speed of fiber optic networks (Fiber Optics Research 2002). Bell Labs Technologies, among others, has recently announced WAN speeds, using fiber as a media, in excess of 10 Gbps (Hecht 2000). In addition, because of the low attenuation of light through glass fiber, repeaters are needed approximately every 20 miles. Copper cabling requires repeaters every mile. This factor allows a significant cost saving. Repeaters, singularly, are not expensive; however, imagine the immense cost savings along an undersea cable of several thousand miles incurred by having to use one-twentieth the number of repeaters. Electrical power surges, corrosive chemicals, and electromagnetic interference do not affect fiber optic cable. Fiber optics is also extremely lightweight compared with copper and is one reason why the military employs it in everything from land vehicles to aircraft carriers. Finally, glass fibers do not leak light and they provide excellent security against potential wiretapping.

Conversely, fiber optic cable can be difficult to install. The connector technology has advanced considerably, but does not approach the simplicity of UTP installations. Incorrect installation will create significant loss of signal. Splicing fiber optic cabling and then recombining the strands also can prove to be challenging, and loss of signal can occur here as well. Fiber optic cable communicates in simplex mode. As mentioned, simplex mode indicates that data can flow in only one direction at a time over a cable. This requires fiber cabling to be installed in pairs, which can increase the cost. Finally, fiber cable and its installation are more expensive than copper. This price differential has narrowed considerably in the past 3 to 4 years, but persists nonetheless.

NETWORK TOPOLOGIES

The topology of a network describes the arrangement of nodes, connectivity devices, and media. Three topologies (bus, ring, and star) are commonly used today in wired networks. Mesh topology is another, but rarely used, type. The following discussion presents a description of each topology. See Figure 1–14 for a comparison of these topologies.

Bus

It is perhaps easiest to understand the bus topology. Imagine a tree trunk, and all major limbs connecting to this trunk, as seen in Figure 1–15.

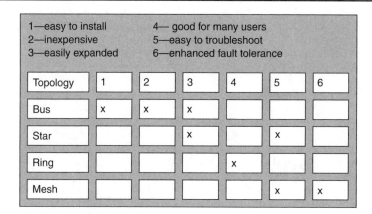

Figure 1–14
Comparison of
Topologies

1—easy to install 2—inexpensive 3—easily expanded			4— good for many users 5—easy to troubleshoot 6—enhanced fault tolerance			
Topology	1	2	3	4	5	6
Bus	x	x	x			
Star			x		x	
Ring				x		
Mesh					x	x

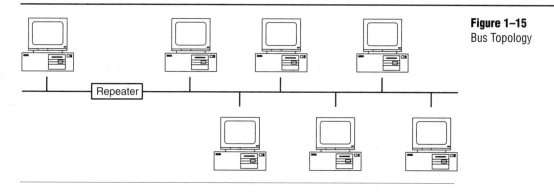

Figure 1–15
Bus Topology

Repeater

That topology is identical to a bus. In a bus topology, all computers are connected to one cable, typically either coax or UTP. Not only is the bus topology the easiest to understand, but it is also the easiest to install. You simply connect your cabling from your network adaptor card to the main bus cabling, add terminators at the end of the main bus cable, and the installation is complete. This method is certainly cost effective from an organizational perspective. As such, the bus has been in existence since the earliest LANs in the 1970s and will be present for some time. Ethernet is typically the protocol seen most in bus topologies; it uses a passive communications approach. If a computer wishes to transmit, it first listens to determine if the bandwidth is in use, and if not, it then transmits its data. Collisions sometimes occur and are handled by detection and retransmission at the data link layer. Each computer then receives the signal and determines if its address

Figure 1–16
Ring Topology

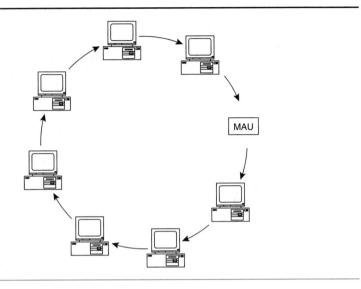

matches that of the recipient. Messages that are not addressed to that computer are disregarded.

Ring

Ring topologies involve cabling that is laid out in a circle. Figure 1–16 depicts this type of connectivity.

Data packets travel from one computer to another around this ring (actually each station on the ring retransmits the signal). A special "token" packet of data continually travels around the ring to each station. A station (computer) may transmit only if in possession of this special token packet. The transmitted message data travel around this ring until reaching a destination. The destination device returns an acknowledgment to the sender. This signal is for the token to now travel to the next station on the ring, thereby allowing it to transmit. This process creates a high-speed, orderly network.

Star

The star topology uses a separate cable for each computer, connecting all nodes to a central switching device. The switching device is typically a hub, as shown in Figure 1–17. This configuration provides a reliable, easily maintained network with extensive expansion possibilities. A hub is a simple device that broadcasts signals to all network nodes attached to it.

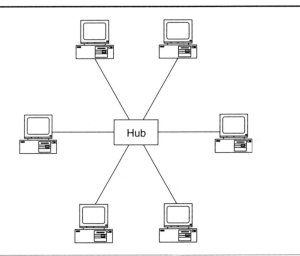

Figure 1–17
Star Topology

For example, an additional computer may be easily added by installing cable from the central hub or switching device to the network adaptor in the new computer. Correspondingly, it is easy to diagnose problems in this network, because each computer has its own cable that is connected to the central switching device. If, for example, only one computer is not connecting correctly, most likely the source of the problem is in that computer or the cable to the central switching device. If no computers are connecting correctly, then the central switching device is probably defective.

Mesh

The mesh topology provides the highest level of **fault tolerance** of all the network topologies, which means that network data are less likely to be lost due to an inactive or inoperative network node or communications line. A mesh network employs distinct cables to connect each device to every other device on the network. This requires an inordinate amount of cabling. This type of topology also creates a network management nightmare for the systems administrator and produces a cost picture that is unreasonable. For these reasons, mesh topologies are rarely seen. Their one advantage, however, is their fault tolerance, so they are seen occasionally in DOD applications. Figure 1–18 shows the mesh topology.

fault tolerance

Figure 1–18
Mesh Topology

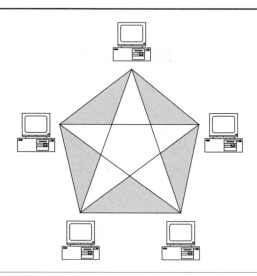

CONNECTIVITY DEVICES

Several devices are utilized in the network realm, depending on other devices, topologies, and purposes for use. Regular and high-speed networking could not transpire without such devices. Examples of connectivity devices are hubs, switches, and routers. Figure 1–19 shows how these devices function relative to the OSI reference model.

Multistation Access Units

Multistation access units (MAUs) are unique to the token ring topology. MAUs function as a central hub on a Token Ring Network. They connect

Figure 1–19
Connectivity Devices and
the OSI Reference Model

Device	Layer 1	Layer 2	Layer 3	Layer 4	Layer 5	Layer 6	Layer 7
Repeaters	x						
Bridges		x					
Routers			x				
Gateways	x	x	x	x	x	x	x
ATM switches		x	x				

computers via a physical star topology, pass the special token packets around the ring, and provide orderly movement of data throughout the network. MAUs can also be either passive or active hubs. Active MAUs amplify the signal, meaning they essentially incorporate a repeater. See Figure 1–16 for a picture of a typical MAU.

Repeaters

Repeaters provide inexpensive solutions to expand a network. The addition of another device, in this case a repeater, is necessary because of signal **attenuation.** Attenuation is the loss of signal over a distance. All media, wired or wireless, succumb to attenuation. For example, a UTP network cannot exceed 100 meters (or 185 meters for 10BASE2 or 500 meters for 10BASE5) without a repeater, because of its relatively poor attenuation characteristics. The repeater amplifies an incoming signal, retimes it, and reproduces it. As such, it can expand network segments, increase the number of nodes beyond the limit of one segment, and of course amplify the signal. See Figure 1–15 for a picture of a typical repeater.

attenuation

Bridges

A bridge attaches two different network segments and passes data from one network to the next. A bridge, unlike a repeater, can filter data traffic. Filtering traffic means that the messages can or cannot be scrutinized and kept out of a network segment. Bridges, therefore, are more intelligent than repeaters in that some rudimentary decision making is performed. This is certainly an advantage for bridges over repeaters, although the decision making takes a modest toll in performance, because the CPU takes time to make these decisions. Until the decision is made, the device is essentially in a "wait" mode. See Figure 1–20 for an example of a bridged network.

Routers

Routers connect networks where complex routing decisions must be performed. Routers use IP addresses in this routing decision and therefore operate at the network layer of the OSI model. The Internet consists of many thousand routers. For example, what if you had a network where each device was connected to several other devices and so on, until your network consisted of hundreds of these devices? In addition, what if this network had communication lines with various capabilities, from 56 K to OC-3? Finally, what if you had to route data along that network, taking into account the different speeds of the communication lines, plus the time of day or month and the rerouting of data when lines become inoperative? A router is the only type of connectivity device with the decision-making capability to accomplish this task. Figure 1–21 provides an example of a routed network.

Figure 1–20
Bridged Network

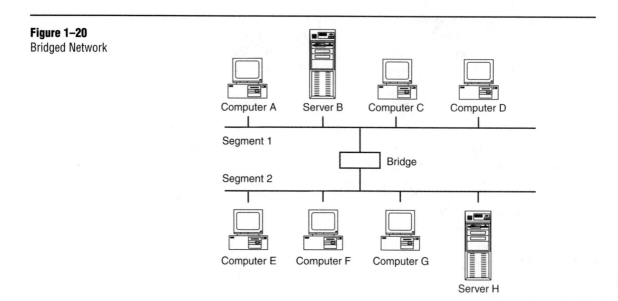

Figure 1–21
Routed Network

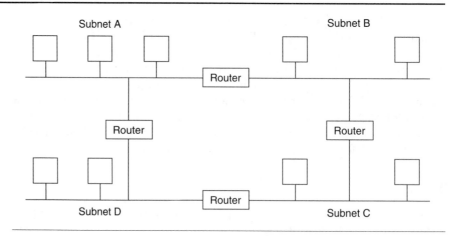

Gateways

Gateways are software or hardware interfaces that enable two differing types of networked systems or software to communicate. As tasks, they convert from one protocol to another (e.g., TCP/IP to Systems Network Architecture [SNA]), translate differing addressing schemes, provide terminal emulation, direct e-mail, and perform various other tasks. Typically, gateways are software constructs comprising many or all seven layers of the OSI reference model.

EVOLUTION OF NETWORKING

Networks have undergone a radical change in both their complexity and their characteristics. Not only have modern networks proliferated, but they have also gotten faster while carrying a rapidly increasing array of multiple media types such as text, voice, and image traffic.

Early Networking

In the earliest days of what we call the telephone network, manually operated switches were used to transfer calls from one line to another until the final destination was reached. This method is called circuit switching. It involves setting up a circuit by essentially joining the source and destination telephone sets. The operators were eventually replaced by mechanical switches and, within the last several decades, electronic switches. Figures 1–22 and 1–23 illustrate this evolutionary process.

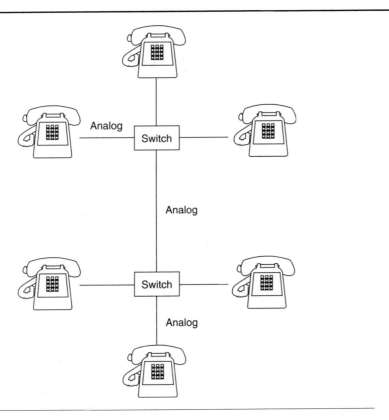

Figure 1–22
Early Telephone Network Design

Figure 1–23
Later Telephone Network
Design

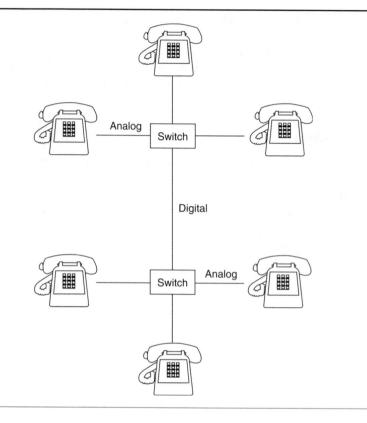

Current Networking

As shown in Figure 1–23, a major development in the 1980s was the updating of the communications between the switches from analog to digital. This update increased throughput while allowing for special functions and low error rates. In fact, most of these links today are fiber optic, thus leaving the local loop from the subscriber home to the neighborhood switch the only segment still copper based. This update to digital technology was not without its complexities, in that now there needed to be a translation process between calls coming in from the local subscriber loop, in analog format, and those being transmitted to digital format, to other switches in the system. Fortunately, computer technology has increased to the point where these devices could be computers and not mere mechanical switches. Control information is sent back and forth between the switches to synchronize and control the network via a new system called common channel signaling (CCS).

The data (typically referred to as bit streams) in today's telephone networks have an organization all their own. The data links are organized into digital carrier systems; and the capacity is divided into a series of logical

Medium	Signal	Rate (in Mbps)			No. of voice circuits
		Europe	United States	Japan	
T=1 Cabling (copper)	DS=1	2.0	1.5	1.5	24
T=3 Cabling (copper or fiber)	DS=3	32	45	34	672

Figure 1–24
Telephone Network Channels

channels, as shown in Figure 1–24. The basic rate was established to be the DS-0 channel, operating at 64 Kbps. This channel is a one-voice circuit. The larger channels use multiplexing techniques to efficiently handle several voice channels.

Figure 1–24 shows that the specified rates vary between the United States, Japan, and Europe. This is a rare instance in which international cooperation in the standards process has not succeeded, and is due to several factors including legacy systems, political pressures, and marketing issues.

X.25

X.25 was developed in 1974 by the CCITT as a network layer protocol centering around the use of the new packet switching technology. For its time, it was cutting edge with a speed of 64 Kbps. X.25 was truly one of the earliest WAN protocols. It was used for decades and still can be found in isolated portions of the world.

The X.25 standard specifies an interface between a host system and an associated packet switching network. X.25 has three layers, corresponding to the bottom three layers of the OSI reference model.

One reason why X.25 was not faster than its counterparts was that it assumed an unreliable underlying network, and employed extensive error control overhead in answering this assumption. In its defense, it could be argued that networks at that time were somewhat unreliable, and thus the error control was a necessary fact.

Frame Relay

In 1988, to further meet the needs of high-volume traffic, frame relay was introduced by CCITT. Nearly 60 percent of Fortune 1000 companies use frame relay or plan to do so.

Unlike X.25 and the **Integrated Services Digital Network (ISDN),** frame relay is designed to interface with modern networks that do their own error checking. It achieves high-speed data transmission by recognizing that

Integrated Services Digital Network (ISDN)

newer network technologies such as TCP/IP have error checking on inter-mediate nodes, so it does not incorporate extensive error checking. This was one downfall of the earlier X.25 standard, in that X.25 performed so much error checking that the real data throughput was drastically reduced from the reported, and expected, rate of 64 Kbps. However, one could argue that perhaps this extensive error checking was necessary as international networks were in their infancy and errors were frequent.

Synchronous Optical Network

Since the mid-1980s, with the advent and successful testing and operation of high-speed fiber optic lines, the transmission links of the telephone have been changing to that of the Synchronous Optical Network (SONET) standard. SONET data rates are organized in the synchronous transfer signal (STS) hierarchy, as shown in Figure 1–25. SONET is so fast because of the superior rates received from fiber optics versus copper-based media. Unlike the DS rates, STS rates are exact multiples of the base STS-1 rate, providing for less complex, and therefore less expensive, multiplexing equipment for these lines. Research is pushing the upper envelope on SONET speeds, as is shown from the figure. It is anticipated that SONET speeds may reach as high as 50 Gbps within 3 years.

   A recent innovation involving the telephone networks (obviously the largest WAN in the world) has been Integrated Services Digital Network (ISDN). ISDN came about because of a perceived need by the telephone com-

Figure 1–25
SONET Data Rates

Carrier	Signal	Rate in Mbps
OC-1	STS-1	51.840
OC-3	STS-3	155.520
OC-9	STS-9	466.560
OC-12	STS-12	622.080
OC-18	STS-18	933.120
OC-24	STS-24	1244.160
OC-36	STS-36	1866.240
OC-48	STS-48	2488.320
OC-192	STS-192	9953.280
OC-240	STS-240	12441.600

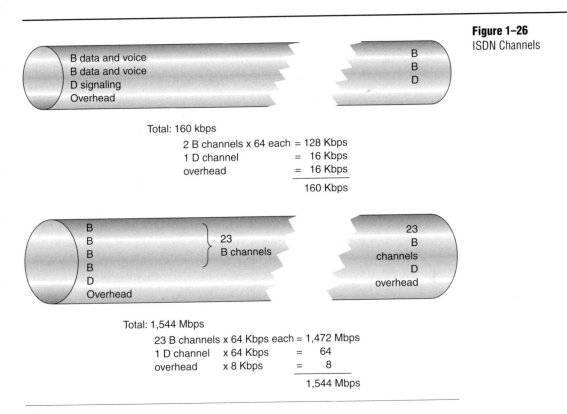

Figure 1–26
ISDN Channels

Total: 160 kbps

2 B channels x 64 each = 128 Kbps
1 D channel   = 16 Kbps
overhead   = 16 Kbps

160 Kbps

Total: 1,544 Mbps

23 B channels x 64 Kbps each = 1,472 Mbps
1 D channel   x 64 Kbps   =   64
overhead   x 8 Kbps   =   8

1,544 Mbps

panies for high-speed *totally digital* service. In 1984, under the auspices of CCITT, the telephone companies collectively created this new technology. The primary goal of ISDN is to integrate voice and nonvoice services. The ISDN consists of two tiers. The first tier, entitled basic rate interface (BRI), is sold as a package to many home and small-office users. BRI uses two B channels to transmit data and one D channel to handle control information. Each B channel is rated at 64 Kbps and the data channel is rated at 16 Kbps for a total bandwidth of 144 Kbps. There are also some overhead bits for a total of 160 Kbps.[3] BRI is actually targeted more toward home and small-business users as it allows the simultaneous use of voice and multiple data applications. The second tier is primary rate service. It consists of twenty-three B channels and one D channel for a total bandwidth of 1,488 Mbps. This service was targeted at high-bandwidth users such as those utilizing a PBX or local network. Figure 1–26 provides a graphical representation of ISDN channels.

---

[3] http://www.press.umich.edu/jep/works/AnaniaFlat.html.

Asynchronous Transfer Mode

Ansynchronous Transfer
Mode (ATM)

Ansynchronous Transfer Mode (ATM), an international standard developed by CCITT, has gained wide acceptance for network interoperability. The acceptance of ATM is related to several factors, most of which other technologies such as ISDN can also claim, such as how it handles data, voice, and video traffic and is rated at high speeds. Conversely, ATM is much faster, beginning at 155 Mbps and increasing in multiples beyond that rate. It also has several unique features that make it favorable among network administrators and the business community; for example, it is a transport method that uses multiple channels and emphasizes efficiency, quality of service, and high-capacity data transport.

Many networks must use separate media for voice, video, and data, because the transmission characteristics are different for each. Voice and video transmissions tend to be continuous streams of signals along the cable, and video signals can occupy large bandwidths. Data signals need less bandwidth but are transmitted in bursts. Because ATM can handle voice, video, and data on a single network medium, it represents a large potential savings in network resources. ATM can be used for both LAN and WAN communications, which eliminates the need for separate short- and long-distance networks (Palmer 2000). Because of these factors, ATM is slowly supplanting other networking technologies, such as frame relay, in the marketplace.

ATM is a streamlined packet transfer–type interface and makes use of fixed-size packets called cells, which allow for faster software processing. These cells are only 53 bytes long (hence the term *streamlined*), of which 5 bytes represent the header and the rest are data. It should be noted that ATM is a star topology network in that one or more interconnected switches form a central hub to which all computers then connect.

Switched Multimegabit Data Service

Switched Multimegabit Data Service (SMDS) is a combination of X.25 and ATM. SMDS was designed by Bell Communications Research over a period of several years, and was released as a product in 1991. It is similar in that it uses cells like ATM and creates virtual circuits like X.25. SMDS is usually offered by long-distance carriers and is explicitly designed to carry data, not voice. It is rated at speeds ranging from 1.544 Mbps to 45 Mbps. Its speed places it at a unique performance position of being faster than frame relay, yet slower than ATM.

Asymmetric Digital Subscriber Line

Asymmetric Digital Subscriber Line (ADSL), developed in the mid-1990s, is slowly replacing ISDN. ADSL is a networking technology that can use existing untwisted or twisted-pair cables for distances up to 4,000 meters. This technology utilizes the existing telephone subscriber lines in a manner similar to ISDN, because of the physics of the cabling involved. Although the telephone voice channel is limited to a bandwidth of 3 kilohertz (kHz), the twisted-pair cable itself, which connects to the central office, has a possible bandwidth of more than 1 megahertz (MHz). This is in theory, however, as noise and attenuation take their toll as well. Some implementations of ADSL have had to resort to replacing large amounts of telephone cabling because of deterioration from the sun, precipitation, and wind. ADSL service that is now offered by the telephone companies can provide up to 1.5 Mbps or more downstream (to the home) and up to 1.5 Mbps upstream (from the home), in addition to regular analog telephone service. It is estimated that 66 percent of subscriber loops in the United States can support the ADSL technology. ADSL is currently being implemented primarily in larger metropolitan areas because of its return on investment considerations.

Local and Metropolitan Area Advances

Research has brought about significant bandwidth increases in the local and metropolitan network arena as well. In the early 1970s, IBM's Token Ring technology was introduced at 4 Mbps. This was soon followed by the Xerox Ethernet product at 10 Mbps. These products subsequently became international standards known as IEEE 802.5 and 802.3, respectively. A newer token ring standard arrived in the 1980s that increased Token Ring Network speeds to 16 Mbps and in the early 1990s, Fast Ethernet arrived, thereby increasing Ethernet LAN speeds from 10 Mbps to 100 Mbps. This was standardized as IEEE 802.3u, also known as 100BASE-X.

Fiber Distributed Data Interface

In addition, as fiber optics became a viable, high-speed technology in the mid-1980s, the Fiber Distributed Data Interface (FDDI) standard was developed. FDDI is rated at 100 Mbps and is a dramatic improvement over earlier Token Ring and Ethernet technologies. FDDI is a token-passing ring network very similar to Token Ring, but it allows several computers to transmit simultaneously. It performs well under a heavy traffic load and allows up to 500 nodes on the network. Two network media are specified: fiber optics and category 5 UTP.

It is typical to find a FDDI two-ring configuration like the one shown in Figure 1–27. These are counterrotating because data flows around the second ring opposite of the direction data flows around the primary ring. Note also that FDDI and Fast Ethernet vie for basically the same tier of business, namely 100 Mbps.

Gigabit Ethernet

A newer arrival, Gigabit Ethernet, is being deployed, with a new standard (IEEE 802.3x). This new technology is rated at 1 Gbps over category 5 or 6 copper cabling.

Figure 1–27
Typical Dual Ring FDDI
Configuration

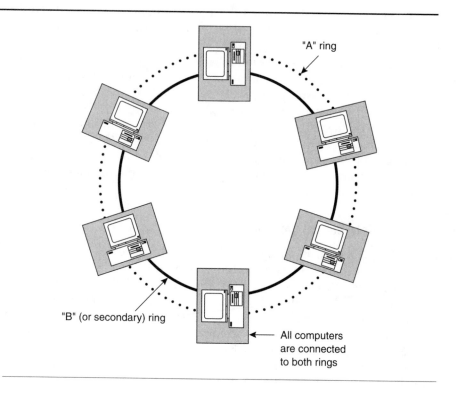

"A" ring

"B" (or secondary) ring

All computers
are connected
to both rings

One major advantage of Gigabit Ethernet is its ability to run over Ethernet networks with little modifications needed. A user simply replaces the network interface cards, upgrades the connectivity devices, and is ready to use Gigabit Ethernet. The standards committee deliberately made Gigabit Ethernet easy to use so it would not require massive infrastructure improvements to be deployed.

It is unclear at this time how the market will respond to Gigabit Ethernet as it competes in the data throughput range of ATM.

Wireless Networking

Wireless communications is certainly the fastest growing segment of the communications industry (Feher 1995). Cellular phones, cordless phones, a host of mobile services, and paging services have experienced incredible growth, especially in the 1990s. In fact, the business community is utilizing wireless communications as a crucial productivity and critical mission tool.

It is certainly true that in today's society, wireless communications allows us to live and work in ways never before possible. Wireless technology has much to offer, particularly flexibility and mobility. Anyone who utilizes cordless or cellular phones, pagers, TV remote controls, keyless car entry, or garage door openers will readily agree. Although telecommunications and computer networking have increased our options on how and what we communicate, we are still physically constrained by physical wire to the network. **Wireless networking** brings us back to a form of communications that is inherently natural to us: as humans we do not like to be physically constrained.

Wireless communications today span an array of different technologies, including cellular, personal communications service (PCS), satellite, infrared, wireless data WANs, specialized mobile radio, global positioning system (GPS), and wireless LANs. The remainder of this textbook will address these concepts in more detail.

■ SUMMARY

Chapter 1 introduces networking basics, from both a local and a wide area networking view. The evolution and current state of the Internet is discussed, as the Internet greatly affects our lives, both at home and at the workplace. We discuss its infant beginnings in the Department of Defense arena to the tremendous growth currently being experienced.

We also discuss the basis upon which all networking protocols are compared, the OSI reference model. The model is described and a sample transaction followed, using this model. In addition, we map several popular networking technologies against the model, including the TCP/IP and ATM models.

For familiarization purposes, we address several network issues such as network types, the underlying media, and the associated topologies. In addition, we discuss several connectivity devices so the reader can grasp how the various networking elements interoperate.

A brief history of the telephone network is presented as well as several popular protocols such as ATM and ISDN.

Finally, various wireless technologies are discussed. The role and characteristics of both wireless LANs and satellite technologies prepare the foundation for the more intensive coverage of wireless technology to follow.

REVIEW QUESTIONS

1. A recent study on the number of pages of unique information on the World Wide Web estimates there now are over _____ pages.
   a. 1 million
   b. 10 million
   c. 100 million
   d. 1 billion

2. The OSI reference model was created by and is a _____ standard.
   a. CCITT de jure
   b. ARPA de facto
   c. ISO de jure
   d. IEEE de jure

3. The Internet's beginnings lie in _____.
   a. ANSNET
   b. ARPANET
   c. ISO
   d. U.S. Department of Commerce

4. The OSI reference model has _____ layers, whereas the TCP/IP reference model has _____ layers.
   a. six, five
   b. seven, four
   c. three, four
   d. four, five

5. Subnet routing is a function of the _____ layer of the OSI reference model.
   a. session
   b. physical
   c. network
   d. application

6. Electronic mail services are found at the _____ layer of the OSI reference model.
   a. application
   b. presentation
   c. transport
   d. physical

7. The TCP/IP reference model has _____ layers.
   a. seven
   b. three
   c. five
   d. four

8. Ethernet is a _____ protocol.
   a. LAN
   b. MAN
   c. WAN
   d. XAN

9. _____ is the most prevalent form of cabling today.
   a. Coax
   b. STP
   c. Fiber optic
   d. UTP

10. The telephone wires are twisted in twisted-pair cabling because of _____.
    a. impedance
    b. crosstalk
    c. behavior modification
    d. 5-4-3 rule

11. _____ is the superior form of fiber optic cabling because of the superior attenuation factor.
    a. Single mode
    b. STP
    c. Multimode
    d. Superior grade

12. _____ are the more powerful light-emitting devices for fiber optic networks.
    a. Lasers
    b. Photon torpedoes
    c. LEDs
    d. Supercharged emitter

13. A major advantage of the star topology is that it _____.
    a. can be expanded easily
    b. uses less cabling
    c. has no single performance throttle point
    d. is difficult to diagnose problems

14. The _____ topology is difficult to manage.
    a. ring
    b. mesh
    c. bus
    d. star

15. The word "open" in the OSI reference model refers to _____.
    a. agreement on standards
    b. vendor-specific solutions
    c. government agreed-upon standards
    d. its approval by the ITU subcommittee of the United Nations

16. The duty of transmitting raw bits over a transmission line is at the _____ layer of the OSI reference model.
   a. physical
   b. data link
   c. network
   d. transport

17. _____ broadcast their information to everyone on the network.
   a. routers
   b. bridges
   c. repeaters
   d. hubs

18. There are now _____ wireless handsets in use worldwide.
   a. 1 million
   b. 100 million
   c. 300 million
   d. 1 billion

19. There are over _____ geosynchronous satellites in orbit today providing wireless services to many nations.
   a. 200
   b. 300
   c. 5,000
   d. 66

20. Mesh topologies are rarely seen because _____.
   a. they are difficult to manage
   b. they are more expensive
   c. they can only employ Ethernet
   d. a and b only

HANDS-ON EXERCISES

1. You are involved in a minor dispute with your professor. You recently submitted your ten-page assignment wherein you described the evolution of the current telephone network. Your professor seems to believe that this report is a bit skimpy and much too simplistic in its content, and he wants you to redo it, this time providing substantive details. Upon seeing your incredulous look, he feels pity on you and points you to the Internet. "Look on the Net and look up the evolution of the Internet and World Wide Web, and then you will see the complexity of this type of technological progress," he says.

   You are that student. Proceed on to the Internet and search. Report your findings in a short overview to your professor to assure him or her you are on the right track.

2. You are a summer intern at Complex Technologies, a local start-up computer firm. The firm is setting up its third office in as many months on the third floor. They already occupy the second and fourth floor, utilizing two 10BaseT LANs running NetBEUI. Being a frugal-minded firm with great aspirations, the administrators want to expand their LAN onto this floor with as little expense as necessary. What switching device, protocol, and media should they utilize? Use a drawing package such as Microsoft Paint, Corel Photo Paint, or Adobe Photoshop to produce a diagram of not only this new floor, but also how it connects to the other two floors.

3. Conduct some private research using the Internet and discover the differences, if any, between the old MILNET and NSFNET. What types of data did they permit users to see? Was there really a need for two separate networks? Were their missions different? Are they existing in any form today? Produce a listing in a word processor of your choice, with Internet sites you visited to locate this information. Is there a similarity in these sites?

4. Use of the Internet has exploded, particularly in the last 5 years. Search the Internet and produce a table describing this increase in graphical and text form using a drawing package such as Microsoft Paint, Corel Photo Paint, or Adobe Photoshop.

5. Produce a table, using Microsoft Word, that lists every connectivity device described in this chapter, and on which layer of the OSI model they operate and a terse description. Save this table for use in later chapters as we will be adding more information to it as we proceed.

6. Use any combination of tools you desire to create a web page that provides a one-page history of the evolution of the Internet. You may wish to use a tool such as Microsoft's Front Page, Macromedia's Dreamweaver, or simply use HTML tags to create it.

7. Research and then produce, using a graphics package, a picture of the seven-layer OSI reference model. List each layer and provide a one-line description of each.

8. Using your browser, go to www.ieee.org and learn about the 802 series of LAN standards. IEEE is the developer and holder of the detailed specifications that describe the concepts and operation of these standards.

9. Using graphic tools available on your system, develop a series of drawings where you link a network topology to its most appropriate purpose, using a fictional setting. For example, you may choose to draw a mesh topology linking systems on a military base (as the military occasionally selects the mesh topology for its increased fault tolerance).

10. Produce a several-page research paper discussing the purposes of, and differences between, the various standards bodies: ANSI, IEEE, OSI, and CCITT. For example, do they overlap in the areas in which they produce standards? Internet sites to explore for this include www.ieee.org, www.osi.org, www.ansi.org, www.cnet.com, and www.zdnet.com.

11. Trace the path of an e-mail message from one peer to another, using the OSI reference model as a foundation.

2 Introduction to Wireless Communications

OBJECTIVES

After reading this chapter and completing the exercises, you will be able to:

- Understand the role of wireless communications in telecommunications technology
- Understand the growth of the wireless market
- Understand the role wireless has in the global telecommunications market
- Understand wireless technologies including satellite, cellular telephony, microwave, paging systems, and the global positioning navigation system

WHY WIRELESS?

Wireless communications operate without a bounded media such as wires or fiber optic cable. As an **unbounded network,** wireless also operates as an unguided system. The transmission disperses through the air in all directions. Instead of relying on electrons running through wires (even though it requires electrical energy to create the transmission), wireless systems make use of radio waves. This unique means of transmission has evolved greatly since **Guglielmo Marconi** successfully initiated the first wireless telegraph transmission in 1895, but the concept has not changed. Marconi proved that electrical waves could be transmitted successfully at a considerable distance through the air. Jean Antoine de Nollet had already observed electric transmission without the use of wire in the eighteenth century. He succeeded in sending an electric discharge, without conducting wires, from one bank of the Seine River to the other. The water in the river carried the electrical charge. Since that time, wireless communication has developed at a feverish pitch as the range and power of wireless technology continues to increase.

Wireless technologies provide inherent benefits in many situations, both in traditional and nontraditional settings. For example, imagine an environmental monitoring station on a ship at sea or a forest ranger station in mountainous terrain. Both need connectivity with the outside world; however, deploying traditional wired technologies such as a copper-based or fiber optics communications line would provide an expensive solution to these communications problems and also may damage the environment. Wireless communications via a series of microwave towers or satellites provides a less expensive alternative for connectivity for these situations and many others.

This scenario plays out in many areas of the world where entire communities rely on wireless technology for communications needs, ranging from telephone to emergency services. In many of these circumstances, wireless networks may offer an advantage of reduced costs. The expense of acquiring the needed right-of-way from landowners, the labor-intensive process of digging trenches for cabling, and the continued maintenance of those cables provide significant obstacles to deploying traditional wired systems. In developed countries, a wire-based telephone network may reduce installation costs and concerns by running a great deal of cabling in railroad right-of-way or along highways, but expenses remain enormous. In most countries, the basic telephone infrastructure does not exist. Thus, wireless technology provides a cost-effective means for electronic communication, and is sometimes the only chance of obtaining such an infrastructure.

unbounded network

Guglielmo Marconi

TYPES AND RANGE OF WIRELESS COMMUNICATIONS

Types of Wireless Communications

Wireless technologies use appliances, or devices, to carry signals. The most prevalent technologies and their associated appliances are listed below. These appliances have proliferated to the extent that we consider them a necessity.

Technology	Appliance
Wireless PBX/LANS	Personal computers
Fixed satellite	Nonconsumer earth-based and satellite-based devices
Mobile satellite	Nonconsumer earth-based and satellite-based devices
Direct broadcast satellite	Consumer (home system/ receiver) and nonconsumer earth-based and satellite-based devices
Aeronautical/maritime (nonsatellite)	Commercial and governmental systems
Military mobile communications (terrestrial)	Governmental systems
Military mobile communications (satellite)	Governmental systems
Microwave relay and distribution systems	Nonconsumer devices
Noncellular telephony	Commercial devices
Wireless emergency/police	Commercial devices
Industrial wireless products (remote controls, etc.)	Commercial devices
Radio and television transmission networks	Consumer and nonconsumer devices

Range of Wireless Services

It is certainly true today that the range of wireless services continues to expand. The use of these services has increased at a phenomenal rate and the number of wireless applications introduced has increased as well. The early services and applications include the following:

- TV broadcast services
- Radio broadcasts (first AM and then FM)
- Microwave relay services
- Satellite communications services

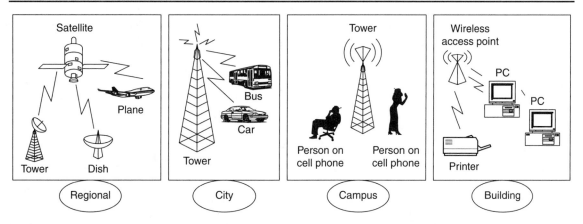

Figure 2–1
Range of Wireless Services

Figure 2-1 provides a pictorial view of the range of services. In the last 20 years, these services have improved dramatically and the devices used to communicate via the services have developed as well.

As an example, the following wireless services and products are now within the range of the wireless spectrum.

- Garage door openers
- Cordless phones
- Cordless keyboards and other peripheral devices
- Remote controls for television
- Walkie-talkies with dramatically enhanced clarity and distance
- Cellular phones (both analog and digital)
- Direct broadcast satellite services
- Wireless personal computer networking for the home
- Satellite-based cellular telephony
- Satellite-based global positioning system
- Automobile speed detection technology
- Wireless local area networking
- Paging services
- Wireless local loop (WLL)—wireless capability within the neighborhood
- Local multipoint distribution services (LMDS)—wireless local high-speed access for small businesses
- **Personal digital assistants (PDAs)**—small, calculator-sized information delivery system providing personal scheduling capabilities

personal digital assistants (PDAs)

Wireless systems have certain limitations. One is they must operate within the electromagnetic wave spectrum. Table 2-1 shows the wireless frequencies in this spectrum and the uses typically applied to them.

Table 2–1

Frequency Spectrum: Band Designations, Frequency Ranges, Wavelengths, and Examples of Communications Applications

Band Designation	Frequency Range*	Wavelength**	Usage
Audible	20 Hz–20 kHz	> 100 km	Acoustics
Extremely/very low frequency (ELF/VLF) radio	3 kHz–30 kHz	100 km–10 km	Navigation, weather, submarine communications
Low frequency (LF) radio	30 kHz–300 kHz	10 km–1 km	Navigation, maritime communications
Medium frequency (MF) radio	300 kHz–3 MHz	1 km–100 m	Navigation, AM radio
High frequency (HF) radio	3 MHz–30 MHz	100 m–10 m	Citizens band (CB) radio
Very high frequency (VHF) radio	30 MHz–300 MHz	10 m–1 m	Amateur (HAM) radio, VHF TV, FM radio
Ultra high frequency (UHF) radio	300 MHz–3 GHz	1 m–10 cm	Microwave, satellite, UHF TV, paging, cordless telephony, cellular and PCS telephony, wireless LAN
Super high frequency (SHF) radio	3 GHz–30 GHz	10 cm–1 cm	Microwave, satellite, wireless LAN
Extremely high frequency (EHF) radio	30 GHz–300 GHz	1 cm–0.1 mm	Microwave satellite
Infrared light	10^3–10^5 GHz	300 μ–3	LAN bridges, wireless LANs
Visible light	10^{13}–10^{15} GHz	1 μ–0.3 μ	Fiber optics
X-rays	10^{15}–10^{18} GHz	10^3 μ–10^7 μ	N/A
Gamma and cosmic rays	> 10^{18} GHz	< 10^7 μ	N/A

*k = Kilo = 1,000 (1 thousand)
*M = Mega = 1,000,000 (1 million)
*G = Giga = 1,000,000,000 (1 billion)
*T = Tera = 1,000,000,000,000 (1 trillion)
**cm = centimeter (1/100 meter)
**mm = millimeter (1/1,000 meter)
**μ = micron (1/1,000,000 meter)

GROWTH OF WIRELESS COMMUNICATIONS

The growth of wireless communications over the previous 10 years indicates a tripling of wireless products and services (Prohm 2000). Some market sectors such as cellular telephony will increase at rates approaching 100 percent every several years, whereas over-the-airwaves radio and television broadcasting and microwave relay services will increase at a steady rate (Dulaney 2000). In fact, the Gartner Group estimates that by 2004, at least 40 percent of business-to-consumer e-commerce transactions outside North America will initiate from wireless devices (Prohm 2000). Wireless firms also are pervasive in the communications sector. For example, the top ten U.S. mobile wireless firms are names easily recognizable to most people.

- Verizon Wireless
- Cingular Wireless
- AT&T Wireless
- Sprint PCS
- Nextel Communications
- Alltel
- VoiceStream Wireless
- Western Wireless
- Dobson Communications
- Powertel

Over-the-airwaves broadcasting will increase due to the advent and implementation of **high definition television (HDTV).** High-saturation markets such as New York City and selected areas in California now have access to HDTV. Secondary and third-tier markets such as Macon, Georgia, or Richmond, Virginia, will soon follow. This implementation of HDTV technology will require over $100 billion in infrastructure investments by the broadcasting industry, because it must replace and/or dramatically enhance transmission and storage facilities due to the bandwidth requirements of HDTV.

high definition television (HDTV)

An additional issue regarding how the wireless market will look in a few years is the impact of fiber optic market penetration into markets previously held by microwave relay services. The use of **low earth orbit (LEO)** satellites for cellular telephony purposes also may capture some of the previously held microwave relay market.

low earth orbit (LEO)

The number of wireless users has grown at an astounding rate, meaning that over 1 billion wireless appliances will be used globally by 2005. The Gartner Group (1999) estimates that by the year 2005, wireless will consume almost 40 percent of the total global telecommunications marketplace.

Table 2-2 provides projections for growth in market revenue of wireless services. It shows the overall annual wireless telecommunications market as of 1994, 2000, and as projected for 2005. As the table also shows, cellular telephony will continue its phenomenal increase in market share.

Table 2–2
Global Wireless Market

Wireless Market Segment	Size of Market in 1994 ($US)	Size of Market in 2000 ($US)	Size of Market in 2005 ($US)
Paging	< 1 billion	> 3 billion	> 10 billion
Personal digital assistants	N/A	< 1 billion	> 50 billion
Position determination/navigation	< 1 billion	> 5 billion	50 billion
Amateur radio	< 1 billion	> 1 billion	> 2 billion
Citizens band/walkie-talkies, etc.	< 1 billion	> 1 billion	> 1 billion
Cellular telephones	35 billion	> 200 billion	> 1 trillion
Specialized mobile radio	3 billion	7 billion	15 billion
Radio and television (transmission networks)	20 billion	26 billion	100 billion
Wireless PBX/LANS	2 billion	8 billion	80 billion
Fixed satellite	14 billion	23 billion	50 billion
Mobile satellite	2.5 billion	10 billion	40 billion
Direct broadcast satellite	1.5 billion	6 billion	30 billion
Aeronautical/maritime (nonsatellite)	2 billion	3 billion	4 billion
Military mobile communications (terrestrial)	10 billion	25 billion	35 billion
Military mobile communications (satellite)	7 billion	20 billion	50 billion
Microwave relay and distribution systems (noncellular telephony)	3.5 billion	6 billion	15 billion
Wireless emergency/police	6 billion	11 billion	20 billion
Industrial wireless products (e.g., remote controls)	< 1 billion	5 billion	50 billion
Total Market Size	113.5 billion	361 billion	1.602 trillion

Source: The Gartner Group, 2000, www.gartnergroup.com.

Factors Affecting Growth

personal communications system (PCS)

frequency

Regardless of whether wireless technologies such as AM radio originated many years ago or recently, such as the **personal communications system (PCS)** in the 1990s, these services have common features and considerations. They share a plethora of technical, operational, business, policy, and health considerations. These typically include an ever-pressing concern for **frequency** allocations, health radiation standards and impacts, electrical interference, attenuation, ghosting problems, **throughput,** reliability, system availability, and quality of service and/or reception.

Frequency is important for wireless communications. The frequency is a unique sliver, or slice, of the airwaves that can be used by a device specially suited to transmit and receive over that unique sliver or slice. It is crucial that devices using a sliver carefully observe the frequency over which they are programmed to operate, or they may cause interference with other devices operating in nearby frequencies.

Throughput is a measure of performance in telecommunications, usually indicated by the amount of data transmitted in a given amount of time. Typically, throughput is measured in terms of bits per second (bps). The typical throughput over a LAN is either 10 million or 100 million bits per second or 10 Mbps. Wide area networks of years ago began at 64,000 bits per second (Kbps) and now exceed 10 million bits per second (Gbps).

In addition, the following present obstacles to the rapid deployment of wireless technologies.

- Governments and/or private enterprises may choose not to deploy new or additional wireless technologies because of the potential health hazards from radio transmissions near humans, cattle, and so forth (GSM 2001). An important characteristic of omnidirectional microwave for people living close to a source is that the strength of exposure is squared as the distance from the source is halved.
- **Latency** or transmission delay in satellite communications, particularly in regard to geosychronous satellites, poses another problem. Geosynchronous satellites are positioned at an orbit that allows them to be over a spot on the earth's surface and remain there. To be positioned this way, geosychronous satellites must be placed high in orbit, typically over 22,000 miles. Therefore, communications to and from these satellites can be delayed due to the sheer distance involved.

- The performance gap between wired and wireless technologies, as seen in Figure 2-2, represents the difference in performance (i.e., speed) between wired and wireless technologies. For example, a wired LAN using Ethernet protocols operates at 100 Mbps. The recently approved wireless LAN standard, IEEE 802.11b, operates at 11 Mbps.
- There is an increasing need for more spectrum and frequency allocations for new and emerging technologies. The list of applications continues to grow, all vying for the limited wireless communications spectrum. Within each spectrum (see Table 2-1) there is only so much bandwidth available. Advanced compression algorithms are being deployed to alleviate this problem, but it is not likely to go away.
- The fiber optic cable market penetration into markets previously held by microwave relay services will also affect growth. Advanced fiber optic networks approach transmission speeds of 20 Gbps.

Figure 2–2
Performance Gap
between Wired and
Wireless Technologies

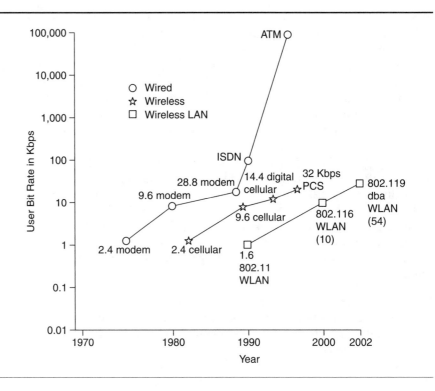

- The lack of cooperation in the wireless standards arena is another factor. In the cellular telephony market, there are over twenty competing cellular telephony systems worldwide, with adjoining European nations deploying different systems. See Figures 2-3 and 2-4.

The question now is whether these obstacles are truly significant in the sense of slowing the rapid expansion of wireless technologies. The answer, at least so far, is a firm NO. The demand simply is too strong for these forces to deter the expansion shown in Figures 2-3 and 2-4. The rate of growth in markets such as cellular telephony is nothing short of miraculous: It was not projected by many, if any, even 5 years ago; and nothing suggests that the rate will slow anytime soon (see Figure 2-5). New markets are merging, and market saturation in existing markets is still immature. In addition, newer developments, such as hybrid networks comprising both bounded and unbounded (i.e., wireless) media, are rapidly increasing.

Analog Cellular Technologies	
AMPS	Advanced Mobile Phone System. Developed by Bell Labs in the 1970s and first used commercially in the United States in 1983. It operates in the 800 MHz band and is currently the world's largest cellular standard.
C-450	Installed in South Africa during the 1980s. Uses 450 Mhz band. Much like C-Netz. Now known as Motorphone and run by Vodacom SA.
C-Netz	Older cellular technology found mainly in Germany and Austria. Uses 450 MHz.
Comvik	Launched in Sweden in August 1981 by the Comvik network.
N-AMPS	Narrowband Advanced Mobile Phone System. Developed by Motorola as an interim technology between analog and digital. It has some three times greater capacity than AMPS and operates in the 800 MHz range.
NMT450	Nordic Mobile Telephones/450. Developed specially by Ericsson and Nokia to service the rugged terrain that characterizes the Nordic countries. Range 25 km. Operates at 450 MHz. Uses FDD FDMA.
NMT900	Nordic Mobile Telephones/900. The 900 MHz upgrade to NMT 450 developed by the Nordic countries to accommodate higher capacities and hand-held portables. Range 25 km. Uses FDD FDMA technology.
NMT-F	French version of NMT900.
NTT	Nippon Telegraph and Telephone. The old Japanese analog standard. A high-capacity version is called HICAP.
RC2000	Radiocom 2000. French system launched November 1985.
TACS	Total Access Communications System. Developed by Motorola, and is similar to AMPS. It was first used in the United Kingdom in 1985, although in Japan it is called JTAC. It operates in the 900 MHz frequency range.

Figure 2–3
Worldwide Cellular Systems

Digital Cellular Technologies	
A1-Net	Austrian Name for GSM 900 networks.
B-CDMA	Broadband CDMA. Now known as W-CDMA. To be used in UMTS.
Composite CDMA/TDMA	Wireless technology that uses both CDMA and TDMA. For large-cell licensed band and small-cell unlicensed band applications. Uses CDMA between cells and TDMA within cells. Based on Omnipoint technology.
CDMA	Code Division Multiple Access. There are now a number of variations of CDMA, in addition to the original Qualcomm-invented N-CDMA (originally just CDMA, also known in the United States as IS-95. Latest variations are B-CDMA, W-CDMA, and composite CDMA/TDMA. Developed originally by Qualcomm, CDMA is characterized by high capacity and small cell radius, employing spread-spectrum technology and a special coding scheme. It was adopted by the Telecommunications Industry Association (TIA) in 1993. The first CDMA-based networks are now operational. B-CDMA is the basis for 3G UMTS (see below).
cdmaOne	First Generation Narrowband CDMA (IS-95). See above.
cdma2000	The new second-generation CDMA MoU spec for inclusion in IMT-2000. It consists of various iterations, including 1xEV, 1XEVDO, and MC 3X.
cdma2000 1XEV	1xEV (Evolution) is an enhancement of the cdma2000 standard of the Telecommunications Industry Association (TIA). The CDMA 1xEV specification was developed by the Third Generation Partnership Project 2 (3GPP2), a partnership consisting of five telecommunications standards bodies: CWTS in China, ARIB and TTC in Japan, TTA in Korea, and TIA in North America. The 1xEV specification is known as TIA/EIA/IS-856 "CDMA2000 High Rate Packet Data Air Interface Specification." It promises around 300 Kbps speeds on a 1.25 MHz channel.
cdma2000 1XEV-DO	1xEV-DO or data-only is an enhancement of the cdma2000 1X standard. It promises around 300 Kbps speeds on a 1.25 MHz channel.
CT-2	A second generation digital cordless telephone standard. CT2 has 40 carriers x 1 duplex bearer per carrier = 40 voice channels.
CT-3	A third generation digital cordless telephone, which is very similar and a precursor to DECT.
CTS	GSM Cordless Telephone System. In the home environment, GSM-CTS phones communicate with a CTS Home Base Station (HBS), which offers perfect indoor radio coverage. The CTS-HBS hooks up to the fixed network and offers the best of the fixed and mobile worlds: low cost and high quality from the Public Switched Telephone Network (PSTN), services and mobility from the GSM.

Figure 2–3
Continued

D-AMPS (IS-54)	Digital AMPS, a variation of AMPS. Uses 3-timeslot variation of TDMA, also known as IS-54. An upgrade to the analog AMPS. Designed to address the problem of using existing channels more efficiently, DAMPS (IS-54) employs the same 30 kHz channel spacing and frequency bands (824-849 and 869-894 MHz) as AMPS. By using TDMA instead of FDMA, IS-54 increases the number of users from 1 to 3 per channel (up to 10 with enhanced TDMA). An AMPS/D-AMPS infrastructure can support use of either analog AMPS phone or digital D-AMPS phones. This is because the Federal Communications Commission mandated only that digital cellular in the United States must act in a dual-mode capacity with analog. Both operate in the 800 MHz band.
DCS 1800	Digital Cordless Standard. Now known as GSM 1800. GSM operated in the 1,800 MHz range. It is a different frequency version of GSM, and (900 MHz) GSM phones cannot be used on DCS 1800 networks unless they are dual band.
DECT	Digital European Cordless Telephone. Uses 12-timeslot TDMA. This started off as Ericsson's CT-3, but developed into ETSI's Digital European Cordless Standard. It is intended to be a far more flexible standard than the CT2 standard, in that it has more RF channels (10 RF carriers x 12 duplex bearers per carrier = 120 duplex voice channels). It also has a better multimedia performance since 32kbit/s bearers can be concatenated. Ericsson has developed a dual GSM/DECT handset.
EDGE	UWC-136, the next generation of data heading toward third generation and personal multimedia environments builds on GPRS and is known as Enhanced Data rate for GSM Evolution (EDGE). It will allow GSM operators to use existing GSM radio bands to offer wireless multimedia IP-based services and applications at theoretical maximum speeds of 384 Kbps with a bit-rate of 48 Kbps per timeslot and up to 69.2 Kbps per timeslot in good radio conditions.
E-Netz	The German name for GSM 1800 networks.
Flash-OFDM	Flash-OFDM (Orthogonal Frequency Division Multiplexing) is a new signal processing scheme from Lucent/Flarion that will support high data rates at very low packet and delay losses, also known as latencies, over a distributed all-IP wireless network. The low-latency will enable real-time mobile interactive and multimedia applications. It promises to deliver higher quality wireless service and better cost effectiveness than current wireless data technologies.
FDMA	Frequency Division Multiple Access.
GERAN	GERAN is a term used to describe a GSM and EDGE (Enhanced Data rates for GSM Evolution)-based 200 kHz radio access network. The GERAN is based on GSM/EDGE Release 99, and covers all new features for GSM Release 2000 and subsequent releases, with full backward compatibility to previous releases.
GMSS	Geostationary Mobile Satellite Standard, a satellite air interface standard developed from GSM and formed by Ericsson, Lockheed Martin, U.K. Matra Marconi Space and satellite operators Asia Cellular Satellite and Euro-African Satellite Telecommunications.
GSM	Global System for Mobile Communications. The first European digital standard, developed to establish cellular compatibility throughout Europe. Its success has spread to all parts of the world and over 80 GSM networks are now operational. It operates at 900 MHz.

Figure 2–3
Continued

iDEN®	iDEN® (Integrated Digital Enhanced Network). Launched by Motorola in 1994, this is a Private Mobile radio system from Motorola's Land Mobile Products Sector (LMPS) iDEN technology, currently available in the 800 MHz, 900 MHz, and 1.5 GHz bands. It utilizes a variety of advanced technologies, including state-of-the-art vocoders, M16QAM modulation, and TDMA (Time Division Multiple Access). It allows Commercial Mobile Radio Service (CMRS) operators to maximize the dispatch capacity and provides the flexibility to add optional services such as full-duplex telephone interconnect, alphanumeric paging, and data/fax communication services.
iMODE	Launched in February 1999, this fast-growing system from NTT DoCoMo uses compact HTML to provide WAP-like content to iMODE phones.
IMT DS	Wideband CDMA, or WCDMA.
IMT MC	Widely known as cdma2000 and consisting of the 1X and 3X components.
IMT SC	Called UWC-136 and widely known as EDGE.
IMT TC	Called UTRA TDD or TD-SCDMA.
IMTFT	Well-known as DECT.
Inmarsat	International Martime Satellite System which uses a number of GEO satellites. Available as Inmarsat A,B,C, and M.
Iridium	Mobile Satellite phone/pager network launched November 1998. Uses TDMA for inter-satellite links. Uses 2 GHz band.
IS-54	TDMA-based technology used by the D-AMPS system at 800 MHz.
IS-95	CDMA-based technology used at 800 MHz.
IS-136	TDMA-based technology.
JS-008	CDMA-based standard for 1,900 MHz.
N-CDMA	Narrowband Code Division Multiple Access, or plain old original CDMA. Also known in the United States as IS-95. Developed by Qualcomm and characterized by high capacity and small cell radius. Has a 1.25 MHz spread spectrum air interface. It uses the same frequency bands as AMPS and supports AMPS operation, employing spread-spectrum technology and a special coding scheme. It was adopted by the Telecommunications Industry Association (TIA) in 1993. The first CDMA-based networks are now operational.
PACS-TDMA	An 8-timeslot TDMA-based standard, primarily for pedestrian use. Derived from Bellcore's wireless access spec for licensed band applications. Motorola supported.
PCS	Personal Communications Service. The PCS frequency band is 1,850 to 1,990 MHz, which encompasses a wide range of new digital cellular standards like N-CDMA and GSM 1900. Single-band GSM 900 phones cannot be used on PCS networks. PCS networks operate throughout North America.

Figure 2–3
Continued

PDC	Personal Digital Cellular is a TDMA-based Japanese standard operating in the 800 and 1,500 MHz bands.
PHS	Personal Handy System. A TDD TDMA Japanese-centric system that offers high speed data services and superb voice clarity. Really a WLL system with only 300 m to 3 km coverage.
SDMA	Space Division Multiple Access, thought of as a component of Third Generation Digital Cellular/UMTS.
TDMA	Time Division Multiple Access. The first U.S. digital standard to be developed. It was adopted by the TIA in 1992. The first TDMA commercial system began in 1993. A number of variations exist.
Telecentre-H	A proprietary WLL system by Krone. Range of 30 km, in the 350-500 MHz and 800-1,000 MHz range. Uses FDD FDM/FDMA and TDM/TDMA technologies.
TETRA	**TE**rrestrial **T**runked **RA**dio (**TETRA**) is a new open digital trunked radio standard which is defined by the European Telecommunications Standardization Institute (ETSI) to meet the needs of the most demanding professional mobile radio users.
TETRA-POL	Proprietary **TETRA** network from Matra and AEG. Does not conform to TETRA MoU specifications.
UltraPhone 110	A proprietary WLL system by IDC. Range of 30 km, in the 350-500 MHz range. Uses FDD FDM/TDMA technologies. The UltraPhone system allows 4 conversations to operate simultaneously on every 25 kMHz-spaced channel. A typical UP 24-channel WLL system can support 95 full duplex voice circuits in 1.2 kHz of spectrum.
UMTS	Universal Mobile Telephone Standard—the next generation of global cellular which should be in place by 2004. Proposed data rates of <2 Mbps, using combination TDMA and W-CDMA. Operates at around 2 GHz.
W-CDMA	One of the latest components of UMTS, along with TDMA & cdma2000. It has a 5 MHz air interface and is the basis of higher-bandwidth data rates.
WLL	Wireless Local Loop limited-number systems are usually found in remote areas where fixed-line usage is impossible. Most modern WLL systems use CDMA technology.

Figure 2–3
Continued

Figure 2–4
Market Shares of Major
Cellular Systems

	GSM	PDC	CDMA	TDMA	GSM-1900
Sep-92					
Dec-92	201,500		1,500		
Mar-93	394,500		3,000		
Jun-93	589,720		24,000		
Sep-93	838,220		48,500	5,000	
Dec-93	1,362,990		78,150	35,000	
Mar-94	1,889,790		154,200	55,000	
Jun-94	2,592,530		276,600	122,000	
Sep-94	3,393,030		396,000	183,500	
Dec-94	4,628,790		565,600	378,000	
Mar-95	5,665,900		800,000	484,000	
Jun-95	7,951,000	1,500,000	0	1,100,000	0
Sep-95	9,884,000	2,100,000	0	1,300,000	0
Dec-95	13,034,000	3,108,000	9,000	2,055,000	40,000
Mar-96	16,343,000	5,227,000	11,300	2,750,000	60,000
Jun-96	21,148,000	7,672,000	180,000	4,500,000	105,000
Sep-96	26,150,000	10,592,500	350,000	6,000,000	150,000
Dec-96	32,878,500	13,920,000	987,000	2,700,000	301,500
Mar-97	40,200,000	16,000,000	1,100,000	3,200,000	500,000
Jun-97	48,900,000	19,000,000	2,500,000	4,000,000	700,000
Sep-97	58,145,570	23,619,000	4,300,000	5,500,000	955,570
Dec-97	71,359,000	26,772,000	5,980,000	6,900,000	1,331,000
Mar-98	83,557,000	30,074,000	10,900,000	8,000,000	1,691,000
Jun-98	98,858,230	33,006,000	12,900,000	10,928,450	2,123,000
Sep-98	114,901,660	35,728,500	16,704,640	13,976,430	2,545,000
Dec-98	138m	38m	23m	18m	2.8m
Mar-99	161m	41m	26m	21m	3.1m
Jun-99	188m	42m	29m	26m	3.3m
Mar-01	460m	65m	34m	61m	7m

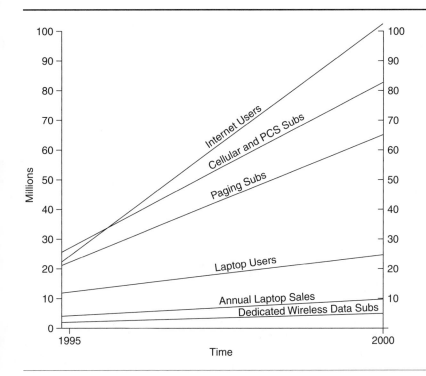

Figure 2–5
Growth of Wireless Services

WIRELESS TECHNOLOGIES

The various wireless technologies are each customized for a specific purpose. Cellular telephony systems are **omnidirectional** in design (broadcast in every direction at once) permitting service for many different users within a specified area. Microwaves involved unidirectional or point-to-point technology. Unidirectional wireless technologies transmit only in one direction at a time. A brief discussion of each of these wireless technologies follows, with more detail appearing later in this text.

omnidirectional

Microwave Technology

Microwave systems are in use every day. For example, they relay data between cellular telephony towers. The cellular telephony towers operate in an omnidirectional fashion while they service all users within their coverage area. Each tower has a microwave system used for transferring data to and from other towers. Microwave transmits within the radio frequencies (RF).

Design and Principles of Operation
Microwave wireless systems are a form of radio transmission and use ultra high frequencies as seen in Table 2-3. This table lists the frequency bands

Table 2–3
Example Microwave
Frequency Bands
(United States)

Frequency Bands	Maximum Antennae Separation	Analog/Digital
4–6 GHz	20–30 miles (32–48 km)	Analog
10–12 GHz	10–15 miles (16–24 km)	Digital
18–23 GHz	5–7 miles (8–11 km)	Digital

set aside by the Federal Communications Commission for conventional microwave. This use of ultra high frequencies makes microwave transmissions harmful to living tissue (Khoundary, 2000, OSHA, 2002). The specific proven harmful effect from exposure to microwaves stems from thermal radiation. RF radiation can enter deep into the body and heat human organs. Therefore, persons in areas containing microwave transmissions must exercise care. Yearly, there are deaths and serious injuries reported because people are too close to high-intensity microwave transmissions (Khoundary 2000).

Microwave transmission systems have been used for over 30 years and continue to be used widely even though microwaves have some inherent physical property limitations. For example, microwaves are highly susceptible to attenuation problems. **Attenuation** is an event during which signal loss occurs due to several factors, including distance and resistance. Repeaters reduce the problem of attenuation. A **repeater** receives a signal on an electromagnetic or optical transmission medium, amplifies the signal, and then retransmits it. Because microwaves are particularly susceptible to attenuation, repeaters are an important part of the decision-making process when selecting and deploying microwave equipment. This extra equipment adds to the cost of microwave service which must be considered when deciding whether to use microwave.

To maximize the strength of microwave transmissions, the microwave beams are tightly focused. Otherwise, due to attenuation, microwave towers would be a necessity every 2 to 3 miles. This highly focused approach allows placement of microwave towers approximately every 30 miles. This makes microwave a **point-to-point communications** technology as opposed to a broadcast technology such as AM or FM radio, which disperses signals in all directions. In addition, microwave is a line-of-sight (LOS) technology because the microwaves are not able to pass through solid objects beyond a distance of a few inches. Each sender and receiver must be able to see each other. Figure 2-6 illustrates this concept. Note the graphical representation of the necessity for line of sight between both towers, or microwave connectivity cannot be secured. The distance limitation of 30 miles between towers is necessary because of the microwave's inability to pass through solid objects such as the earth. Figure 2-7 presents this issue of the earth obstructing con-

attenuation

repeater

point-to-point communications

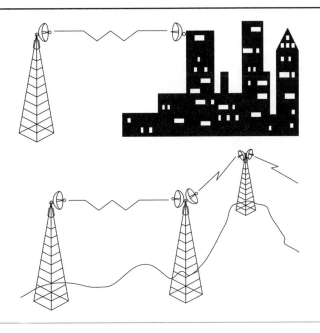

Figure 2–6
Point-to-Point
Microwave

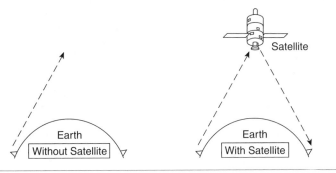

Figure 2–7
The Earth's Curvature
and Line of Sight

nectivity due to its natural curvature. Again, each sender and receiver in a microwave network must be in direct line of sight to one another or connectivity is not possible.

The two types of microwave systems are terrestrial and satellite. Terrestrial microwave systems are primarily used for long-distance telecommunications service. They also carry voice and TV transmission. Terrestrial microwave typically employs the use of a parabolic dish. The parabolic dish focuses a narrow, high-intensity beam in a line-of-sight fashion toward its intended receiving dish, as seen in Figure 2-8.

Figure 2–8
Parabolic Dish

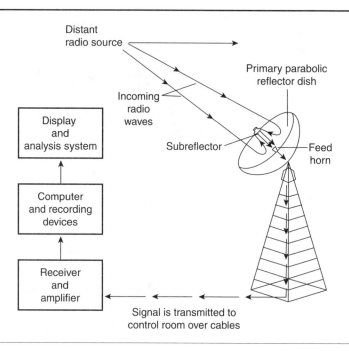

Satellite microwave systems have a dish mounted on a satellite. The satellite serves as a relay platform and is linked with ground-based microwave stations (also known as earth stations). This process can be described as follows: The satellite receives transmissions on a selected frequency band, amplifies this signal, then retransmits it to a ground-based station on another selected frequency. The primary applications of this technology are TV distribution, private networks, and long-distance telephone transmission.

Strengths

Microwave technology has two major advantages over copper or fiber optic cables.

- Microwave technology requires fewer amplifiers and repeaters than copper or fiber optic cable. Thus, microwave communication can cost less because amplifiers and repeaters placed every few miles can prove to be less expensive. This is particularly evident in rough terrain or metropolitan areas where space is at a premium.
- Microwave technology can be transmitted at a more rapid rate than most copper. Microwave transmission rates may exceed 5 Gbps, whereas most copper-based networks transmit at up to 1 Gbps. Fiber-based networks only recently have exceeded microwave speeds.

Weaknesses

As mentioned, microwave technology requires line-of-sight communications. This aspect can present problems in highly congested places such as major metropolitan areas or rapidly developing locations where building construction is frequent. Buildings that can block the necessary line of sight may rise after installation of the microwave system. In addition, the curvature of the earth limits the distance between microwave towers needed to receive the line-of-sight communications.

Cellular Telephone Systems

Cellular telephony systems are commonplace worldwide. Millions of users join a cellular provider each year and the service providers provide connectivity for the users. Sometimes these providers are the firms directly owning the cellular equipment such as Cingular, or providers such as PowerTel that execute corporate agreements with infrastructure providers. They provide a middleman arrangement where they "lease" equipment from cellular firms that own it and then service the cellular user community. For example, over twenty cellular network providers offer service in the Atlanta, Georgia, area.

Design and Principles of Operation

Cellular telephone systems provide two-way voice communications among moving and/or stationary targets and have widespread coverage of up to 5 miles. Originally, cellular systems operated from mobile terminals inside vehicles with large antennas mounted on the vehicle roof. Today these systems have rapidly evolved to support lightweight hand-held mobile terminals operating inside and outside buildings at both pedestrian and vehicle speeds. As recently as 5 years ago, the typical cellular phone was referred to as a "bag phone" and was roughly the same size as a stationary in-house telephone. Cellular phones are now the size of a small calculator.

A cellular telephony network consists of multiple, low-power transmit/receive antennae that are distributed throughout a selected geographical area, as shown in Figure 2-9. The geographical area is divided into hexagonal **cells.** Each cell has an antenna that governs communications within that cell **cells** and includes accepting handoffs from those users entering and leaving the cell. Actually, the coverage area of an individual cell does overlap adjoining cells. Each cell typically has a coverage area spanning 1 to 5 miles in diameter. Major metropolitan areas usually have smaller cells. These areas need more antennae because the proximity of tall buildings limits the effective distance and clarity of cellular telephony. In addition, cellular traffic density limits the effectiveness of cellular telephony networks in major metropolitan areas.

Figure 2–9
Typical Analog Cellular
Design

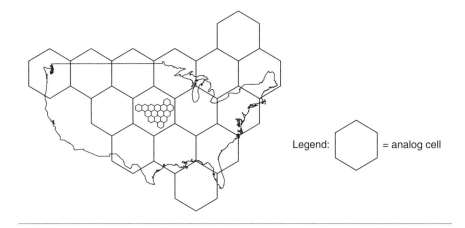

Legend: = analog cell

To provide smooth communications among cells, cellular transmitters perform handoffs. As the cellular telephone user moves out of one cell area and into another, the first cell passes off the transmission to the adjoining cell. Unfortunately, in the earlier days of cellular telephony, users could discern several seconds of uncomfortable silence, or static, during this handoff. Today, due to improvements in speed in switching technology, handoffs are no longer evident.

Cellular transmission towers have several synonymous names, including:

- Mobile traffic switching offices (MTSOs)
- Mobile traffic switching exchanges (MTSXs)
- Mobile switching centers (MSCs)

mobile traffic switching office (MTSO)

These terms identify the same capability, but the names differ depending on the firm installing them. For simplicity purposes, we will use the term **mobile traffic switching office (MTSO).** These MTSOs link each other and connect to the Public Switching Telephone Network (PSTN) through either microwave or leased-line facilities, as shown in Figure 2-10.

Several standards exist in the cellular telephony network. They range from analog to digital, with digital being predominant due to the advantages of clarity (clearer signal), performance (higher transmission rates), and cost. These advantages will be discussed in depth later.

The following cellular standards are currently used.

Advanced Mobile Phone System (AMPS)

- **Advanced Mobile Phone System (AMPS)** is used primarily in the United States. It operates in analog transmission mode and was the first cellular telephony standard adopted by any country. AMPS does not handle data well, as it was designed to be a voice-carrying system

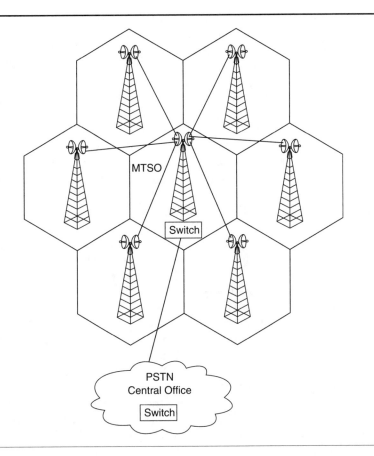

Figure 2–10
MTSOs Connected the PSTN

and it operates at only 6,800 bps. AMPS is widely deployed in the United States, Australia, the Philippines, and several other countries. The market share for AMPS continues to decline as digital systems become more prevalent. As an example of this trend toward digital, the Australian government mandated a switchover to digital beginning December 31, 1999. This transition period is expected to take approximately 12 to 24 months to complete (Budde 2001). Expansion of cellular telephony is rapid in Australia as can be evidenced by the recent announcement of yet another nationwide mobile phone network. The digital network will cover areas outside of Sydney and Melbourne, although roaming agreements are in place to allow users total coverage of Australia. Coverage included 7.7 million people in 2001 and is expected to reach 9.3 million by 2010 (Budde 2001).

- **Total Access Communications System (TACS)** is an analog system similar to AMPS. This system is used in the United Kingdom and provides

Total Access Communications System (TACS)

twice as much capacity as AMPS. TACS was the first analog network system launched in Europe; it was launched jointly by the European firms Vodafone and Cellnet. TACS has attempted to expand its market base, but generally is considered just a stopgap to true digital cellular networks (TACS Analogue Systems 2001).

Nordic Mobile Telephone (NMT)

- **Nordic Mobile Telephone (NMT)** is a cellular telephony system deployed in the Scandinavian countries. NMT is optimized for use in areas of low population density (it operates in the somewhat unique 450 MHz range), which is ideal for this area of Europe. A version of NMT called NMT900 has been developed for higher population density areas, but is also considered just a temporary technology (Forschungszentrum Julich 2001).

Personal Digital Cellular (PDC)

- **Personal Digital Cellular (PDC)** is used in Japan and is a digital cellular technology (Nokia 2001). PDC is based on TDMA technology.[1] If you have a European standards-based phone or a wireless technology from America, for instance, your phone will not work in Japan (Wireless Rentals 2001). In fact, Japan has the second largest mobile telephone network in the world with over 62 million subscribers (Eurotechnology.com 2001).

Digital-AMPS (D-AMPS)

- **Digital-AMPS (D-AMPS)** is also known as US TDMA. It is a North American digital cellular standard. Paradoxically, it operates in the same frequency as the earlier AMPS networks and can coexist with an analog-based AMPS network. D-AMPS supports many more channels (i.e., more calls can be handled) than AMPS and is faster than AMPS in its transmission speed for voice traffic. TDMA supports data communications at speeds up to 28.8 Kbps. Group 3 facsimile (the latest, most efficient fax capability) can also be supported in D-AMPS. Figure 2-11 shows a typical digital cell layout.

Global System for Mobile Communications (GSM)

- **Global System for Mobile Communications (GSM)** is essentially the standard for European cellular systems. An interesting characteristic of GSM is the use of a subscriber identification module (SIM). The SIM plugs into a card slot in the GSM handset and provides additional security protection. As hackers become more active in this arena, this security protection rises in importance. GSM was developed to be the international digital cellular system of choice. Currently, over 100 countries use the GSM standard. GSM dominates the European market, and much of the Asian market, and actually supports full roaming privileges throughout these countries. GSM is not compatible with the United States' dominant digital standard, the PCS.

Personal Communications System (PCS)

- **Personal Communications System (PCS)** uses digital communication and is predominant in the United States. Due to advanced technology features, PCS can exist in the same network as AMPS. PCS can oper-

---

[1] See www.nokia.com.

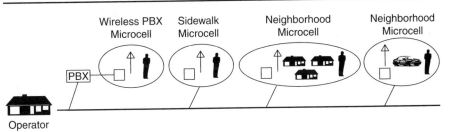

Figure 2–11
Typical Digital Microcell
Structure

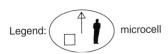

ate in both analog and digital modes and can operate with AMPS even though AMPS uses analog communications. PCS also offers improved bandwidth utilization compared with AMPS at a 15-to-1 advantage. It also supports both voice and data at an aggregate channel bandwidth of 1.152 Mbps. Finally, PCS provides additional enhancements such as soft handoff technology (a more efficient method of handing off calls between two MTSOs), the capability of effectively deploying smaller cells which is excellent for frequency reuse, and improved phone battery times due to precise power control mechanisms. Figure 2-12 shows the growth of PCS.

Cellular systems permit **frequency reuse,** which means that cellular systems use the same frequency spectrum at spatially separated locations and that fewer unique frequencies are needed to cover a geographical area. Specifically, the coverage area of a cellular system is divided into cells, where a set of channels is specifically assigned to each cell. This is similar to **medium access control (MAC)** addresses on Ethernet-based network interface cards, where each card is stamped with a unique identifier or address.

Frequency reuse permits identical channel (i.e., frequency) use by cells separated by distance, as shown in Figure 2-13. As you can see, the key in frequency reuse is separating cells using the same channel from each other, typically by placing a cell of another channel between the two.

Early cellular system designs reflect the massive infrastructure costs involved, typically from $1 million to $2 million each, for an individual cell transmitting and receiving antennae setup. These devices, called base stations, handle all calls within a cell (area) as well as the performing handoffs to and from neighboring cells (areas). This station typically includes the necessary antennae, power generating and management equipment, and a

frequency reuse

medium access control (MAC)

Figure 2–12
Growth of PCS

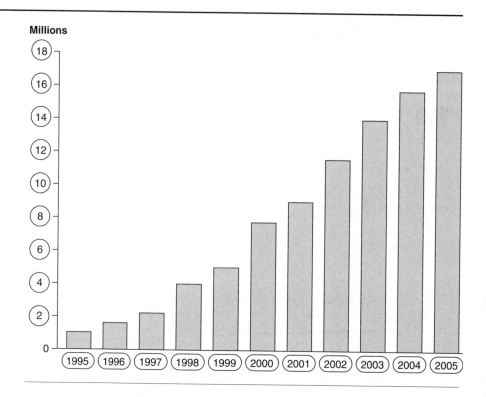

Millions

sophisticated computer system (of various levels of complexity). The computer provides the complex call management required on a 24-hour-a-day 365-day-per-year schedule of service. Historically, it is for this financial reason that early cellular systems, in the early 1980s, used a relatively small number of cells (called macrocells) to cover an entire city or region. Usually these cell base stations were built and operated on skyscrapers or tall topographical features such as hills, outcroppings, plateaus, or mountains. This provided cell coverage areas of several square miles.

microcells
Today, cellular systems typically use a network design utilizing smaller cells called **microcells.** This design of allowing better utilization of frequencies is possible due to several characteristics such as the inherent advantages of digital cellular telephony and better costing factors. The smaller cells allow the base stations to be built closer to ground level, thus reducing costs significantly. Digital systems also allow cheaper, faster components. This cell design using digital technology requires less power to operate. The digital systems also have much higher capacity than analog systems and can take advantage of sophisticated data compression algorithms. This allows these digitally based cellular systems to provide voice mail, paging services, and e-mail services in addition to voice.

Legend: ◯ Frequency A
⬡ Frequency B

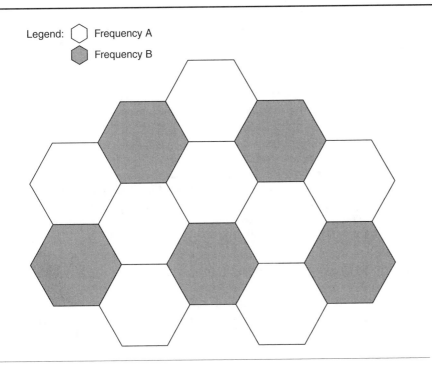

Figure 2–13
Cellular Systems with
Frequency Reuse

WIRELESS LANS

Wireless local area networks (WLANs) are useful in many situations. For example, a firm that is rapidly expanding would commonly install a WLAN. This firm adds personnel on a frequent basis, so it is constantly adding, moving, and reconfiguring offices. Instead of having to frequently relocate LAN wiring, the firm selects a wireless LAN that will transmit to all workstations within range, regardless of the physical location of workstations. Thus, WLANs provide a less expensive approach and save many labor hours otherwise spent in installing LAN wiring.

wireless local area networks (WLANs)

Design and Principles of Operation

The typical WLAN configuration utilizes wireless network interface cards on each workstation. These special network interface cards have a low-power transmit/receive radio antenna installed. A hub antenna is located at a central point such as the center of a room or even the corner near the ceiling (to be unobtrusive). This configuration provides line-of-sight transmission to all other wireless devices in the room. Line of sight is not required, but is a desirable trait, especially for higher frequency (i.e., high performance) LANs, because wireless LANs can penetrate solid objects, but with reduced performance. Thus, many solid objects mean reduced performance, whereas fewer solid objects mean better performance.

Some WLANs use unlicensed frequencies to avoid expensive and lengthy licensing procedures by national regulatory authorities. The problem with this approach, however, is unwanted interference from other such systems in proximity, such as other WLANs, walkie-talkies, garage door openers, and bar code scanners, because these devices typically use the same frequencies. The WLANs that use licensed frequencies avoid this problem, but the implementer must wade through the massive paperwork and lengthy delays present in applying to the FCC for a license.

de jure standards

de facto standards

There is a de jure standard for WLANs. **De jure standards** are those standards approved by a nationally and/or internationally recognized organization. These differ from **de facto standards,** which are approved by organizations or individuals who have no national and/or international recognition. The de jure WLAN standard, the IEEE 802.11 standard, began at theoretical maximum transmission rates of 2 Mbps, with optional fallback to 1 Mbps in extremely noisy environments such as factory floors. The average transmission rate was around 1.6 Mbps. The newer IEEE 802.11b WLAN standard approved in 2000 operates at 11 Mbps. At the present time, a limited number of products adhere to this new standard, but this is expected to change as WLAN technology improves to a point where market penetration is sizable. Figure 2-14 shows a typical WLAN design.

Strengths

Wireless LANs offer the sizable advantage of avoiding cabling issues such as costs, installation problems and delay, and aesthetics of cabling lying about in offices. This is especially important in rapidly changing environments where reconfigurations are necessary. In addition, wireless LANs have proved to make an excellent choice in older buildings where cabling installation issues can become a nightmare. Deploying cabling in older buildings has proven more difficult than first imagined throughout the country. College campuses are discovering this problem as they attempt to provide Internet and/or LAN access to students living in older dormitories. Installation costs can easily exceed ten times the amount necessary to wire a new building.

Weaknesses

Given the advantages of WLANs, you may ask why they have not become commonplace. The main reason is performance factors. For example, most WLANs operate at an effective throughput of between 1 and 2 Mbps, which lies well below Ethernet's 10 Mbps or Token Ring's 16 Mbps. Secondly, wireless networks are susceptible to effects from construction materials and ob-

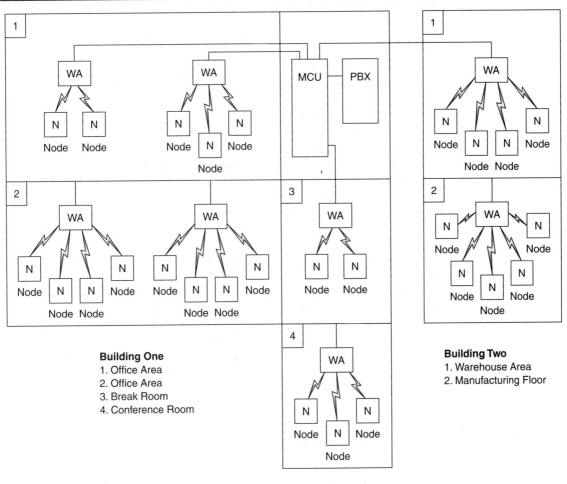

Building One
1. Office Area
2. Office Area
3. Break Room
4. Conference Room

Building Two
1. Warehouse Area
2. Manufacturing Floor

Legend: WA = wireless access point
N = wireless node

Figure 2–14
Typical WLAN Design

jects. As an example, a WLAN in a large relatively open area of a building will operate much more efficiently than one in a high-density, compartmented one. Third, the somewhat controversial issue of health effects on living tissue may cause some to prefer wired networks. At higher transmission rates, there is at least anecdotal evidence of cell damage; however, at lower rates, there appears to be little or no damage (Khoundary 2000).

SATELLITE COMMUNICATIONS

Satellite technology is important for several reasons. Satellite transmission rates are fast, with bandwidth exceeding 3 GHz. Satellites also are impervious to terrestrial interference factors such as precipitation and EMI. They can be placed without regard to obstructions such as buildings or other man-made objects. Remote parts of the world benefit, where communications would be almost nonexistent without this technology, including areas of South America, Asia, Africa, and the polar regions.

Globalstar, whose primary investors are Loral and Qualcomm, is an example of a satellite system. It uses an LEO satellite-based telecommunications system that offers wireless digital telecommunications services worldwide (Kary 2001). Digital communications provide more efficient communications than analog. Globalstar also provides low-cost, high-quality telephony and other services such as data transmission, messaging, facsimile, and position location to both mobile and fixed markets worldwide.

Globalstar's forty-eight-satellite LEO constellation provides wireless service to virtually every area of the world between the latitudes of 70° North to 70° South. It performs this function as follows: All telephone transmissions enter the service provider's existing land-based network from a local gateway. Each ground station or gateway has the ability to provide connections for more than 1,000 simultaneous circuits into the PSTN, the local telephone network. The overall configuration has the capacity to serve as many as 9 million subscribers on a demand usage basis. Gateways are distributed around the world to connect users with their local PSTN. Figure 2-15 shows this architecture.

Design and Principles of Operation

Satellite communications function as a terrestrial transmission system that happens to use a nonterrestrial relay space station. Figure 2-16 shows the three components necessary to use satellite communications. First, an earth station is required to transmit data to the satellite. This can be a significant investment, particularly in the case of geosynchronous satellites. Because geosynchronous satellites are further away, it can require a larger antenna setup for high-bandwidth data transfer to/from these types of satellites. The second element in this system is the satellite. Geosynchronous satellites are those satellites in high earth orbit in a stationary position in which they remain over the earth in the same location relative to the earth. Geosynchronous satellites require a large dish at the earth station to transmit data to their high orbit. Lower earth orbit satellites (nongeosynchronous) require a smaller dish because the distance is not as great. Third, a receiving dish or station is required. This may be either the large earth stations used by NASA or the 18-inch parabolic dishes used by consumers.

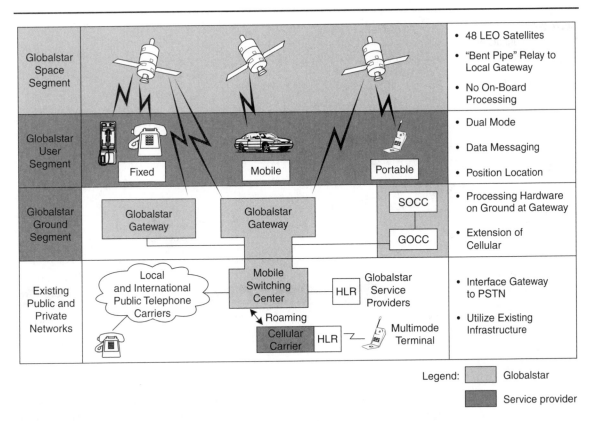

Figure 2–15
Globalstar Architecture

All satellites require an approved frequency in which to transmit. Both national governments and the International Telecommunications Union (ITU) govern this approved frequency. In addition, all orbital paths are registered and approved at the national and international levels by the United Nations. This approval is critical in order to avoid collisions or, more likely, interference from one satellite to another in transmissions and receptions. For example, if one satellite takes a position in front of another, difficulties can arise. Apparently, this problem occurred recently when the Chinese launched the APSTAR1 satellite in close proximity to a Russian and a Japanese satellite. Frantic international negotiations ensued and the Chinese government repositioned the satellite (Apstar1 2001).

Interestingly, satellites come in a variety of sizes and weights, as shown in Figure 2-17. These sizes and weights vary depending on the specific application of the satellite (e.g., weather, geographical survey, communications,

Figure 2–16
Typical Components of a
Satellite System

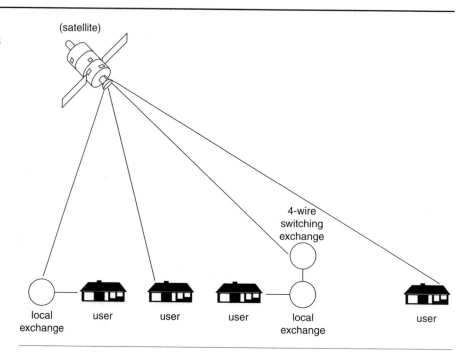

surveillance, orbital distance, funding issues). In addition, the range of telecommunications services they perform varies as well. Some are capable of simply storing and then forwarding messages, whereas others carry hundreds of broadcast-quality TV channels at real-time speeds. This latter capability of transmitting at real-time speeds can be seen when TV networks talk with correspondents overseas on newscasts. Satellites also provide a wide range of other services including multimedia, voice, imaging, fax, electronic mail, and television.

geosynchronous orbit (GEO)

Satellites occupy three different orbits. First, in a **geosynchronous orbit (GEO),** a satellite is roughly positioned over a portion of the earth's surface and remains there. This orbit is the highest at approximately 22,500 miles above the surface of the earth. The satellites in these orbits typically serve various purposes from broadcast television to weather. Because they are so far above the earth's surface, only eight satellites are needed to cover the entire globe. Secondly, low earth orbit satellites are typically used for intelligence gathering activities or the recent advance in satellite telephony, such as the Iridium system of sixty-six satellites that is now financially bankrupt or the more financially successful Globalstar network. At this low orbit, it requires sixty-six satellites to cover the earth's surface. Finally, there are **medium earth orbit (MEO)** satellites that perform a wide variety of tasks from data transfer to broadcasting (such as DirecTV). MEOs provide a compromise between the

medium earth orbit (MEO)

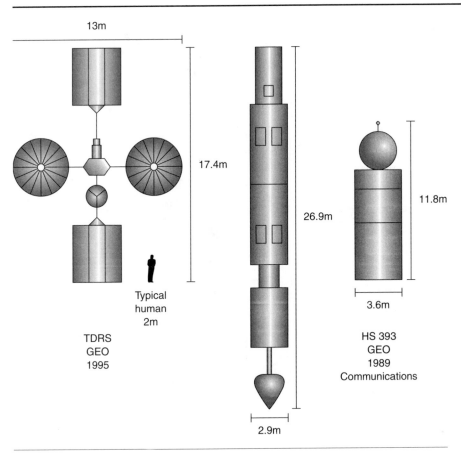

13m

17.4m

26.9m

11.8m

Typical
human
2m

TDRS
GEO
1995

2.9m

3.6m

HS 393
GEO
1989
Communications

Figure 2–17
Relative Sizes of Satellite
Spacecraft

high and low orbits. Thirty-two MEOs can provide full earth coverage. Figure 2-18 is a graphical look at this coverage issue. Figure 2-19 provides a perspective view of a GEO satellite's orbit.

One of the most used satellite systems is the **Global Positioning System (GPS).** GPS is an advanced satellite navigation system created and managed by the U.S. government. It provides precise earth positions for both military and civilian sectors. Figure 2-20 shows a pictorial of this system. GPS performs tasks such as tracking military targets (its initial purpose) and various vehicles in the trucking industry as they transport goods around the country. A wide variety of people (hikers, boaters, automobile drivers, and anyone else who wants to pinpoint his or her location) use the GPS system. Future uses may include a GPS card in desktop or laptop computers giving its exact location as part of the IP address. The IP address is used in many networks, including the Internet, to route data. The source computer places its unique

**Global Positioning
System (GPS)**

Figure 2–18
Satellite Coverage of the
Earth's Surface

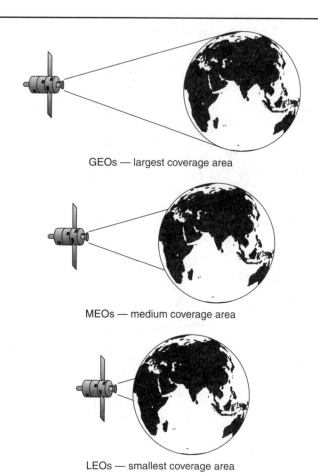

GEOs — largest coverage area

MEOs — medium coverage area

LEOs — smallest coverage area

IP address and the destination IP address in the header that is then affixed to data. Each router on the Internet then looks at that IP address to determine its route along the network.

The DOD launched the first Navstar GPS satellite in 1978. This satellite reached full operational capability in 1995 with the ability to track military vehicles and to accurately aim military weapons, such as missiles, at enemy targets. In addition, the satellites can detect nuclear explosions anywhere on earth using the onboard nuclear detection software/hardware suite. The GPS system was never intended for civilian use, and parts of the signal were encrypted for military reasons. Other navigation systems have been in use, but most are no longer practical or are phasing out because they are land based, providing limited access by users. GPS is satellite based, providing access to users virtually anywhere on the planet.

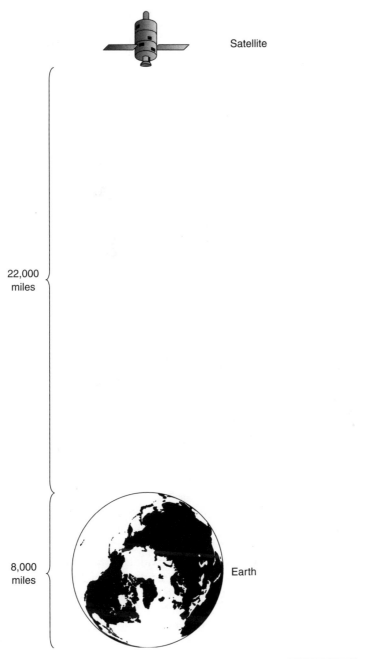

Satellite

22,000
miles

8,000
miles

Earth

Figure 2–19
Relative Height of a GEO
Satellite

Figure 2–20
The GPS

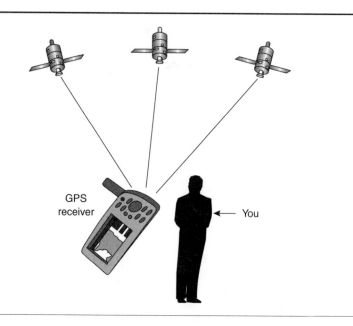

GPS receiver

You

The GPS was not accurate enough for many civilian applications, partly because of the DOD's encryption and because of atmospheric conditions. As various commercial GPS applications developed, such as using GPS in surveying, investors remained skeptical about relying on a system that was both encrypted and operated by the government. In 1996, President Clinton signed the Presidential Decision Directive on GPS, guaranteeing that GPS signals, although government controlled, would be permanently available for civilian use.

Other nations have quickly realized the importance and practicality of such a navigation system. Russia has developed its own satellite navigation system, GLONASS. The Europeans also object strongly to using the U.S. system, and are in the process of creating their own system rather than using a system controlled by our government. Specifically, European countries are building national GPS-based information networks that provide surveyors and navigators with 24/7 access to precise position data. These GPS information networks are proving invaluable for the construction of new buildings and roads, increasing safety for commercial marine operations, making vehicle tracking more efficient, and monitoring the natural environment (Europe 2001).

The GPS contains twenty-four satellites, maintaining an altitude 12,500 miles above the earth. The clocks on board each satellite are synchronized with an accuracy of 3 billionths of a second (SNS). These clocks work in the process of location determination to ensure accuracy, which ensures almost perfect location accuracy, to within a few feet.

A minimum of four GPS satellites are required to calculate a receiver's three-dimensional position. The twenty-four satellite configuration ensures that there is a minimum of six satellites within view of a receiver at any point on the globe at any given moment. It is possible for twelve satellites to actually be within range of a single receiver. The greater the number of satellites in range, the greater the accuracy of the position solution calculated by the GPS receiver. Handheld receiving units are quite compact—the size of a cell phone or smaller. Major U.S. manufacturers of GPS receivers include Trimble Navigation, Ltd., Ashtech-Magellan Corp., and Novatel. A wide variety of GPS software is also available for the personal computer and Palm Pilot. Chapter 11 explores the issue of GPS in more detail.

Although GPS and the technology to use it have been available for some time, the costs, inaccuracy, government ownership, and privacy issues have prevented this useful and practical service from being fully utilized. Future prospects are more promising and there is not another system of its type likely to take its place for personal, commercial, and military use in the predictable future (Horak 2000).

If GPS cards become commonplace technology, will consumers object to the government or to commercial enterprises knowing their exact location? If your computer's IP address includes the precise geographic location, it would be available to anyone on the Internet. Is this good business or is it an invasion of privacy? The courts will have to determine the legal issues involved with this technology.

GPS provides many positive effects. The safe and speedy navigation of ships and boats of all sizes, the ability to track a package from the distributor to the consumer, or the pinpointing of yield effects of fertilizer application to the square meter (thus reducing fertilizer pollution runoff) are practical applications of the GPS today. Many new cars can combine the information from its GPS card, cell phone connections with a central assistance location, and on board maps provided on CD-ROM to plot directions to the next location. If the car is stolen, the GPS locator can relay its location to assist the police in returning the stolen vehicle.

Strengths

Satellite communication has many strengths. Satellites are efficient transmitters of data, transmitting at high speeds exceeding 3 Gbps. They are impervious to signal obstacles such as buildings and mountains. These obstacles limit the effectiveness of other high-speed technologies such as microwave systems or wireless LANs. It is extremely difficult to intercept data going to and from satellites because vertical transmissions are inherently more secure (Tech TV 2001). For this reason, the military heavily relies on this form of wireless communications. Satellites also can be configured for a myriad of applications from voice to data to multimedia.

Weaknesses

Satellite communications can be expensive because of the cost involved in launching it. For example, it is typical to spend at least $200 million to launch a low orbit or medium orbit communications satellite and over $300 million to launch a geosynchronous satellite. This cost must be recouped over a period of a few years, because the average lifespan of one of these satellites is 10 years. Satellite communications also are susceptible to solar activity such as solar flares, which can limit communications typically from a few minutes to several days.

WIRELESS APPLICATION PROTOCOL

Design and Principles of Operation

Most of the technology developed for use on the Internet has been designed for desktop and larger computers, and medium to high bandwidth, which are generally reliable data networks. Fiber-based networks with reliability rates exceeding 98 percent are now becoming common. More fiber-based T3 networks are being ordered than copper-based ones. These networks use sophisticated software algorithms to provide additional stability. Router and switching technology has advanced to the point where errors, when they do occur, are handled and ported more efficiently than ever before. These factors contribute to a stable and growing technology base for web-based applications, services, and products.

This environment is proving to be a challenge for low-power hand-held access devices used in wireless networks. Easily purchased, hand-held wireless devices such as cellular phones and pagers present a more constrained computing environment compared with desktop computers. Because of basic limitations of power and size, these hand-held devices tend to have severely restricted power consumption, less powerful CPUs, less memory (ROM and RAM), and smaller displays and input devices such as a phone keypad or voice recognition rather than a keyboard.

Similarly, wireless data networks present a more constrained communication environment compared with wired networks. Because of fundamental limitations of power, available spectrum, and mobility, wireless data networks tend to have less bandwidth, more latency, less connection stability, and less predictable availability.

The Wireless Application (WAP) Standard attempts to emulate this very successful World Wide Web (WWW) (Internet) model, except with the assumption that users are not at a workstation, but are using a hand-held device to access the Internet.

The WAP model, as shown in Figure 2-21, is similar to the WWW model. This model provides several benefits to the application developer community, including a familiar programming model, a proven architecture, and the

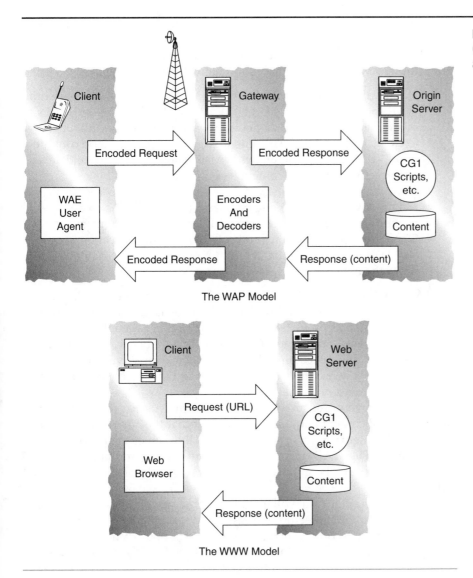

Figure 2–21
Comparison of the WAP and WWW Models

ability to leverage existing tools (e.g., Web servers, XML tools, Javascript, URLs). Over the past several years, the WWW standard has added extensions to match the characteristics of the wireless environment. Wherever possible, existing standards have been adopted or have been used as the starting point for the WAP technology.

WAP content and applications are specified in a set of well-known content formats based on the familiar WWW content formats. Content is transported using a set of standard communication protocols based on the WWW communication protocols. A micro browser in the wireless terminal coordinates the user interface and is comparable to a standard Web browser.

The WAP model defines a set of standard components that enable communication between mobile devices and network servers, including:

- Standard naming model. WWW-standard URLs are used to identify WAP content on origin servers. WAP uses WWW-standard URLs to identify local resources in a device, for example, call control functions.
- Content typing. All WAP content is given a specific type consistent with WWW typing. This allows WAP user agents (a hand-held device) to correctly process the content based on its type.
- Standard content formats. WAP content formats are based on WWW technology and include display markup, calendar information, electronic business card objects, images, and scripting language.
- Standard communication protocols. WAP communication protocols enable the communication of browser requests from the mobile terminal to the network web server.

The WAP protocols have been optimized for mass-market, hand-held wireless devices. This provides functionality to as large a group of users as possible. WAP utilizes proxy technology to connect between the wireless domain and the WWW, similar to MTSOs in cellular networks that connect to the PSTN. WAP devices communicate with their own servers and wireless switching stations, which then communicate with the PSTN network, which then route calls to users on the PSTN.

This infrastructure ensures that mobile terminal users can browse a wide variety of WAP content and applications, and that the application author is able to build content services and applications that run on a large base of mobile terminals. The WAP proxy allows hosting content and applications on standard WWW servers and using proven WWW technologies such as CGI scripting.

The typical use of WAP will include a web server, WAP proxy, and WAP client; but the WAP architecture can easily support other configurations. It is possible to create an origin server—a server with both WAP and non-WAP functionality—that includes the WAP proxy functionality. Such a server facilitates end-to-end security solutions or applications that require excellent access control or a guarantee of responsiveness (e.g., WTA). Figure 2-22 shows how a WAP network would look.

In the example, the WAP client communicates with two servers in the wireless network. The WAP proxy translates WAP requests to WWW requests, thereby allowing the WAP client to submit requests to the web server. The proxy also encodes the responses from the web server into the compact binary format understood by the client.

If the web server provides WAP content (e.g., WML), the WAP proxy retrieves it directly from the web server. However, if the web server provides WWW content (such as HTML), a filter is used to translate the WWW con-

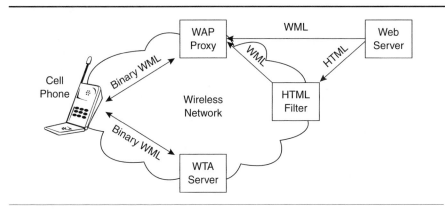

Figure 2–22
An Example WAP
Network

tent into WAP content. For example, the HTML filter would translate HTML into WML.

The WAP architecture provides a scalable and extensible environment for application development for mobile communications devices. This is achieved through a layered design of the entire protocol stack as shown in Figure 2-23. Each layer of the architecture is accessible by the layers above, as well as by other services and applications.

The WAP layered architecture enables other services and applications to utilize the features of the WAP stack through a set of well-defined interfaces.

Strengths

The Wireless Application Protocol provides many benefits. It brings Internet content and advanced data services to wireless phones and other wireless terminals. WAP also creates a global wireless protocol specification that works across all wireless network technologies. WAP enables the creation of content and applications that scale across a wide range of wireless bearer networks and device types. WAP also embraces and extends existing standards and technology to be cost effective.

Weaknesses

Deployment of the WAP standard for hand-held Internet devices involves additional costs. All devices must be fully compliant, therefore requiring the retrofitting of current noncompliant devices. The WAP standard, because it is designed to be universal, exacts a minimal performance "hit." As with a majority of standards-based software, the process of providing a "service to all" can make the software somewhat larger than a proprietary solution. This typically means it will execute more slowly than a proprietary solution. WAP devices also are more expensive to manufacture due to the sheer size of the standardized features that must be present.

Figure 2–23
WAP Architecture

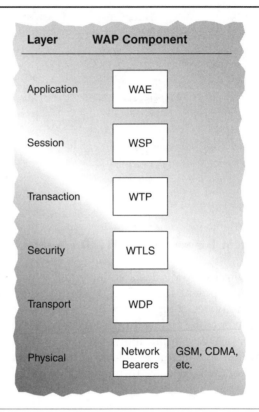

OTHER WIRELESS SYSTEMS

Other than the better-known wireless technologies such as microwave and satellite, there are many other lesser-known systems. For example, RF communications play a crucial role for most North American cities. Without this technology, ambulances could not be sent to assist the injured, fire trucks would not arrive at their assigned destinations, trash would not be picked up, sewer personnel could not be sent to trouble spots throughout the city in a moment's notice, and policemen would not be able to be at locations of unrest or to intercede in a domestic quarrel.

In times of floods or other natural disasters, RF communications allow resources to be diverted as needed to lessen nature's devastating effects on an area that has been victimized. In 1994, many cities along the Mississippi River during a well-documented and dangerous multi-state flood attribute the use of RF communications to saving stranded citizens who were in danger.

Other wireless systems being deployed today we take for granted, such as garage door openers or TV channel remotes are often taken for granted.

One of these wireless systems is well known and widely used. In 1950, New York City introduced the first paging system covering a large metropolitan area. Users carried small, lightweight devices called pagers or beepers that received signals from a centralized antenna base station. The signal contained the identification number of a particular pager (the one to whom the message is destined). Once the pager received the broadcast signal, verifying the pager was the correct page, it beeped to alert the user of an incoming message. At that point in the evolution of this technology, the user had to call a central telephone number and talk to an operator to receive the actual content of the message. Now, however, paging systems provide a wealth of services such as:

- Two-way messaging
- Data transmission
- LED readout

Figure 2-24 provides a look at typical paging services.

Paging systems use both terrestrial and satellite communications, with today's most common practice being a combination of both, forming a hybrid network. Estimates of users of paging technology vary widely and the advent and proliferation of cellular telephones is encroaching into this lucrative market. Table 2-4 provides a glimpse into the user base for this technology.

Pagers connect to centralized antennae on a wireless basis. These centralized antennae interconnect via either microwave or standard dedicated or leased-line networks. Some paging systems such as SkyTel provide satellite links to these terrestrial networks.

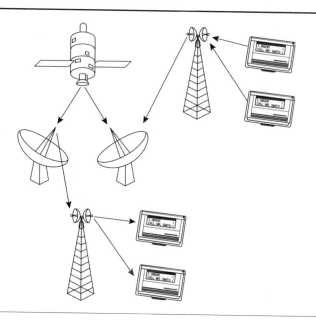

Figure 2–24
Terrestrial Paging
Environment

Table 2–4
Paging Subscribers in
the United States

Year	Number of Subscribers
1995	34 million
1997	60 million
2000	120 million
2002	160 million
2005	200 million

Source: The Strategies Group, 2001.

The data in paging systems originate in several ways. The most common approach involves the sender dialing a telephone number. That number may be dedicated to one pager or be one of many numbers associated with the service provider. The telephone number terminates in a voice processor at the central location of the service provider. The voice processor prompts the paging party to use the touchtone keypad of the telephone set to enter a return number. If the telephone number is not dedicated to that one pager, the paging party first receives a prompt to enter a pager identification number (PID), typically called a personal identification number (PIN), assigned to that one pager. To send a text message, the paging service provider has the option of accessing a human attendant who will answer the call and enter the message (usually limited to no more than 500 characters) for transmission to the alphanumeric pager.

In addition, many service providers now support web-based messaging. This approach enables any paging party to access the service provider's website via the Internet, and to enter a message of limited length without the intervention of an attendant. Additionally, large organizations with great numbers of employees equipped with pagers may have direct access to the paging system to send alphanumeric messages to their employees. Such access may be on either a dedicated or a dial-up basis, and typically involves proprietary software, which turns a PC into a paging dispatch terminal. These direct links from the user organization to the service provider also may permit the redirection of e-mail and facsimile transmission from corporate servers, once those message formats are converted (the types of message formats vary depending upon vendor). As not to overwhelm both the paging network and the terminal equipment, filters must limit the size of the transmitted file.

Paging systems offer the ability to notify a user of an incoming message, regardless of terrain or locality. This is possible because paging systems are not a line-of-sight technology like the microwave systems. They also require very little power in the receiving unit (i.e., pagers), which allows for them to be extremely small, typically the size of a small calculator. No studies have indicated any tissue damage by the use of these devices.

A host of other applications and services in the wireless arena are not as well known as paging systems. One recent announcement involves a product titled HomeRF. This product is an application of wireless networking that will provide wireless connections throughout the home to all computer devices such as desktop personal computers and laptops, and includes devices as diverse as sophisticated refrigerators, other kitchen appliances, and a wide range of audio and video devices such as DVD and stereos.

Another example is **switched mobile radio (SMR),** which has been in existence since the 1920s. One of its primary purposes is dispatching emergency or police vehicles. Applications of this technology have become commonplace as trucking firms and industrial firms use this wireless technology for everyday operations.

switched mobile radio (SMR)

Cordless phones, another widely used wireless technology, first appeared in the late 1970s. Some estimates indicate that one in four households now possess at least one cordless phone (The Strategies Group 2001). Originally, cordless phones lacked clarity and were severely limited in distance away from their base station. Now, with the advent of digital cordless phones, clarity and distance from the base station are largely issues of the past. In addition, cordless phones use sophisticated scrambling technologies to provide more security; however, this security should be taken with a grain of salt. Remember that a cordless phone is nothing more than a transceiver that broadcasts your voice, and the person with whom you are talking, on radio frequencies that are easily accessed.

Additional wireless systems include remote sensor networks that collect data for a variety of purposes, climate control devices, and several devices for remote control activities such as garage doors and TV channel switching.

■ SUMMARY

This chapter explains how the wireless market is evolving over a 10-year period (1995–2005). Wireless is defined and the advantages and disadvantages are discussed. After all, why select wireless as a choice unless definite advantages can be obtained?

In addition, we discuss the cellular standards developed and those currently in use, from AMPS to PCS, and various aspects of cellular telephony such as the allocation of cells, MTOS, and frequency reuse.

We conclude with other wireless technologies such as microwave, the GPS, satellite systems, and wireless LANs.

In the next chapter we look into some of the fundamental technical underpinnings of wireless technology, such as how wireless signals travel (i.e., propagate) and are absorbed by the surrounding environment.

REVIEW QUESTIONS

1. The major advantages of all forms of wireless are _____.
   a. mobility, flexibility, cost efficiency
   b. flexibility, stability, clarity
   c. cost efficiency, mobility, standards-adherence
   d. mobility, flexibility, throughput

2. Select the major obstacles to an even faster deployment of wireless technologies (select all that apply):
   a. health concerns
   b. cost
   c. performance gap
   d. spectrum issues

3. Cells are _____.
   a. imaginary boundaries in space for satellite positioning
   b. boundaries used in cellular telephony
   c. components in a wireless local area network
   d. on Alcatraz Island, currently unoccupied

4. Digital cellular telephony is superior to analog due to _____.
   a. enhanced clarity, performance, cost
   b. longer transmission distances, enhanced clarity, cost
   c. less towers required, clarity, cost
   d. cost

5. _____ is the earliest cellular telephone network.
   a. PCS
   b. MTSO
   c. AMPS
   d. PCS-II

6. PCS is a _____ cellular telephone network.
   a. digital
   b. analog
   c. macrocell-based
   d. semi-esoteric

7. The standard for pan-European cellular system is _____.
   a. PCS
   b. PDC
   c. GSM
   d. AMPS

8. Frequency reuse is _____.
   a. using the same frequency in the same cell
   b. using the same frequency in an adjoining cell
   c. using the same frequency in nonadjoining cells
   d. none of the above

9. Microcells are used in _____ cellular telephone networks.
   a. analog
   b. digital

10. Microwave systems use _____ frequencies.
    a. low
    b. high
    c. nonradio
    d. thermal

11. Wireless LANs typically operate at _____.
    a. 10 Mbps
    b. 100 Mbps
    c. 5 Mbps
    d. none of the above

12. Will refer to _____.
    a. wireless logical link
    b. wireless local loop
    c. wireless link loop
    d. wireless logical link

13. Satellites that perform imaging functions are typically _____.
    a. LEOs
    b. MEOs
    c. GEOs
    d. WEOs

14. Satellite frequencies are governed by _____.
 a. ITU
 b. CCITT
 c. EIA/TIA
 d. ANSI

15. It requires approximately _____ LEOs to cover the earth's surface.
 a. 66
 b. 8
 c. 32
 d. 100

16. _____ is the official standard for wireless LANs.
 a. IEEE 802.6
 b. IEEE 802.11
 c. IEEE 802.3
 d. ANSI XT39

17. Paging systems first arrived on the scene in the _____.
 a. 1920s
 b. 1950s

 c. 1970s
 d. 1990s

18. _____ is a satellite-based navigational positioning system.
 a. GSM
 b. PCS
 c. GPS
 d. Iridium

19. The first wireless telegraph transmission was made by _____.
 a. Jonathan Hertz
 b. Guglielmo Marconi
 c. Albert Einstein
 d. Clifford Stoll

20. GPS was developed by the _____.
 a. U.S. Department of Commerce
 b. U.S. Coast Guard
 c. U.S. Department of Defense
 d. United Nations Orbital Registration Agency

HANDS-ON EXERCISES

1. An offshore drilling firm has links to all 230 platforms it operates throughout the world. The firm just bought another firm that does approximately the same thing. Your problem is to devise a connectivity strategy to present to the board of trustees of the combined firm on how you will connect to the additional 415 platforms the old firm operated. Which wireless technologies would be appropriate for this task, in order of preference, and why?

2. You are brought in as a consultant to a firm that operates wireless networks throughout the South. Their technical people are on strike and may be for some time, and the load is increasing throughout the network. The IS director of the firm would like you to explain the advantages and disadvantages of how wireless LANs usually operate (his technical people know, but he obviously does not).

3. You have received an inheritance of $120 million from a long-lost uncle and want to get into the cellular telephony business. You invite specialists in this arena to come to your new estate and brief you on cellular telephony. Now imagine you are one of these specialists. How would you describe the current state of cellular standards/systems?

4. You are the new owner of a satellite telecommunications firm. Describe which type of satellite (LEO, MEO, or GEO) you would deploy for connectivity to your five offices throughout Georgia.

5. Using your browser and a search engine, locate three businesses that use satellites for communications. Prepare a report on these firms, explaining how and why they use satellites and which type of satellites they use (LEOs, MEOs, and/or GEOs).

6. Using your browser, discover how many satellite launches have occurred since the first satellite was launched many years ago. Also, how many of these satellite launches were successful? As an extra credit, can you discover the amount of LEOs, MEOs, and GEOs?

7. Using your browser, go to www.gsmworld.com. This important website provides a significant amount of information on various satellite systems. Prepare a spreadsheet in which you list the following:
   • Satellite system title
   • Origin
   • Primary purpose
   • Current status (Example: The Iridium system is now defunct, or is it?)

3

Technical Foundation of Wireless Technologies

OBJECTIVES

After reading this chapter and completing the exercises, you will be able to:

- Learn where and when wireless technologies originated
- Learn about the domain in which wireless technologies operate in the electromagnetic spectrum
- Learn how the various wireless technologies operate

WHAT IS WIRELESS?

In 1873, the British physicist James Clerk-Maxwell announced the theory of electromagnetic waves. In 1888, Heinrich Hertz produced the first electromagnetic waves by supplying an electric charge to a capacitor and then short-circuiting it. The energy from the resulting spark was radiated in the form of electromagnetic waves, and Hertz was able to measure the wavelength and velocity of these so-called hertzian waves. In 1895, the Italian scientist and inventor Guglielmo Marconi became interested in electricity, particularly the work of Hertz and Clerk-Maxwell, through the efforts of his neighbor, the distinguished physicist Professor Righi.

Marconi built the equipment and transmitted electrical signals through the air from one end of his house to the other, and then from the house to the garden, using these waves. Figure 3–1 shows Marconi's first radio transmitter. Hertzian waves were produced by sparks in one circuit and detected in another circuit a few meters away. Marconi could soon detect signals over several kilometers, which led him to try, although unsuccessfully, to interest the Italian Ministry of Posts and Telegraphs. Independently and a little earlier that same year, Alexandrovich Popov, a Russian, successfully transmitted Morse code signals from ship to shore. The Russian government considered this new technology a military secret and kept it under tight wraps. Marconi, however, having no such governmental constraints and clearly seeing the commercial value, let the world know.

Marconi's achievement arose from the fact that he was able to produce and detect the waves over long distances, laying the foundations for what we know today as radio. Historians consider these experiments the dawn of practical wireless telegraphy or radio.

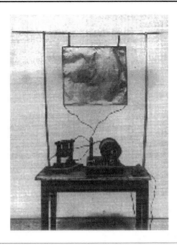

Figure 3–1
Marconi's Radio
Transmitter

Following his successful home-based experiments, Marconi became obsessed with the idea of sending messages across the Atlantic Ocean. He built a transmitter that was 100 times more powerful than any previous station in November 1901, at Poldhu, England. He also built a receiving station at St. John's, Newfoundland. On December 12, 1901, Marconi received signals from across the ocean. News of this achievement spread around the world, and he was acclaimed by outstanding scientists, including Thomas A. Edison. Thus, Marconi became known as the "father of radio."

So, what is wireless? Simply, it is an unbounded technology discovered in an era when wired or bounded technology, in the form of the telegraph and telephone, was already in existence and being used. Wireless technology continues to develop and proliferate worldwide, as we will discuss in this chapter.

ANTENNAS

antenna

An **antenna** is an object that carries (transmits) electrical energy and that receives this energy. The function of the antenna is to convert electrical signals flowing down a conductor (the metal inside) into airborne waves (in a transmitter) or to convert airborne waves into electrical signals flowing down a conductor (in a receiver), or both.

In radio transmission, a radiating antenna converts an electric current into an electromagnetic wave or field, which freely propagates through a nonconducting medium such as air or space. In the broadcast radio channel arena, an omnidirectional antenna radiates a transmitted signal over a wide service area. This can be illustrated by your local radio station transmitting in an omnidirectional pattern from a radio tower with a radio transmitter mounted at the summit, as opposed to point-to-point radio channel transmission in which a directional transmitting antenna is used to focus the wave into a narrow beam. This beam is then specifically directed toward a single receiver. In either case, the electromagnetic wave transmitted through the air is picked up by a remote receiving antenna and reconverted to an electric current.

Antenna Characteristics

Sizes, Shapes, and Strength

Antennas come in many shapes and sizes, from the tall towers transmitting broadcast radio signals, to the 18-inch backyard dishes receiving satellite TV signals. The size and shape of the antenna depends on the frequency and the direction of the signal.

First, as a rule of thumb, the lower the frequency the antenna must handle, the larger the antenna must be. This reasoning explains why broadcast radio stations that transmit at 530 kHz have antennas several hundred feet high, whereas cellular phones, which operate at 900 MHz, have antennas less than 6 inches long. Direction of the airborne signal is the second aspect. If the objective is to transmit or receive signals in an omnidirectional pattern, then the antenna will have a different shape than one transmitting in only one direction.

In general, **directional antennas** require more power for users to receive signals within the same area than omnidirectional antennas, because the **omnidirectional antenna** must divide the energy into many "slices," to cover the same distance as a directional antenna which transmits in only one direction. This distance limitation explains why broadcast radio stations begin to fade: Broadcast radio uses omnidirectional antennas. Therefore, a directional antenna with its higher power over a smaller area will have a greater signal range than an omnidirectional antenna with the same output power (Weisman 2000).

directional antennas

omnidirectional antenna

Antennae and Frequency

A relationship exists between the velocity, the wavelength, and a quantity called frequency. Frequency measures the quantity of waves passing a given point in 1 second. The relationship is given by the following simple equation.

$$\text{frequency} = \text{velocity of light} / \text{wavelength}$$

For light, the frequency is 3×108 m/s divided by 0.000001 m, which gives $3 \times 1{,}014$ waves (or cycles) per second. The unit per second is called a **hertz (Hz).** So, we can say light has a frequency of 300 million million hertz (or 300 million MHz).

hertz (Hz)

Frequency is what separates one **radio frequency (RF)** signal from another and distinguishes one wireless application from another. Figure 3–2 contains a sample of different wireless activities at different frequencies.

radio frequency (RF)

There also is a grouping of frequencies called bands (Figure 3–3). They are defined by the Federal Communications Commission (FCC). As you look at the figure, a few facts may help to better understand it. First, the greater the channel allotment given by the FCC, the more information that can be broadcast on that frequency. Compare AM radio and television. The channel allotment for AM radio is only 10 kHZ, which is why you never hear AM stereo (stereo requires a larger allotment). Television, however, can broadcast in stereo and broadcast video, which requires a high amount of channel allotment. The 6 MHz allotment for television may sound high, but high definition television (HDTV) requires at least twice that amount. So, are the broadcasters to transmit only half the number of HDTV channels as they do in regular television? No. The answer lies in advanced compression techniques to "shrink" the allotment necessary to transmit HDTV into the same amount as regular television today (6 MHz).

A relationship also exists between wavelength and antenna size. Antennas radiate or receive radio signals. Higher frequencies, with their associated shorter wavelengths, require smaller antennas. The so-called perfect antenna is typically one-half the length of the wavelength it receives. In the case of

Frequency (Hz)	Application
60	Electrical wall outlet
2,000	Human voice
530,000	AM radio
54,000,000	TV Channel 2 (VHF)
88,000,000	FM radio
746,000,000	TV Channel 60 (UHF)
824,000,000	Cellular phone
1,850,000,000	PCS phone
2,400,000,000	Wireless LAN
4,200,000,000	Satellite big dish
9,000,000,000	Airborne radar
11,700,000,000	Satellite small dish
500,000,000,000,000	Visible light
1,000,000,000,000,000,000	X-files

Figure 3–2
Daily U.S. Frequencies

Service	Frequency Band(s)	Channel Allotment
AM radio	535–1,605 kHz	10 KHz
FM radio	88–108 MHz	200 KHz
TV (VHF)	54–72 MHz	6 MHz
TV (VHF)	76–88 MHz	6 MHz
TV (VHF)	174–216 MHz	6 MHz
TV (UHF)	470–890 MHz	6 MHz

Figure 3–3
Frequency Bands

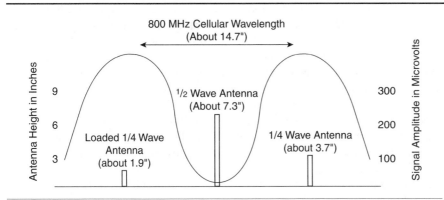

Figure 3–4
Antenna Size and
Wavelength

lower frequencies, however, an antenna the size of Maine would be ludicrous. To address this issue, you can successfully reduce the size of the antenna, by using an electrical device called an **inductor.** Figure 3–4 provides a look at the relationship between signal frequency and antenna size.

inductor

NARROWBAND AND SPREAD SPECTRUM TECHNOLOGY

Most conventional radio transmitters, whether amplitude modulated (AM) or frequency modulated (FM), operate in a mode known as narrowband. What this means is that the transmitted signal contains all of its power in a very narrow RF **bandwidth.** Because of this narrow operating frequency, these radios are prone to suffer from more interference than other types of radios. Sometimes it is possible for a single interfering signal at or near the radio's tuned frequency to render that radio unusable. The answer to this dilemma is **spread spectrum,** a relatively new technology of radio systems. The spread spectrum system is a somewhat distinct brand of RF technology. It is a relatively new concept that entails taking a radio signal, expanding its bandwidth via complex coding schemes (a highly mathematical endeavor), transmitting that expanded signal, and then recovering the desired signal by reversing the process.

bandwidth

spread spectrum

Technically, spread spectrum technology has been in existence since the hertzian waves were generated in 1888. When the notable Hertz created his first spark gap, he also unknowingly created the first spread spectrum transmitter. When a crude transmitting device like a spark gap operates, it generates other radio frequencies in addition to the fundamental frequency desired by the operator. Soon after, scientists such as Marconi also realized this fact. At that time, scientists were striving for maximum transmitted power of the desired frequency. They thought that the additional frequencies generated above and below the desired frequency were a nuisance. In fact, they were more than a nuisance. Precious power was being robbed from the fundamental frequency at the transmitter; and at the receiver, these additional nuisance signals that were so close above and below the fundamental frequency made clear, strong reception more difficult. In fact, this knowledge and the need for more efficient and more powerful RF transmitting devices eventually led to the development of the radio valve, better known as the vacuum tube. One of Thomas Edison's scientists discovered this principle of electromagnetic inductance when he noticed that electricity flowed in a wire held near an incandescent light. Years later he improved that device and patented it as the vacuum tube. Even with today's ideal equipment, however, operators are encumbered by atmospheric and man-made interference.

Over 50 years after Hertz demonstrated radio waves and many years after radio was refined, another major wireless technological breakthrough occurred. Musician George Antheil invented and patented the concept of spread spectrum radio technology in 1942. This significant technological breakthrough was originally intended as a radio guidance system for controlling navy torpedoes. Antheil developed the concept of frequency hopping, to quickly shift the radio signals of control devices and thus make them invulnerable to radio interference or jamming. When the United States entered World War II shortly thereafter, little development work was done on this technology and it was not used in the war.

Although it was available for many years, spread spectrum radio was employed almost exclusively for military use. Then in 1985, FCC rule changes allowed spread spectrum's unlicensed commercial use in three radio frequency bands—902 to 928 MHz, 2.4000 to 2.4835 GHz, and 5.725 to 5.850 GHz—for transmission under 1 watt of power. This power limit prevents interference within the band over long distances. The FCC rule changes, combined with the continuing evolution of digital technology, have greatly accelerated the development of spread spectrum communications equipment. The key to commercializing spread spectrum is overcoming its complexity and cost. Most of the complexity relating to the technology lies in the digital signal processing (DSP) circuitry required to operate direct sequence or frequency hopping radios. Today, all kinds of complex processing circuits are available in a variety of forms in everyday products and at ever-decreasing prices.

The term *unlicensed,* as applied by the FCC, means that a single license is granted to the device manufacturer. That manufacturer, in turn, can produce and sell the various products covered under that license and can do so anywhere in the United States. This enforces the FCC standards at the source, while making possible the unlicensed operation at the application end. Two other common characteristics of unlicensed spectrum devices are as follows:

- Very specific and narrow bandwidth
- Low RF power emissions of their transmitters

Spreading

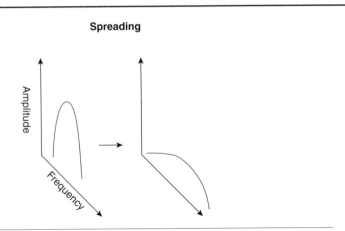

Figure 3–5
Direct Sequencing

In theory, this enables the most efficient usage of these spectrum segments by the maximum number of users.

Spread spectrum radio differs from other commercial radio technologies because it spreads, rather than concentrates, its signal over a wide frequency range within its assigned bands. The two main signal-spreading techniques are direct sequencing and frequency hopping. **Direct sequencing** continuously distributes the data signal across a broad portion of the frequency band. This technique modulates a carrier by a digital code with a bit rate that is much higher than the information signal bandwidth, as seen in Figure 3–5.

direct sequencing

Frequency hopping moves a radio signal from frequency to frequency in a fraction of a second. Signals hop to each frequency, then carry the encoded information at the maximum capacity as determined by the bandwidth and frequency of the carrier, as seen in Figure 3–6.

frequency hopping

Let us look at some of the advantages of using unlicensed spread spectrum technology. First, no FCC site license is required. The FCC will grant a one-time license on the radio product. After that license is granted, the product can be sold anywhere in the United States. Second, spread spectrum radios are inherently more immune to noise and interference than conventional radios; thus, they operate with a higher efficiency than conventional technology. Conventional radios operate on a specific frequency controlled by a matched crystal oscillator. The specific frequency is allocated as a part of the FCC site license, and the equipment must remain on that frequency (except for very low power devices such as cordless phones). Lastly, spread spectrum data radios offer the opportunity to have multiple channels which can be dynamically changed. This allows for many applications such as repeaters, redundant base stations, and overlapping antenna cells.

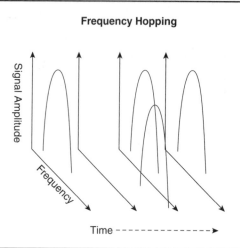

Frequency Hopping

Figure 3–6
Frequency Hopping

ACCESSING CHANNELS: SPREAD SPECTRUM AND CELLULAR TELEPHONY

The FCC Frequency Allocation Chart, which lists all the frequencies and how they are allocated, shows a significant portion devoted to what is called "land mobile." This frequency band between 806 MHz and 902 MHz is more commonly referred to as the cell phone region. Until about 1984, this was an infrequently used area of the public spectrum; however, starting that year, advances in integrated circuits and other hardware technology enabled analog wireless communications device usage to grow at a rapid rate. More recently, the advent of digital cellular phone service, concurrent with improvements in the power, performance, and features of cellular phones, has caused a veritable explosion in phone proliferation and usage. As could be expected, the age-old problem of limited bandwidth has resurfaced. As the number of users exceeds the amount of bandwidth available in a given geographical area, instantaneous communication is no longer possible. Callers must continue to redial until they access an available channel.

Several techniques have been developed to deal with the important issues in wireless communications systems of multiple and random access. These techniques are **frequency division multiple access (FDMA), time division multiple access (TDMA),** and **code division multiple access (CDMA).** Each of these will be explained briefly in the following sections.

First we need to understand the nature of the problem in order to understand the significance of the fixes. They will be explained in more detail in Chapter 6. The wireless communications link activation is random and unpredictable, because people make calls at any and all times of the day. In ad-

frequency division multiple access (FDMA)

time division multiple access (TDMA)

code division multiple access (CDMA)

dition, one or many links may be activated at any moment, while several links can be active simultaneously. Let us now turn to the three types of multiaccess communication systems in use today.

FDMA

Frequency division multiple access is the division of the frequency band allocated for wireless cellular communication into thirty channels, each of which can carry a voice conversation or, with digital service, digital data. FDMA is a basic technology in the analog Advanced Mobile Phone Service (AMPS), the most widely installed cellular phone system in North America, although PCs are climbing in popularity. With FDMA, each channel can be assigned to only one user at a time. FDMA is also used in the Total Access Communication System (TACS). See Figure 3–7 for a graphical representation of FDMA.

The Digital-Advanced Mobile Phone Service (D-AMPS) also uses FDMA, but adds time division multiple access to get three channels for each FDMA channel, thus tripling the number of calls that can be handled on a channel. This obviously provides a large performance improvement.

This is how conventional public switched telephone systems work. Here, every user is assigned a certain frequency band and can use that part of the spectrum to communicate. The system capacity is directly related to the number of channels the allocated bandwidth is capable of carrying. If only a small number of users are active, then only part of the allocated frequency spectrum is used. Assignment of the channels is done either by carrier sensing in the mobile unit (random access) or by centrally controlled assignment.

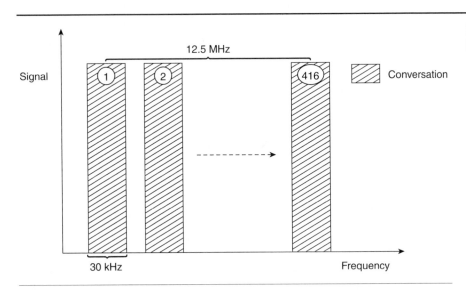

Figure 3–7
FDMA

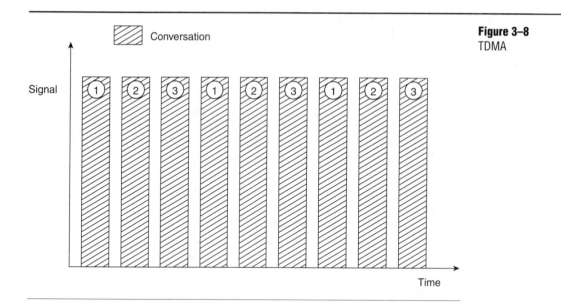

Figure 3–8
TDMA

TDMA

Time division multiple access is digital transmission technology that allows a number of users to access a single RF channel without interference by allocating unique time slots to each user within each channel. TDMA digital transmission design multiplexes three signals over a single channel. The current TDMA standard for cellular divides a single channel into six time slots, with each signal using two slots, providing a 3-to-1 gain in capacity over AMPS. Each caller is assigned a specific time slot for transmission. See Figure 3–8 for a graphical representation of TDMA.

TDMA is used in current mobile phone systems. Here, every user is assigned a set of time slots. Transmission of data is only possible during this time slot; after that, the transmitter must wait for another time slot. As can be imagined, synchronization of all active users is vital. Because of the importance of the timing issue, this technique uses a central base station that controls the synchronization and the assignment of time slots. Thus, TDMA is difficult to apply to random access systems.

CDMA

Code division multiple access is a digital cellular technology that uses spread spectrum techniques. CDMA does not assign a specific frequency to each user. Instead, every channel uses the full available spectrum. Individual conversations are encoded with a pseudo-random digital sequence. CDMA was first used during World War II by the English allies to foil German attempts

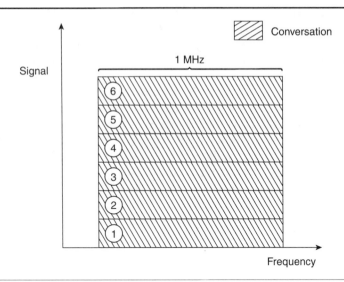

Figure 3–9
CDMA

at jamming transmissions. The allies decided to transmit over several frequencies, instead of one, making it difficult for the Germans to pick up the complete signal.

CDMA assigns a unique code to each user. This code is used to encode the data message. The codes are designed so that the possibility of one user's code correlating with or matching another user's code is unlikely. Thus, all users can transmit simultaneously in the same frequency channel while a receiver can still recover the desired signal. Because synchronization between links or cells is not a requirement, random access is possible. Figure 3–9 shows a graphical representation of CDMA.

These three multiple access/random access techniques are the ones currently in use worldwide; but the spread spectrum world does not revolve around cellular communications. In fact, this powerful technology is being applied to many other areas of radio spectrum usage. Several categories of devices in the unlicensed segment include appliances such as cordless telephones, RF and infrared (IR) computer keyboards and mice, wireless intrusion detection and alarm systems, and television or video cassette recorder (VCR) remote control units. Both the licensed and the unlicensed spectrums include voice, PCS, and data applications.

As mentioned, the FCC defines the wireless, or radio, spectrum as that band of frequencies starting at 3 KHz and ending at 300 GHz. This vast spectrum contains one unallocated and many suballocated frequency bands. In 1985, the FCC changed the rules to make spread spectrum applications practical and the so-called unlicensed spectrum available to the public for the first time. Since then, the theoretical and practical development of spread spectrum has advanced rapidly. The usage of the two kinds of spread spectrum technology—direct sequence and frequency hopping—depends on the application involved. Spread spectrum transmits the same amount of power as a single frequency transmitter; the difference is that it is spread across a spectrum of frequencies.

In 1994, additional FCC rule changes instituted the auctioning off of RF spectrum. Since then, twenty-seven blocks of the wireless spectrum have been auctioned off, thus enabling the current explosion in the use of these frequencies. Usage includes the traditional applications such as broadcast radio and television as well as state-of-the-art applications such as WAN/LAN, pagers, PCS, cellular, IVDS, and cordless telephones. Licensed and unlicensed devices alike use spread spectrum technology, including unlicensed computer keyboards and sophisticated long-range military data and communications gear.

PROPAGATION

propagation

A major issue in this transmission arena is propagation. **Propagation** involves deviations from a state of rest or equilibrium. Although surface waves on water are the most familiar, both sound and light travel as wavelike disturbances, and the motion of all subatomic particles exhibits wavelike properties.

Electromagnetic waves represent oscillations of the electromagnetic field. These include radiant heat, light, radio, microwave, ultraviolet, and x-rays. Electromagnetic waves are produced by moving electric charges and varying currents, which can travel through a vacuum. Unlike sound waves, they are not disturbances in any medium.

Because radio wave propagation is not constrained by any physical conductor (such as network cabling or fiber optic cabling), radio is an excellent choice for mobile communications, satellite and deep-space communications, broadcast communications, and other applications in which the laying of physical connections may be impossible or costly. However, atmospheric conditions can vary widely due to storm activity, solar activity, and various sources of reflection and diffraction.

The range of a radio communications link is typically the maximum distance that the receiver can be from the transmitter and still maintain a sufficiently high signal-to-noise ratio (SNR) for dependable signal reception. The received SNR is negatively affected by a combination of two factors: beam divergence loss and atmospheric attenuation. Beam divergence loss is caused by the geometric spreading of the electromagnetic field as it travels through space. As the original signal power is spread over a constantly growing area, only a fraction of the transmitted energy reaches a receiving antenna. For an omnidirectional radiating transmitter, which broadcasts its signal as an expanding spherical wave, beam divergence causes the received field strength to decrease by a factor of $1/r^2$, where r is the radius of the circle, or the distance between the transmitter and receiver. The other cause of SNR degradation, atmospheric attenuation, depends on the propagation mechanism, or the means by which unguided electromagnetic waves travel from transmitter to receiver. Attenuation is a serious factor that is difficult to solve.

Radio propagation involves the following issues.

- *Reflection.* This involves the issue of a propagating radio wave which impinges on an object that is large as compared with the wavelength. Examples include buildings, mountains, walls, heavy equipment such as in a factory environment, and the earth itself.
- *Scattering.* Objects that are smaller than the wavelength of the propagating wave include trees, large plants, and street signs and lights.
- *Diffraction.* This is an effect where the radio path between the transmitter and the receiver is obstructed by a surface with sharp and irregular edges such as certain rock formations or even statues.
- *Mobility effects.* The channel can vary with the location of the user (e.g., where you are on an interstate roadway), time, and shadowing effects from dominant objects and multipath scattering from objects nearby. All this causes an aggravating effect of quick fluctuations in received power.
- *Attenuation effects.* Attenuation is the effect of signal loss over distance, even with clear, unobstructing environments. For example, the signal will degrade in strength over distance until it is no longer recognizable. The rate of signal loss due to attenuation can vary, depending on environmental conditions and equipment issues, as shown in Figure 3–10. This figure provides a typical attenuation curve. In free space (vacuum), power degradation is much less (by $1/d^2$).
- *Movement effects.* Not only does movement by the transmitter and receiver affect power and reception, but also the received signals may fade due to the movement of surrounding objects.
- *Indoor propagation.* The indoor receiving signal decays much faster because of objects such as walls, furniture, and even people. In fact, the received signal can be affected by issues such as the density of the people moving around in a building and the type of construction materials. Concrete and steel flooring produce less attenuation than full steel plate flooring, with wood materials offering even less resistance to attenuation. This textbook is not oriented toward formulas, but if you are curious, the formula for determining indoor path loss is as follows:

$$loss = unit\ loss + 10 \times log(dist) = k\ L + N\ Wa + P$$

where:

x => power delay index

dist => distance between the transmitter and the receiver

k => number of floors in which the signal must travel

Figure 3–10
Outdoor Propagation

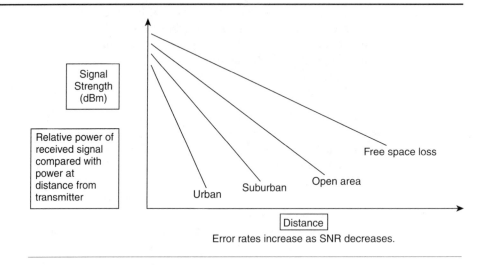

Error rates increase as SNR decreases.

L => loss per floor
N => number of walls the signal must travel
Wa => loss per indoor wall
P => people moving around (varies)

- *Solar activity.* Severe activity by the sun also can affect radio-based communications. This solar activity can affect communications not only in North America, but also at all latitudes. This solar activity produces severe fluctuations in signals, and propagation may take unexpected paths. As an example, television and FM radio stations are only slightly affected by solar activity, whereas typical radio communications can bring to life major propagation issues. Figure 3–11 provides a glimpse of solar activity as discussed in the textbook.

FREQUENCIES AND SPECTRUM

As mentioned, to understand wireless transmission you must understand characteristics of electromagnetic waves. Figure 3–12 shows a representation of a wave as a curved line. The number of oscillations per second of a wave is the frequency, measured in hertz. One oscillation is presented in the figure as movement from one high point of the curved line (the wave) to the next lowest point. Thus, the *time* required for the wave to change from the high point to the low point is the frequency. In Figure 3–13, the wave has a frequency of 2 Hz, or two oscillations per second. The *distance* between the highest point of one oscillation and the highest point of the next oscillation in the wave is called **wavelength.** Note the relationship between wavelength and frequency in Figure 3–13.

wavelength

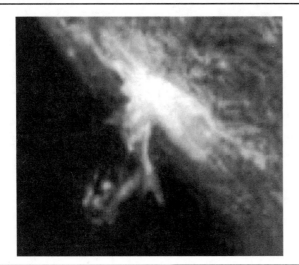

Figure 3–11
Solar Activity

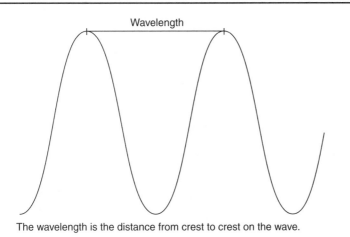

Wavelength

The wavelength is the distance from crest to crest on the wave.

Figure 3–12
Wavelength

Wavelength greatly impacts the use of electromagnetic waves for communication. Figure 3–14 depicts the range of electromagnetic waves, known as the **electromagnetic spectrum,** grouped into bands. These band names derive from conventional use and have no real meaning for communications.

Figure 3–15 depicts the electromagnetic spectrum used for telecommunications. The band names are those proposed by the International Telecommunications Union (ITU). The ITU is an organizational element of the United Nations that is responsible for establishing international standards and procedures designed to facilitate cooperation between nations.

**electromagnetic
spectrum**

Figure 3–13
Frequency and
Wavelength

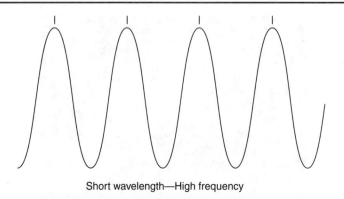

Short wavelength—High frequency

Long wavelength—Low frequency

Figure 3–14
Electromagnetic
Spectrum

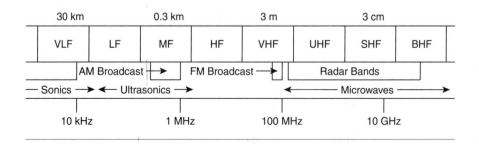

Most often wireless communications uses wavelengths between 10^{24} Hz and 10^{12} Hz. This range is the radio frequency or wireless spectrum. The actual radio frequency or wireless spectrum spans a huge range from approximately 3 kHz to 300 GHz, and covers many frequency bands, as shown in Figure 3–14. As noted, the term *wireless* usually refers to only one or two narrow bands of frequencies of this entire spectrum (which is shown in Figure 3–15).

Table 3–1 presents the wavelength, frequency, and energy of the electromagnetic spectrum. The boundaries between the different radiations are artificial. As you progress from radio waves to gamma rays, the wavelength

Figure 3–15
Spectrum Chart

Radio Frequency Bandwidth
The Allocated Radio Spectrum is located between 9 KHz and 300 GHz

BANDWIDTH DESCRIPTION	FREQUENCY RANGE		
Extremely Low Frequency (ELF)	0	to	3 KHz
Very Low Frequency (VLF)	3 KHz	to	30 KHz
Radio Navigation & maritime/aeronautical mobile	9 KHz	to	540 KHz
Low Frequency (LF)	30 KHz	to	300 KHz
Medium Frequency (MF)	300 KHz	to	3,000 KHz
AM Radio Broadcast	540 KHz	to	1,630 KHz
Travellers Information Service	1,610 KHz	to	
High Frequency (HF)	3 MHz	to	30 MHz
Shortwave Broadcast Radio	5.95 MHz	to	26.1 MHz
Very High Frequency (VHF)	30 MHz	to	300 MHz
Low Band: TV Band 1—Channels 2–6	54 MHz	to	88 MHz
Mid Band: FM Radio Broadcast	88 MHz	to	174 MHz
High Band: TV Band 2—Channels 7–13	174 MHz	to	216 MHz
Super Band (mobile/fixed radio & TV)	216 MHz	to	600 MHz
Ultra High Frequency (UHF)	300 MHz	to	3,000 MHz
Channels 14–70	470 MHz	to	806 MHz
L-band	500 MHz	to	1,500 MHz
Personal Communications Services (PCS)	1,850 MHz	to	1,990 MHz
Unlicensed PCS Devices	1,910 MHz	to	1,930 MHz
Superhigh Frequencies (SHF) (Microwave)	3 GHz	to	30.0 GHz
C-band	3,600 MHz	to	7,025 MHz
X-band	7.25 GHz	to	8.4 GHz
Ku-band	10.7 GHz	to	14.5 GHz
Ka-band	17.3 GHz	to	31.0 GHz
Extremely High Frequencies (EHF) (Millimeter Wave Signals)	30.0 GHZ	to	300 GHZ
Additional Fixed Satellite	38.6 GHz	to	275 GHz
Infrared Radiation	300 GHz	to	810 THz
Visible Light	810 THz	to	1,620 THz
Ultraviolet Radiation	1.62 PHz	to	30 PHz
X-Rays	30 PHz	to	30 EHz
Gamma Rays	30 EHz	to	3,000 EHz

gets shorter (so they become more penetrating), the frequency gets higher (so the oscillation needed to produce them gets faster), and the energy gets higher (so it takes more energy to produce x-rays than it does to produce radio waves).

A brief description of these waves follows. More extensive information can be gathered from many other sources, including a good high school

Table 3–1
Electromagnetic
Spectrum: Frequencies
and Wavelengths

Name	Wavelength (m)	Frequency (Hz)
Radio waves	10^4–10^{-3}	10^3–10^{10}
Infrared	10^{-3}–10^{-6}	10^{10}–10^{14}
Visible	10^{-6}	10^{14}
Ultraviolet	10^{-6}–10^{-8}	10^{14}–10^{16}
X-rays	10^{-8}–10^{-10}	10^{16}–10^{19}
Gamma rays	10^{-10}–10^{-14}	10^{19}–10^{24}

physics textbook. Figure 3–16 also provides a graphical representation of the electromagnetic spectrum.

Radio waves are produced when free electrons are forced to move in a magnetic field, or when electrons change their spin in a molecule. They are used for a wide range of communications purposes, for example, transmission of radio and television signals. Radio waves consist of electromagnetic radiation which has the lowest frequency, the longest wavelength, and is produced by charged particles moving back and forth. Radio waves have wavelengths that range from less than a centimeter to tens or even hundreds of meters. FM radio waves are shorter than AM radio waves. The atmosphere of the earth is transparent to radio waves, with wavelengths from a few millimeters to about 20 meters.

All electric appliances generate radio waves. Fortunately or unfortunately, we cannot detect radio waves because the wavelength is beyond our means of detection.

infrared radiation

Infrared radiation (including microwaves) is produced by the vibrations of molecules. Human skin feels this radiation as heat. Microwave ovens work by using infrared radiation of the correct frequency to make water molecules vibrate faster. A faster vibrating molecule is a hotter molecule. Only the food that contains water is affected. The plate, which is a dry mineral, is unaffected.

visible and ultraviolet light

Visible and ultraviolet light is produced by chemical reactions and ionizations of outer electrons in atoms and molecules. Electromagnetic radiation at wavelengths that are visible to the human eye are perceived as colors ranging from red (longer wavelengths: ~ 700 nanometers) to violet (shorter wavelengths; ~ 400 nanometers).

There are many chemical reactions instigated by this radiation: the chemical retinal in animal eyes, chlorophyll in plants, silver chloride in photography, the chemical melanin in human skin, and silicon involved in electricity. Light is the most familiar electromagnetic radiation, because the earth's atmosphere is transparent to it. Light (with a bit of infrared and ultraviolet on either side of it) can pass through the atmosphere. Ultraviolet is electromagnetic radiation at wavelengths shorter than the violet end of visible light; the atmosphere of the earth effectively blocks the transmission of

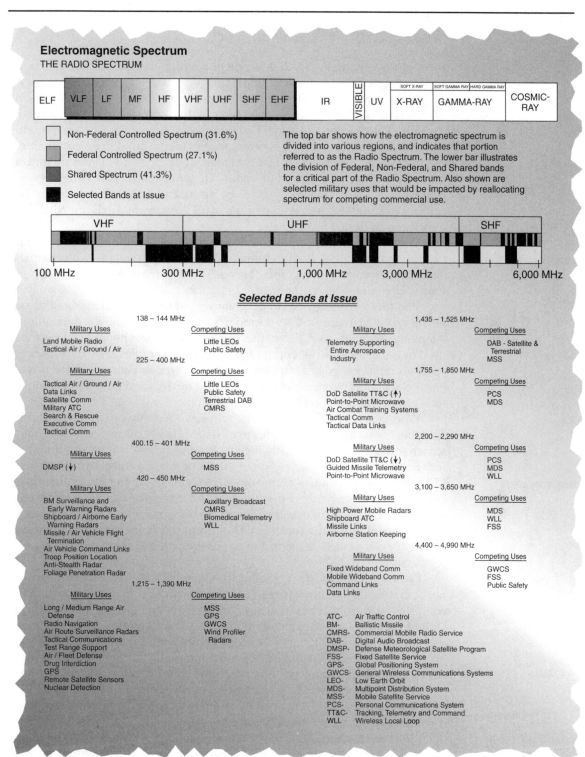

Electromagnetic Spectrum

THE RADIO SPECTRUM

| ELF | VLF | LF | MF | HF | VHF | UHF | SHF | EHF | | IR | VISIBLE | UV | X-RAY | GAMMA-RAY | COSMIC-RAY |

SOFT X-RAY · SOFT GAMMA RAY · HARD GAMMA RAY

☐ Non-Federal Controlled Spectrum (31.6%)

☐ Federal Controlled Spectrum (27.1%)

☐ Shared Spectrum (41.3%)

■ Selected Bands at Issue

The top bar shows how the electromagnetic spectrum is divided into various regions, and indicates that portion referred to as the Radio Spectrum. The lower bar illustrates the division of Federal, Non-Federal, and Shared bands for a critical part of the Radio Spectrum. Also shown are selected military uses that would be impacted by reallocating spectrum for competing commercial use.

| VHF | UHF | SHF |

100 MHz   300 MHz   1,000 MHz   3,000 MHz   6,000 MHz

Selected Bands at Issue

138 – 144 MHz

Military Uses	Competing Uses
Land Mobile Radio	Little LEOs
Tactical Air / Ground / Air	Public Safety

225 – 400 MHz

Military Uses	Competing Uses
Tactical Air / Ground / Air	Little LEOs
Data Links	Public Safety
Satellite Comm	Terrestrial DAB
Military ATC	CMRS
Search & Rescue	
Executive Comm	
Tactical Comm	

400.15 – 401 MHz

Military Uses	Competing Uses
DMSP (↓)	MSS

420 – 450 MHz

Military Uses	Competing Uses
BM Surveillance and Early Warning Radars	Auxillary Broadcast
Shipboard / Airborne Early Warning Radars	CMRS
Missile / Air Vehicle Flight Termination	Biomedical Telemetry
Air Vehicle Command Links	WLL
Troop Position Location	
Anti-Stealth Radar	
Foliage Penetration Radar	

1,215 – 1,390 MHz

Military Uses	Competing Uses
Long / Medium Range Air Defense	MSS
Radio Navigation	GPS
Air Route Surveillance Radars	GWCS
Tactical Communications	Wind Profiler Radars
Test Range Support	
Air / Fleet Defense	
Drug Interdiction	
GPS	
Remote Satellite Sensors	
Nuclear Detection	

1,435 – 1,525 MHz

Military Uses	Competing Uses
Telemetry Supporting Entire Aerospace Industry	DAB - Satellite & Terrestrial
	MSS

1,755 – 1,850 MHz

Military Uses	Competing Uses
DoD Satellite TT&C (↑)	PCS
Point-to-Point Microwave	MDS
Air Combat Training Systems	
Tactical Comm	
Tactical Data Links	

2,200 – 2,290 MHz

Military Uses	Competing Uses
DoD Satellite TT&C (↓)	PCS
Guided Missile Telemetry	MDS
Point-to-Point Microwave	WLL

3,100 – 3,650 MHz

Military Uses	Competing Uses
High Power Mobile Radars	MDS
Shipboard ATC	WLL
Missile Links	FSS
Airborne Station Keeping	

4,400 – 4,990 MHz

Military Uses	Competing Uses
Fixed Wideband Comm	GWCS
Mobile Wideband Comm	FSS
Command Links	Public Safety
Data Links	

ATC-	Air Traffic Control
BM-	Ballistic Missile
CMRS-	Commercial Mobile Radio Service
DAB-	Digital Audio Broadcast
DMSP-	Defense Meteorological Satellite Program
FSS-	Fixed Satellite Service
GPS-	Global Positioning System
GWCS-	General Wireless Communications Systems
LEO-	Low Earth Orbit
MDS-	Multipoint Distribution System
MSS-	Mobile Satellite Service
PCS-	Personal Communications System
TT&C-	Tracking, Telemetry and Command
WLL	Wireless Local Loop

Figure 3–16
Graphical Display of the Electromagnetic Spectrum

most ultraviolet light, which is excellent for humans, because too much ultraviolet light can cause severe skin damage.

Living organisms have evolved to use these waves. Visible light is simply the part of the electromagnetic spectrum that reacts with the chemicals in the eyes. For example, research has shown that most bees can see more ultraviolet than human beings, whereas snakes use infrared to determine prey.

X-rays are produced by fast electrons stopping suddenly, or by ionization of the inner electrons of an atom. They are produced by high energy processes in space, such as by gases being sucked into a black hole and becoming compressed and by exploding stars. X-rays are used in medicine to look through flesh.

gamma rays

Very high energy processes, usually involved with the nucleus of atoms, produce **gamma rays.** Radioactivity and exploding stars are two examples. Gamma rays are very dangerous, because if they strike atoms and molecules they will incur extensive damage.

Wireless technology has undergone a rapid transformation from the earliest days of Marconi in 1895, to today's sophisticated satellite and cellular telephone technologies. From the earliest frequency hopping algorithms used in the telephone network to today's highly complex code multiplexing schemes, this change has affected industry as well as common citizens who use their cellular phones and remote control devices and who watch a multitude of programming on broadcast television.

When we hear a reference to the wireless spectrum or wireless frequencies, most of us have no real concept of what it really means, or for that matter what the term *wireless* really means. Most often the reference is to only a tiny fraction of a particular region of the RF or wireless spectrum. The actual RF or wireless spectrum is a huge range, covering many **frequency bands.** (A frequency band is a group of frequencies used for similar purposes.) When we discuss wireless we are usually referring to only one or two narrowbands of frequencies of this entire spectrum. Usage and technology standards of these narrowbands, as well as the rest of the entire wireless spectrum, are closely regulated by the FCC.

frequency bands

Overseas, the FCC has no authority. Outside of the United States, most if not all national governments regulate their own spectrum. Often the various countries' spectrum usage policies conflict with one another and with the United States. Some forward-thinking individuals foresaw these conflicts and formed the International Telecommunications Union. The ITU, headquartered in Geneva, Switzerland, is an international organization within which governments and the private sector coordinate global telecom networks and services. The ITU is comprised of 189 countries and 650 firms. Today, this respected organization is directly involved in all international regulatory matters involving wireless technology and other wired or bounded media standards.

The FCC licensed the so-called unlicensed frequency spectrums, including some practical present-day uses of those frequencies. All data encoding techniques are covered. Special emphasis is given to spread spectrum technology, due to its importance in current and future wireless communications. Practical applications of wireless technology are also included.

The wireless spectrum starts at 3 KHz and extends beyond 300 GHz. In the United States, the FCC actively regulates all frequencies from 3 KHz to 300 GHz, as shown in Figures 3–14 through 3–16. The ITU plays the same role on the international scene.

The majority of this chapter concerns only a small segment of the entire regulated radio frequency spectrum; however, for the sake of a more complete understanding of the entire physical radio spectra, let us briefly cover the areas from 3 KHz to 9 KHz and 300 GHz and beyond. These areas are currently unallocated to any organization, private or commercial. The first region, 3 KHz to 9 KHz, is considered part of the spectrum known as the infrasonic region in the very low frequency (VLF) band. This frequency range is part of the human audible or sonic range, from 20 Hz through 20 KHz. It stands to reason that any equipment designed to transmit information in the 3 KHz to 9 KHz range would have some major obstacles to overcome, including environmental and atmospheric noise interference. These problems not withstanding, successful military research and experimentation has been carried out in this and other slightly different frequency ranges. The navy currently operates VLF communications equipment located in the Midwest. This equipment was once used to communicate with submarines at sea. Frequencies in this range, if given enough power, travel through the earth and oceans. The antenna for this type of wireless equipment is several kilometers long.

The region starting at 300 GHz and beyond is not considered part of the RF spectrum. This region contains the subareas of infrared, starting at 300 GHz, then proceeds to visible range, at 10^{14} Hz. Ultraviolet, x-ray, gamma ray, and cosmic ray follow in order.

Frequency Hopping Spread Spectrum (Random Hop Sequence)

Like its military predecessor, commercial spread spectrum technology camouflages data by mixing the actual signal with a spreading code pattern. Code patterns shift the signal's frequency or phase, making it extremely difficult to intercept an entire message without knowing the specific code used.

Imagine trying to predict which frequency to listen to next and at what time! As in the military's case, adding secret encoding schemes and encrypting the data—including the encryption of digitized voice transmissions—would make it almost impossible for the enemy's decoders to successfully extract any intelligence from the spread spectrum "noise." That is assuming they are listening to the proper frequency at the right interval. Considering this type of technology, the U.S. Air Force's ability to fly over enemy territory during Desert Storm while totally impervious to its defenses is not as amazing as first believed. Transmitting and receiving radios must use the same spreading and despreading code, so only they can decode the true signal. Logic would tell us that spread spectrum radio for voice is neither the only nor always the best wireless technology for the situation; however, in specific applications, its inherent attributes make it the technology of choice over traditional microwave radio or the optical technologies such as infrared and laser transmission, particularly in environments where wires cannot be utilized.

Spread Spectrum in Area Networks

The most recent spread spectrum developments have come in wide area network and local area network WAN/LAN technology. These developments have come through the integration of the spread spectrum radio transmitter and receiver with a full-function Ethernet and other types of networking bridges. An Ethernet bridge, in this case, is a device that converts the information packets from the local network node into a form suitable for transmitting via a spread spectra radio transmitter. The intended receiver will be a spread spectrum receiver running identical spreading algorithms as the originating transmitter. Its Ethernet bridge will reconvert the transmitted RF signal back to the 802.3-compliant datagram signal for insertion back onto the receiver-side network. The spread spectrum techniques are the same as those used in cellular phone applications with a few minor exceptions. First, the area of the wireless spectrum utilized for indoor WAN/LAN applications is in the unlicensed frequency bands, such as the 2.4 to 2.4835 GHz band. Transmitters utilizing this frequency band are limited to 1 watt maximum output power. Cellular phone repeater transmitters have much higher power limitations. Second, the spread spectrum encoding technique can be much simpler. Most likely, the transceiver (a combination of a transmitter and a receiver) will use direct sequence spread spectra with no encryption, because the limited range of the transmitted signal coupled with the attenuating properties of the structure precludes having any additional complex, and therefore more expensive, security technology.

Outdoor LAN/WAN applications utilize the same spread spectrum technology except that other higher powered licensed equipment can be purchased if needed. Using special parabolic dishes or broadband linear array antennas, the 1 watt unlicensed signal can be sent and received reliably at ranges up to an incredible 14 miles. If a longer range is to be spanned, other more powerful licensed technology can be utilized to reliably connect sites separated by as much as 25 miles. In either indoor or outdoor WAN/LAN configurations, obstacles are a main concern when establishing a wireless link. Therefore, planning includes determining the optimum zone (where the wireless signal can be picked up reliably) and the forbidden zones (where it cannot be picked up at all). Another concern is a phenomenon called multipath or ghosting. This occurs when the transmitted signal is reflected off an obstacle and arrives at the receiver slightly after the signal that arrived directly. In traditional receiver systems, ghosting can seriously hamper or even prevent the receiver from picking up the intended signal. Spread spectrum greatly reduces—and in most cases eliminates—this problem.

As practical commercial applications become better understood, wireless in general and spread spectrum in particular will play an increasingly critical role in a world destined to depend on wireless technology. Because the wireless spectrum is a finite natural resource, debate over and competition for spectrum will become ever more intense. Technology will continue to advance, providing some relief, but we as users must carefully conserve and use these resources efficiently. It is far better that we all have some limited use of the spectrum than have the spectrum be so overcrowded and overtaxed that it is rendered useless for everyone.

SIGNALS

Because wireless involves the exchange of signals, let us examine this issue a little further. Electrical energy (either current or waves) can actually store information if it is made to vary (in intensity) over time. The key to this process is to control the signal. This energy, controlled over time, is called a **signal**. The two types of signals are analog and digital.

signal

Analog Signals

A **sine wave,** as illustrated in Figure 3–17, is an example of an **analog** signal. An ideal wireless signal has a sine waveform, with a frequency usually measured in either megahertz or gigahertz. Household utility current has a sine waveform with a frequency of 60 Hz in most countries including the United States. In other countries the frequency is 50 Hz.

sine wave, analog

How did this wave come to be called a sine wave? In the twelfth century, the Arabic word for the half chord of the double arc was confused with another word and translated as *sinus* (*sine*). This word first appeared in the *Almagest,* a thirteen-volume work on astronomy by Ptolemy of Alexandria in the mid-second century.

As shown in Figure 3–17, a sine wave signal varies in **intensity**. Why is intensity important? The intensity of a signal is typically a measure of power.

intensity

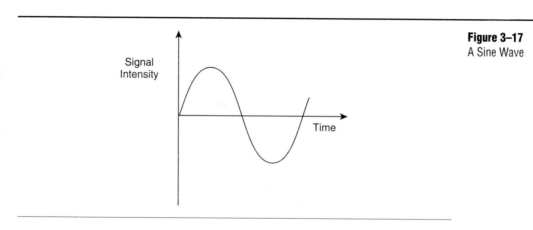

Figure 3–17
A Sine Wave

Figure 3–18
A Digital Signal

The number of times a signal goes through a complete up and down cycle in 1 second is the signal's frequency (measured in hertz). For example, an 800 MHz signal utilized in cellular telephony goes through 800 million ups and downs in a single second.

Digital Signals

digital

Unlike the (analog) sine wave signal, which varies gradually between its high and low points, a **digital** signal is one that varies instantaneously between two electrical values. For all practical purposes, there are no values between the high and low levels in a digital signal (see Figure 3–18). Note the two signal levels: up and down (high and low). Digital signals can represent information in the pattern of highs and lows. For example, a certain pattern of highs and lows can be used to represent your voice as you talk on a cellular phone.

Absorption

absorption

When signals travel through the air, they are changed by their environment—usually the signal is reduced. This process is called insertion loss or **absorption,** because the signal is absorbed by whatever it comes in contact with. The type of contact material is also an important factor: Water absorbs more RF signal than metal, for example.

Wavelength

Wavelength measures the length of an RF signal. The wavelength is inversely proportional to the signal's frequency, which means the higher the frequency, the shorter the wavelength. For example, a cellular phone's RF signal (900 MHz) is higher in frequency than an AM radio's signal (530 kHz), and therefore the cellular phone's signal has a shorter wavelength. Antennas begin to radiate RF energy (as waves) when the wavelength of the RF signal it is carrying becomes similar to that of the antenna itself. The electrical current flowing into the antenna begins to radiate out of the antenna as invisible waves (Weisman 2000).

Digital transmission clearly involves another order of complexity: (1) The conversion of the analog waveform into thousands of digital samples every second, (2) the transmission and regeneration of these samples within a system that must be tightly synchronized end to end, and (3) the faithful recreation of the voice signal from a stream of abstract 1s and zeros, all involving elaborate processing at both ends and extremely challenging hardware requirements. Although the average person thinks the computer is the marvel of digital electronics, communications applications are much more demanding than most data processing applications, because of the complexity of communications signals and the need to handle huge amounts of data in real time. Ten seconds of T-1 transmission encompasses almost 2 megabytes of data; a DS-4 digital transmission processes more than 300 megabytes in the same interval. Historically, the availability of fast electronic components was the limiting factor on the implementation of digital voice systems. The first PCM systems were experimentally developed during World War II—in fact, the principles of PCM had been elaborated even before the war, but the inadequacy and high cost of digital processing circuits held back the implementation of commercial digital systems until the mid-1960s.

The cost of digital circuitry still determines which applications are suitable for digital techniques. For example, digital microwave radio was unable to compete economically with analog systems until the late 1970s in some niches. Realizing the potential of digital techniques in applications such as mobile radio will call for the development of more highly sophisticated designs, more advanced materials, more complex software, faster processors, and more ingenious mathematics—by comparison with which even the T-carrier may look primitive.

Analog versus Digital

An analog transmitter such as a telephone microphone or a radio transmitter is designed to launch a signal through a transmission medium such as a copper wire or a radio channel. An analog receiver is designed to capture this signal from the medium. Particularly in radio, the transmitter and receiver do not actually function as a system after the initial call setup. They do not exercise control over one another. Both operate more or less independently; the transmitter transmits, and the receiver receives anything within range.

Digital, however, is more complex. It is important to realize that the digital network is not a physical entity, but a logical entity that is superimposed upon a highly heterogeneous set of physical transmission facilities. In the digital network, one telephone call may encompass a copper wire local loop to the local central office, a fiber optic interoffice trunk to the toll switch, a microwave or satellite link from the toll switch to the destination toll switch, and a mobile radio-telephone link to the called party traveling in a car. Yet the digital network possesses the ability to link these disparate facilities into a single digital circuit, the performance parameters of which can be precisely monitored and controlled. It is embodied in an interrelated system of processors which originate, transmit, detect, and decode the messages at different stages in the transmission path. Such a network offers several advantages over the analog system that it is gradually replacing.

Quality Issues and Digital

A digital transmission system allows for communication to be established and maintained at uniform high-quality levels in environments where analog techniques are costly and ineffective. The more difficult the communications link, the more digital communications will stand out over analog. One of the chief areas of digital application today is upgrading the quality of the long-distance segment of the network using digital techniques, particularly optical fiber. Long-distance circuits have always been subject to an assumption of reduced quality. With optical fiber systems, this situation is changing. Another particularly difficult communication environment is the field of mobile radio, where, until very recently, digital techniques have not been applied.

Digital communications is superior because of the following factors.

- *Resistance to noise.* This is due to the inherent resistance of the digital signal to degradation due to channel noise.
- *Resistance to crosstalk.* Crosstalk in cable pairs occurs when the electrical energy from one circuit is magnetically superimposed upon an adjacent circuit. Many analog systems are not noise limited as much as crosstalk limited. In a digital system, crosstalk is largely eliminated by the threshold effect.
- *Error correction.* In more modern digital systems, powerful error-correction techniques have been developed which can be applied to scrub a signal and remove a large percentage of errors from any source. The application of such advanced techniques becomes a matter of weighing costs (the additional digital circuitry, plus, in most cases, a bandwidth penalty) against the benefits (the ability to operate in even noisier channels or in the presence of even more severe crosstalk or interference).
- *Signal regeneration.* By inserting regenerative repeaters at proper intervals, the effects of noise and other transmission degradations can be controlled to any level desired. Many transmission problems involve time variant or statistical sources of signal degradation against which an analog system is often largely helpless. Such problems can be controlled in a properly designed digital system. A digital system can be designed to monitor and adapt to changing channel conditions.
- *Echo control.* This is a reflection of the transmitted signal back to the transmitter from some discontinuity in the transmission path. Analog systems address this issue crudely and with little improvement. Digital systems are more effective. By storing the transmitted speech for a period of time equal to the round-trip delay of the circuit, attenuating the stored signal to the proper level of the returning echo, and then

subtracting it from the incoming return signal, the echo is completely eliminated.

- *Flexibility.* One of the most important characteristics of digital systems is they are usually under tight software control, which greatly increases user and operator flexibility. An operation as simple as changing the telephone number associated with a particular telephone line is impossible for many analog central offices still in operation in the United States today.

Security

The first real work on digital-based radio systems was stimulated during World War II by the need for a truly secure radio-telephone system. Digital communications systems were utilized in military and sensitive government applications long before they became economical in commercial settings, precisely because of the benefit of security. The problem of communications privacy is a serious issue in mobile radio systems, where the ease of interception greatly compromises even routine cellular radio-telephone calls. Analog scrambling systems exist, but they are not accepted as having a high level of security, and they typically degrade voice quality.

By contrast, a digital radio system lends itself readily to intensive encryption, which can be designed to provide almost any level of security desired and has a rather small impact on system performance or economics. Even without encryption, the digitization process itself provides a fair degree of privacy by making inexpensive interception devices harder for amateurs to construct.

PERSONAL COMMUNICATIONS SYSTEM

The personal communications system (PCS) is the current state of cellular telephony. PCS uses the cellular concept, but in comparison with cellular communications, which utilizes frequencies in the 800 MHz range and has both analog and digital capabilities, PCS is purely digital and operates in the 1,900 MHz range. PCS came into existence after it became clear that the explosive growth of cellular telephony would require the FCC to allocate another frequency range for wireless communications to avoid overloading the 800 MHz range. PCS has developed as a digital tool that not only carries voice communications but also provides paging service, fax service, and mobile connection to the vast Internet.

HOW AN ORGANIZATION GETS A PIECE OF SPECTRUM

Usage and technology standards of these narrowbands as well as the rest of the entire wireless spectrum are closely regulated in the United States by the FCC. This regulation is required because radio transmission is omnidirectional; thus, cellular telephones must use a frequency different from that used by broadcast radio or else the two transmissions will interfere with each other.

As mentioned, the FCC has no authority outside the United States. In most countries, national governments regulate the electromagnetic spectrum, although policies may conflict with one another and with the United States. This situation is particularly real in Europe and Japan, where European Union wireless transmission has developed standards differing from those in this country, thus making compatibility of internet appliances impossible in both the European Union and the United States. Today the ITU handles such matters. ITU radio regulations are decided during World Radio Communications conferences.

Prior to the establishment of the FCC in 1934, if users wanted spectrum they simply took it. When commercial broadcast radio was becoming popular in the 1930s and as increasingly powerful radio stations were built, problems started to occur. In 1934, the FCC was established and charged with the allocation and administration of the wireless spectrum, among other things. Since 1994, the FCC has instituted new policies and now most spectrum is allocated by auction bid. Approximately thirty auctions have been conducted as of this writing.

FCC auction rules appear to be numerous and complicated. To enable a fair share of the potential bidders an opportunity to buy spectrum, each auction goes through as many rounds as necessary until the bidding stops. Auctions involve a band or bands of frequencies with one or more licenses being offered depending on the technology involved and the geographical regions covered. For example, if you look at auction 26 in Table 1, you will see that 2,499 licenses were sold in that pager spectrum auction. Compare this with auction 27, which sold FM broadcast spectrum in the 88 MHz to 108 MHz range, and you will see that a single license was granted. It appears that a great deal of time and effort went into making the bidding rules fair overall. Over $32 million was bid in one auction where multiple additional licenses were available; yet, that organization did not get all the licenses. The rules are set up so that no one individual or organization can bid on all licenses.

Cooperation between the FCC and ITU is both beneficial and frequent. Under the provisions of the ITU treaty, the United States is obligated to comply with the spectrum allocations specified in ITU Radio Regulations Article 8 (Table of Frequency Allocations). U.S. domestic spectrum uses may differ from the international allocations, provided these domestic uses do not conflict with other nations' spectrum uses that comply with ITU regulations or agreements.

■ SUMMARY

In this chapter, we discussed the history of radio, from Marconi's first radio transmitter to recent U.S. Navy transmitters. We also learned about the various types of antennas and their characteristics, information about frequencies, and all about the electromagnetic spectrum. We also learned how this electromagnetic spectrum relates to cellular telephony. We also learned about how the basic underlying mechanics such as frequencies, wavelengths, and spectrum relate to a wide range of wireless applications.

REVIEW QUESTIONS

1. Wireless is _____.
   a. communication without wires
   b. unbounded communications
   c. subject to propagation
   d. a and b
   e. all of the above

2. _____ developed the theory of electromagnetic waves.
   a. Marconi
   b. Clerk-Maxwell
   c. Hertz
   d. Antheil

3. _____ developed the first radio.
   a. Marconi
   b. Clerk-Maxwell
   c. Hertz
   d. Antheil

4. The wireless spectrum begins at _____.
   a. 300 MHz
   b. 30 kHz
   c. 3 kHz
   d. none of the above

5. True/False: An antenna transmits electrical energy.

6. The size and shape of an antenna is determined by _____ (select all that apply).
   a. direction of signal
   b. wavelength frequency
   c. manufacturer's unique address
   d. distance of target
   e. directional versus omnidirectional use

7. FDMA stands for _____.
   a. frequency division multiplexing association
   b. frequency division multiple addition
   c. frequency division modulation attenuation
   d. none of the above

8. True/False: CDMA uses spread spectrum technology.

9. The technology that came first, chronologically, is _____.
   a. FDMA
   b. TDMA
   c. CDMA
   d. none of the above

10. True/False: Propagation effects do not affect wireless communications.

11. SNR stands for _____.
    a. signal-to-noise ratio
    b. signaling not radio
    c. single noise ratio
    d. none of the above

12. True/False: The effects of reflection are the same regardless of substance.

13. True/False: Digital technology produces clearer signals in cell phones than analog technology.

HANDS-ON EXERCISES

1. Using a drawing program such as Corel's Photopaint, Adobe's Photoshop, or Microsoft Paint, produce a graphical representation of analog and digital signals.

2. Conduct research into omnidirectional and directional antennas. Using Word table formatting, list the wireless technologies that use each of these antennas.

3. Using your browser, research the history of wavelength and frequency. Discuss the roles of scientists who figured prominently in the study of wavelength and frequency.

4. How did Marconi blaze the trail in wireless communications? Use your browser to conduct research in this issue.

5. Conduct research on the entire electromagnetic spectrum. What are those areas above and below the RF spectrum discussed in this chapter?

6. List the propagation effects and an example of each.

4 Wireless Application Protocol

OBJECTIVES

After reading this chapter and completing the exercises, you will be able to:

- Learn the importance of Wireless Application Protocol in the wireless world
- Learn the various components of Wireless Application Protocol
- Learn how Wireless Application Protocol is applied over various protocol suites and bearer (network) services

KDDI TO LAUNCH NEXT GENERATION SERVICES ACROSS JAPAN IN DECEMBER

SALES OF NEW GPS KEITAI AND MOVIE KEITAI TO START SIMULTANEOUSLY

KDDI Corp. and Okinawa Cellular Telephone Company are due to launch three new cellular phone services across Japan from December this year Eznavigation, a GPS navigation system, and ezmovie, a moving image transmission system, will be the first services of their kind available in Japan through mobile phone networks. The third new service, an all-new cellular phone service compatible with the global mobile Internet standard WAP2.0, will be the first of its kind anywhere in the world.

Accompanying the new services will be two new cellular phone units. The GPS Keitai will give users access to the services and functionality required to use the GPS to determine their exact location. For its part, the Movie Keitai will give users access to the moving image service outlined above. Movies will come complete with stereo sound, and will be available anyplace, anytime, for extremely reasonable prices.

Sales of the new phone units are to commence in early December. Customers will be able to choose from two different models of GPS Keitai, the C3001H and the C3002K, though only one model, the C5001T, will initially be available for the Movie Keitai. At the start of the services, GPS services will offer customers approximately 20 different contents, with moving image services totaling approximately 60.

Source: Wireless Application Protocol Forum Ltd. Retrieved November 12, 2001, www.wapforum.org/new/20020612433new.htm.

BACKGROUND

Most of the technology developed for the Internet has been designed for desktop and larger computers with medium-to-high bandwidth and generally reliable data networks. Fiber optic cable networks with reliability rates exceeding 98 percent have become common. More fiber-based T3 networks are being ordered than copper cable-based networks. The fiber-based networks employ sophisticated software algorithms to provide additional stability. **Connectivity devices** such as routers and switches also have advanced to the point where errors, when they do occur, are handled and

connectivity devices

reported more efficiently than ever before. These factors contribute to a stable and growing technology base for fiber-based network applications, services, and products.

Wireless data networks, however, present a more constrained communications environment compared with wired networks. Because of fundamental limitations of power, available spectrum, and mobility, wireless data networks tend to have the following:

- Less bandwidth
- More latency
- Less connection stability
- Less predictable availability

Problems with the environment have provided a challenge for low power hand-held access devices. More information on this issue can be found in Chapter 3.

Mass-market hand-held wireless devices present a more constrained computing environment compared with desktop computers. Because of fundamental limitations of power and form factor (the proper size and shape of a device), mass-market hand-held devices tend to have the following:

- Less powerful CPUs
- Less memory (ROM and RAM)
- Restricted power consumption
- Smaller displays (sometimes as small as 2 inches)
- Different input devices (e.g., a phone keypad, voice input)

HISTORY

Wireless Application Protocol (WAP)

The **Wireless Application Protocol (WAP)** is a product of the WAP Forum, a consortium of over 400 firms. (Please note that the WAP Forum, in late 2002, changed its name to the Open Mobile Alliance.) How the WAP Forum began provides valuable information into the formation of WAP. The WAP Forum was founded in June 1997, by four firms: Phone.com (formerly Unwired Planet), and Ericsson, Motorola, and Nokia (the world's three largest wireless handset manufacturers). This consortium provides a worldwide standard for the delivery of Internet-based services to mass-market mobile phones. The four cofounders initially published the architecture for this new standard in September 1997.

In January 1998, the founding companies established the WAP Forum, Ltd. to administer the WAP specification process. Membership in the WAP Forum was subsequently opened to firms across the wireless industry to work on the new WAP specification process and to drive the continuing evolution of WAP. The WAP Forum membership of over 400 companies represents all sectors of the industry, including leading wireless network operators, infrastructure providers, handset manufacturers, and software providers (WAP Forum 2001).

The WAP, and the organization that defines and promotes it, the WAP Forum, originally grew out of efforts by cellular phone manufacturers to work with the World Wide Web Consortium (W3C) on a specification for the wireless transmission of data on the Internet. This fit in well with the WAP Forum's mission of creating global wireless protocol specifications that work across the different wireless network technologies.

The WAP Forum and the W3C work in cooperation. The WAP Forum is dedicated to enabling advanced services and applications on **mobile** (wireless) **devices,** such as cellular telephones, and the W3C is dedicated to leading and advancing the development of the World Wide Web. Although the WAP Forum and the W3C have different organizational goals, they share similar goals for the future of the Internet. As an example, W3C's long-term goals for the Web are as follows:

mobile devices

- *Universal access.* To make the Web accessible to all by promoting technologies that take into account the vast differences in culture, education, ability, material resources, and physical limitations of users on all continents
- *Semantic Web.* To develop a software environment that permits each user to make the best use of the resources available on the Web
- *Web of trust.* To guide the Web's development with careful consideration for the novel legal, commercial, and social issues raised by this technology.

Specific goals of the WAP Forum are as follows (WAP Forum 2000):

- To bring Internet content and advanced data services to wireless phones
- To create a global wireless protocol specification that works across all wireless network technologies
- To facilitate the development of content and applications that scale across a wide range of wireless bearer networks and device types
- To embrace and extend existing standards and technology when it is both possible and appropriate

To accomplish these goals, the WAP Forum has developed WAP specifications according to the following design principles (WAP Forum 2000):

- To create license-free standards for the entire industry to use for product development
- To know that the best technology standards can only come about with full industry participation
- To work optimally with all air interfaces in order to best address the needs of the widest possible population of end users
- To maintain WAP specifications independent of any particular device
- To encourage easy, open interoperability between its key components

Part of the challenge of both the WAP Forum and the W3C stems from the different wireless technologies in use worldwide, for example:

- AMPS is used primarily in the United States. It is analog and is the earliest cellular telephony standard.
- TACS is a spin-off of AMPS. This system is used in the United Kingdom and provides more capacity than AMPS. TACS has attempted to expand its market base, but is generally considered a stopgap to true digital cellular networks.
- Nordic Mobile Telephone (NMT) is deployed in the Scandinavian countries. NMT is optimized for use in areas of low population density (it operates in the somewhat unique 450 MHz range), which is ideal for this area of Europe.
- Personal Digital Cellular (PDC) is used primarily in Japan and is a digital cellular technology.
- Digital-AMPS (D-AMPS) is also known as US TDMA. It is a North American digital cellular standard.
- Global System for Mobile (GSM) communications is essentially the standard for pan-European cellular systems. GSM was developed to be the international digital cellular system of choice and is currently deployed in over 100 countries. It is particularly predominant in Europe and much of Asia and actually supports full roaming privileges throughout these countries. GSM is not compatible with the United States–dominant digital standard, PCS.
- PCS is digital and is predominant in the United States. Due to advanced technology features, PCS can exist in the same network as AMPS (it can even operate in both analog and digital modes on the same network), even though AMPS is analog.

Chapter 6 addresses this issue in more detail.

DESIGN AND PRINCIPLES OF OPERATION

CDPD

WAP consists of a series of layers on top of each of the transport technologies used by cellular telephony networks such as **CDPD,** IDEN, and CDMA. It is a protocol similar to HTTP that works over different kinds of wired networks. If one looks at how it relates to the OSI model, we see the corresponding protocol stacks, as shown in Figure 4–1. Devices that use WAP include hand-held digital wireless items such as mobile phones, pagers, two-way radios, smart phones, and communicators. WAP works with most wireless networks.

WAP is both a communications protocol and an application environment. It can be built on top of any operating system, for example:

PalmOS—the operating system for Palm's series of PDAs
EPOC—the operating system for Ericsson, Psion, and Nokia hand-held devices

OSI Model	WAP Model
Layer	
7 HTTP, FTP, SMTP (e-mail), Telnet	WAP
6	WAP
5	WAP
4 TCP, UDP, etc.	WAP
3 IP, IPX, etc.	Cellular networks protocols
2 Ethernet, HDLC, etc.	Cellular networks protocols
1 Bits over wires	Bits over airwaves

Figure 4–1
OSI versus WAP
Protocol Stacks

Windows CE—Microsoft's operating system for small devices such as
notebooks

FLEXOS—an operating system designed for palm devices and smart-
phones [The developer is Symbian, a joint venture of Psion, Ericsson, Nokia, Motorola, and Matsushita (Panasonic).]

OS/9—MAC's operating system with hand-held extension capabilities

JavaOS—a highly compact operating system designed to run Java
applications directly on microprocessors in anything from net computers to pagers (JavaSoft, an operating company of Sun Microsystems, Inc., is the developer.)

In short, a WAP phone is like a miniature web browser that can interact
with any Internet-based application. This means that WAP phones can access
much of the Internet's text content. Graphical content, however, is clearly an
area with limitations, because WAP devices are small, although plans are in
progress to bring larger color screens to a wide range of WAP-enabled devices within the next few years (Fowlie 2001).

The WAP Forum, in order to promote the standardized design and use
of a wireless protocol, needed first to develop a specification outlining the details of the approach and the underlying protocols.

The WAP specification is designed to bring Internet access to the wireless mass market. By building open specifications and encouraging communication and technical exchanges among the industry players, the WAP
Forum encourages and fosters market development.

Key elements of the WAP specification include the following:

- *A definition of the WAP programming model.* This model will provide
  guidance for programmers on the design and proposed operation of
  the WAP.
- *A markup language based on XML.* The language is designed to enable
  powerful applications within the constraints of hand-held devices. A

extensible markup language (XML)

markup language is a software language that allows operation of software using a web-centered approach. **Extensible markup language (XML)** is an adaptable way to create common information formats and to share both the format and the data on the World Wide Web or any other network.

- *A specification for a microbrowser.* Found in the wireless terminal that controls the user interface, a microbrowser is similar to a standard web browser such as Netscape Navigator or Microsoft's Internet Explorer.
- *A thin (i.e., not much overhead) protocol stack.* This will act to lessen bandwidth requirements, thus guaranteeing that a wide variety of wireless network types can run WAP applications.
- *A secure connection.* The connection is between the **client** and the application **server.**

client

server

The WAP specification is a truly open standard that enables public content, corporate intranet, and carrier-specific solutions to reach wireless subscribers. The WAP specification leverages and extends existing internet standards, enabling application developers to tailor their content to the special needs of wireless users. The ultimate beneficiaries are wireless subscribers who can be more productive than ever before.

To meet the requirements of mobile network operators, solutions must meet the following requirements.

1. *Interoperability.* Terminals from different manufacturers communicate with services in the mobile network.
2. *Scalability.* Mobile network operators are able to scale services to customer needs.
3. *Efficiency.* Mobile network operators provide quality of service suited to the behavior and characteristics of the mobile network.
4. *Flexibility.* Mobile network operators provide for maximum users for a given network configuration.
5. *Reliability.* Mobile network operators provide a consistent and predictable platform for deploying services.
6. *Security.* Mobile network operators enable services to be extended over potentially unprotected mobile networks while still preserving the integrity of **user data;** protects the devices and services from security problems such as denial of service, which has happened recently at several popular sites such as Yahoo.com and eBay.

user data

Wireless network service providers (e.g., Cingular, Verizon, and Cellular One) operate under several fundamental constraints, which place restrictions on the type of protocols and applications offered over the network. The following constraints are both similar and different from traditional bounded network issues.

- Power consumption is a constraint that traditional bounded network operators do not suffer. However, in a cellular network, one must always be concerned with the limited power of the devices utilized, such as cellular phones.

- Cellular network economics is a similar issue to bounded media networks such as the PSTN. In both cases, the economics are in a constant state of flux due to changing market conditions, customer expectations, and technology advances.
- Latency is an issue present in all networks, but particularly in wireless networks using satellite transmissions. As an example, during the Desert Storm conflict, communications with correspondents in Baghdad required significant delays due to latency time required to reach geosynchronous satellites.
- Bandwidth technology at this stage of development is limited because of a series of environmental, health, and power availability and generation issues. More information on these issues can be found in Chapter 3.

Many wireless devices (e.g., cellular phones and pagers) are consumer devices used in a wide variety of environments and under a wide range of use scenarios. For example, the user interface must be extremely simple and easy to use. Many mobile devices, in particular cellular telephones, are mass-market consumer-oriented devices. Wireless also must utilize single-purpose devices. The goal and purpose of most mobile devices is highly focused (e.g., voice communication), which is in contrast with the general-purpose tool-oriented nature of a personal computer. This motivates a specific set of use cases, with simple and focused behavior (for example, "place a voice call" or "find the nearest ATM"). Finally, wireless devices should offer hands-free, heads-up operation. Many mobile devices are used in environments where the user should not be unnecessarily distracted (e.g., driving and talking).

The WAP specification addresses the limitations of wireless networks and devices by utilizing the existing standards and developing new extensions where needed. It also enables industry participants to develop solutions that are interface independent, device independent, and fully interoperable. The WAP solution encompasses the tremendous investment private industry has in web servers, web development tools, web programmers, and web applications. In addition, it solves many of the problems associated with the wireless domain. The specification further ensures that this solution is fast, reliable, and secure. It enables developers to use existing tools to produce sophisticated applications that have an intuitive user interface.

The WAP architecture specification meets the needs of the WAP Forum through the development and implementation of a stable, flexible, and efficient system architecture for the wireless web community. To enable operators and manufacturers to meet the challenges in advanced services, differentiation, and fast and flexible service creation, the WAP Forum defines a set of protocols in transport, security, transaction, session, and application layers. These are similar to those found in the OSI or TCP/IP reference models.

Mobile networks are growing in both size and complexity, and the costs of providing efficient services are increasing. This certainly complicates matters for both the public sector and private industry, because solutions must meet a multitude of requirements to be useful. For example, to meet the requirements of mobile network operators, solutions must be all of the following:

1. *Interoperable.* Devices from different manufacturers must communicate with services in the mobile network.
2. *Scalable.* Mobile network operators must be able to scale services to a wide range of customer needs.
3. *Efficient.* Quality of service must be suited to the behavior and characteristics of the mobile network.
4. *Flexible.* Solutions must conform to meet changeable conditions in environment, technology, and/or user demands.
5. *Reliable.* A consistent and predictable platform must be provided for deploying services (at best, as can be done in the changing environment).
6. *Secure.* Services are to be extended over vulnerable mobile networks while still preserving the integrity of user data. Devices and services must be protected from security problems such as denial of service.

In short, the WAP specification addresses mobile network characteristics and operator needs by adapting existing network technology to the special requirements of mass-market, hand-held wireless data devices, and by introducing new technology when appropriate.

WAP ARCHITECTURE REQUIREMENTS

Requirements of the WAP Forum architecture are as follows:

1. Leverage existing standards when possible, such as HTTP and URLs.
2. Define a layered, scalable, and extensible architecture.
3. Support as many diverse wireless networks as possible.
4. Optimize for narrowband bearers with potentially high latency.
5. Optimize for efficient use of device resources (low memory/CPU usage/power consumption).
6. Provide support for secure applications and communication.
7. Enable the creation of human–computer interfaces with maximum usability, flexibility, and vendor control.
8. Provide access to local handset functionality, such as logical indication for incoming call and/or voice mail indication.
9. Support multivendor interoperability by defining the optional and mandatory components of the specifications.
10. Provide a concise programming model for telephony services and integration.

WAP ARCHITECTURE OVERVIEW

World Wide Web (WWW) usage increases every day (although estimates vary, depending on the source), as shown in Figure 1–1. It is the logical model in which WAP should interoperate and emulate. In addition, the Internet WWW architecture provides a flexible and powerful programming model, as shown in Figure 4–2. This WWW model presents applications and content in standard data formats, which are browsed by applications known as web browsers. The web browser is a networked application; in other words, it sends requests for named data objects to a network server and the network server responds with the data encoded using the standard formats. The operable term here is *standard*. Because this work had already been done, and is a stable and productive model, the WAP architecture attempts to emulate and interface with it.

The existing WWW standards specify the following mechanisms which are necessary to build a general-purpose, flexible, and stable application environment.

- *Standard naming model.* All servers and content on the WWW are named with an Internet-standard uniform resource locator (URL) with the domain naming service (DNS) translating the URLs to site addresses.

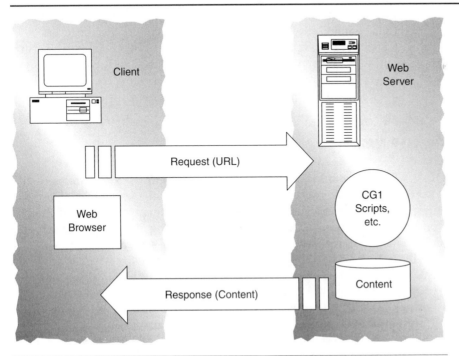

Figure 4–2
WWW Model

JavaScript

user

proxy

- *Content typing.* All content on the WWW is given a specific type, thereby allowing web browsers to correctly process the content based on its type.
- *Standard content formats.* All web browsers support a set of standard content formats, including the hypertext markup language, **JavaScript** scripting language, and others.
- *Standard protocols.* These protocols allow any web browser to communicate with any web server. The most commonly used protocol on the WWW is the HyperText Transport Protocol (HTTP).

This infrastructure enables users to easily reach a large number of third-party applications and content services. It also allows application developers to create applications and content services for a large community of clients. This approach has proven effective in meeting these goals, even with an expanding and diverse **user** and application base.

WWW protocols typically refer to three classes of servers.

1. Origin server, the server on which a given resource (content) resides or is to be created.
2. **Proxy,** an intermediary program that acts as both a server and a client for the purpose of making requests on behalf of other clients. The proxy typically resides between clients and servers that have no means of direct communication, for example, across a firewall. Requests are either serviced by the proxy program or passed on, with possible translation, to other servers. A proxy must implement both the client and server requirements of the WWW specifications.
3. Gateway, a server that acts as an intermediary for some other server. Unlike a proxy, a gateway receives requests as if it were the origin server for the requested resource. The requesting client may not be aware that it is communicating with a gateway.

THE WAP MODEL

The WAP programming model, as shown in Figure 4–3, is similar to the WWW programming model. The WAP model provides several benefits to the application developer community, including a familiar programming model, a proven architecture, and the ability to use existing tools (e.g., web servers, XML tools, JavaScript, and URLs). Over the past several years, extensions have been introduced to match the characteristics of the wireless environment. When possible, existing standards have been adopted or used as the starting point for the WAP technology.

WAP content and applications are specified in a set of well-known content formats based on the familiar WWW content formats. Content is transported using a set of standard communication protocols based on the WWW communication protocols. A microbrowser in the wireless terminal coordinates the user interface and is comparable to a standard web browser.

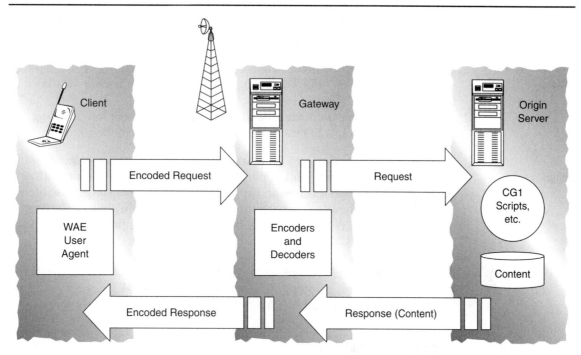

Figure 4–3
WAP Model

WAP defines the following set of standard components which enable communication between mobile devices and network servers.

- *Standard naming model.* WWW-standard URLs are used to identify WAP content on origin servers and local resources in a device; for example, call control functions.
- *Content typing.* All WAP content is given a specific type consistent with WWW typing. This allows WAP user agents to correctly process the content based on its type.
- *Standard content formats.* WAP content formats are based on WWW technology and include display markup, calendar information, electronic business card objects, images, and scripting language.
- *Standard communications protocols.* WAP communications protocols enable the communication of browser requests from the mobile terminal to the network web server.

The WAP content types and protocols have been optimized for mass-market hand-held wireless devices to provide functionality to as large a group of industry and, by definition, users as possible. WAP utilizes proxy technology to connect the wireless domain to the WWW. This is similar to the use of mobile telephone switching offices (MTSOs) in cellular networks which connect to the Public Switching Telephone Network (PSTN). The WAP proxy typically consists of the following functionality.

- The protocol gateway translates requests from the WAP stack (WSP, WTP, WTLS, and WDP) to the WWW protocol stack (HTTP and TCP/IP).
- Content encoders (decoders) translate WAP content into compact encoded formats to reduce the size of data over the network.

author

This infrastructure ensures that mobile terminal users can browse a wide variety of WAP content and applications, and that the application **author** is able to build content services and applications that run on a large base of mobile terminals. The WAP proxy allows content and applications to be hosted on standard WWW servers and to be developed using proven WWW technologies such as CGI scripting.

Although the typical use of WAP will include a web server, WAP proxy, and WAP client, the WAP architecture can easily support other configurations. It is possible to create an origin server that includes the WAP proxy functionality. Such a server might be used to facilitate end-to-end security solutions

Example WAP Network

To fully understand how a WAP network may look, refer to Figure 2–2.

In the example, the WAP client communicates with two servers in the wireless network. The WAP proxy translates WAP requests to WWW requests, thereby allowing the WAP client to submit requests to the web server. The proxy also encodes the responses from the web server into the compact binary format understood by the client.

If the web server provides WAP content (e.g., WML), the WAP proxy retrieves it directly from the web server. However, if the web server provides WWW content (such as HTML), a filter is used to translate the WWW content into WAP content. For example, the HTML filter would translate HTML into WML.

WAP ARCHITECTURE COMPONENTS

The WAP architecture provides a scalable and extensible environment for application development for mobile communications devices, through a layered design of the entire protocol stack, as shown in Figure 4–4. Each layer of the architecture is accessible by the layers above, as well as by other services and applications.

The WAP layered architecture enables other services and applications to utilize the features of the WAP stack through a set of well-defined interfaces.

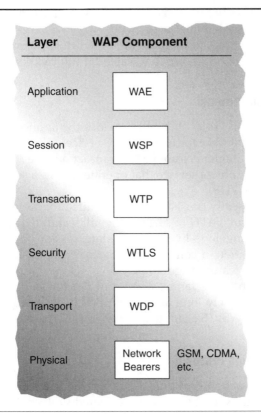

Figure 4–4
WAP Architecture

External applications may access the session, transaction, security, and transport layers directly. The following sections provide a description of the various elements of the protocol stack architecture.

Wireless Application Environment

The wireless application environment (WAE) is a general-purpose application environment based on a combination of WWW and mobile telephony technologies. The primary objective of the WAE effort is to establish an interoperable environment that will allow operators and service providers to build applications and services that can reach a wide variety of different wireless platforms in an efficient and useful manner. WAE includes a microbrowser environment containing the following functionality.

- **Wireless markup language (WML)**—a lightweight markup language, similar to HTML, but optimized for use in hand-held mobile terminals
- **WMLScript**—a lightweight scripting language, similar to JavaScript

wireless markup language (WML)

WMLScript

- Wireless telephony application (WTA, WTAI)—telephony services and programming interfaces
- Content formats—a set of well-defined data formats, including images, phone book records, and calendar information

Wireless Session Protocol

The Wireless Session Protocol (WSP) provides the application layer of WAP with a consistent interface for two session services. The first is a connection-oriented service that operates above the transaction layer protocol WTP. The second is a connectionless service that operates above a secure or nonsecure datagram service (WDP).

The WSPs currently consist of services suited for browsing applications (WSP/B), which provide HTTP/1.1 functionality and semantics in a compact over-the-air encoding, long-lived session state, session suspend and resume with session migration, a common facility for reliable and unreliable data push, and protocol feature negotiation.

bearer networks

Protocols in the WSP family are optimized for low-bandwidth **bearer networks** with relatively long latency. WSP/B is designed to allow a WAP proxy to connect a WSP/B client to a standard HTTP server.

Wireless Transaction Protocol

The Wireless Transaction Protocol (WTP) runs on top of a datagram service and provides a lightweight transaction-oriented protocol that is suitable for implementation in "thin" clients (mobile stations). WTP operates efficiently over secure or nonsecure wireless datagram networks and provides the following features.

transaction

- Three classes of **transaction** service: unreliable one-way requests, reliable one-way requests, and reliable two-way request-reply transactions
- Optional user-to-user reliability: WTP user to trigger the confirmation of each received message
- Optional out-of-band data on acknowledgments

protocol data unit (PDU)

- **Protocol data unit (PDU)** concatenation and delayed acknowledgment to reduce the number of messages sent
- Asynchronous transactions

Wireless Transport Layer Security

Wireless transport layer security (WTLS) is a security protocol based on the industry standard transport layer security (TLS) protocol, formerly known as secure sockets layer (SSL). WTLS is intended for use with the WAP transport protocols and has been optimized for use over narrowband communication channels. WTLS provides the following features.

- *Data integrity.* WTLS contains facilities to ensure that data sent between the **terminal** and an application server are unchanged and uncorrupted.   **terminal**
- *Privacy.* WTLS contains facilities to ensure that data transmitted between the terminal and an application server are private and undecipherable by any intermediate parties that may have intercepted the data stream.
- *Authentication.* WTLS contains facilities to establish the authenticity of the terminal and application server.
- *Denial-of-service protection.* WTLS contains facilities for detecting and rejecting data that are replayed or not successfully verified. WTLS makes many typical denial-of-service attacks harder to accomplish and protects the upper protocol layers.

WTLS may also be used for secure communication between terminals, such as for authentication of electronic business card exchange. Applications are able to selectively enable or disable WTLS features depending on their security requirements and the characteristics of the underlying network (e.g., privacy may be disabled on networks already providing this service at a lower layer).

Wireless Datagram Protocol

The transport layer protocol in the WAP architecture is referred to as the Wireless Datagram Protocol (WDP). The WDP layer operates above the data-capable bearer services supported by the various network types. As a general transport service, WDP offers a consistent service to the upper layer protocols of WAP and communicates transparently over one of the available bearer services.

Because the WDP protocols provide a common interface to the upper layer protocols, the security, session, and application layers are able to function independently of the underlying wireless network. This is accomplished by adapting the transport layer to specific features of the underlying bearer. By keeping the transport layer interface and the basic features consistent, global interoperability can be achieved using mediating gateways.

Network Operators (Bearers)

WAPs are designed to operate over a variety of different bearer services, including short message, circuit-switched data, and packet data. The bearers offer differing levels of quality of service with respect to throughput, error rate, and delays. The WAP protocols are designed to compensate for or tolerate these varying levels of service.

Because the WDP layer provides the convergence between the bearer service and the rest of the WAP stack, the WDP specification lists the bearers that are supported and the techniques used to allow WAPs to run over each bearer. The list of supported bearers will change over time, with new bearers being added as the wireless market evolves.

Other Services and Applications

The WAP layered architecture enables other services and applications to utilize the features of the WAP stack through a set of well-defined interfaces. External applications may access the session, transaction, security, and transport layers directly. This allows the WAP stack to be used for applications and services not currently specified by WAP but deemed to be valuable for the wireless market. For example, applications (electronic mail, calendar, phone book, notepad, and electronic commerce) or services (white and yellow pages) may be developed to use WAP.

Sample Configurations of WAP Technology

WAP technology is expected to be useful for applications and services beyond those specified by the WAP Forum. Figure 4–5 depicts several possible protocol stacks using WAP technology.

The leftmost stack represents a typical example of a WAP application (i.e., WAE user agent) running over the complete portfolio of WAP technology. The middle stack is intended for applications and services that require transaction services with or without security. The rightmost stack is intended

Figure 4–5
WAP Protocol Stacks

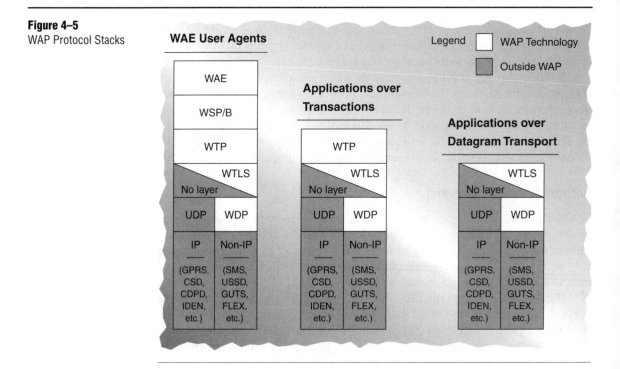

for applications and services that only require datagram transport with or without security.

COMPLIANCE AND INTEROPERABILITY

The WAP Forum views multivendor interoperability as an important element to the success of WAP products. To provide as high a probability as is technically possible that two WAP products developed independently by two different vendors will successfully interoperate, a rigorous definition of conformance, compliance, and testing must be developed. To this end, the WAP Forum has created a WAP conformance specification and is working to maintain current information relating to all issues of WAP interoperability.

Successful interoperability can only be achieved by testing products. Testing can be divided into the two broad categories of static testing and dynamic testing. Static testing is a manufacturer's statement of the capabilities and functions of a product. Static testing will identify obvious areas of incompatibility between two products, in other words, where one implements a feature that the other does not support. All WAP specifications will provide a means for static testing in the form of a protocol implementation conformance statement (PICS).

Dynamic testing is the real form of testing that leads to a high degree of confidence that two products will successfully interoperate. Dynamic testing involves the execution or exercising of a product in a live environment, ultimately proving that the product meets the stated claims given in the static test (i.e., PICS). There are three general approaches to dynamic testing: pair-wise testing or bake-offs; use of a reference implementation against which all products are measured; and definition of formal test suites containing test cases to be run against a product in a testing laboratory. Each of these approaches to dynamic testing has cost trade-offs and some technical pluses and minuses. The cost of each approach is related to the total number of products to be tested. The WAP Forum will promote the most cost-effective method that leads to the greatest degree of confidence for successful interoperability, given the total number of WAP products available in the market at a given time. This evolutionary approach will change over time as the WAP industry matures. As a starting point, the WAP Forum is promoting pair-wise testing in a laboratory environment for new WAP products. As the WAP industry evolves, reference implementations may be identified, followed by the definition of formal test suites for WAP specifications.

WAE OVERVIEW

Wireless application environment is a result of the WAP efforts to promote industrywide standards and specifications for developing applications and services that operate over wireless communications networks. WAE specifies an application framework for wireless devices such as mobile telephones, pagers, and PDAs. The framework extends and leverages other WAP technologies, including WTP and WSP, as well as other Internet technologies such as XML, URLs, scripting, and various content formats. The effort is aimed at enabling operators, manufacturers, and content developers to meet the challenges of implementing advanced differentiating services and applications in a fast and flexible manner.

Background

The WAE effort is an undertaking to build a general-purpose application environment based fundamentally on WWW technologies and philosophies. It is part of the overall WAP effort. The primary objective of the WAE effort is to establish an interoperable environment that will allow operators and service providers to build applications and services that can reach a wide variety of different wireless platforms in an efficient and useful manner.

The output of the WAE effort is a collection of technical specifications that are either new or based on existing and proven technologies. Among the existing technologies leveraged by the WAE effort are the following:

- Unwired Planet's hand-held markup language (HDML)
- W3C's hypertext markup language (HTML)
- ECMA-262 Standard ECMAScript Language Specification, based on JavaScript
- IMC's calendar data exchange format (vCalendar) and phonebook data exchange format (vCard)
- Various WWW technologies such as URLs and HTTP
- Various mobile network technologies such as GSM call control services, and generic IS-136 services such as send flash

The resulting WAE technologies are not fully compliant to all of the motivating technologies. When necessary, modifications were made to better integrate the elements into a cohesive environment and to better optimize the interaction and user interface for small-screen, limited-capability terminals that communicate over wireless networks.

WAE Objectives

The main objectives of the WAE effort are:

1. To define an application architecture model:
   - that fits within the WAP architecture and meets WAP's overall objectives.
   - that is suitable for building interactive applications that function well on devices with limited capabilities including limited memory, small screen size, limited battery life, and restricted input mechanisms.

- that is suitable for building interactive applications that function well in narrowband environments with medium-to-high latencies.
- that employs appropriate security and access control features to allow safe execution of anonymous and third-party content.
- that leverages established and common standards and technologies that make WAE implementation simpler, and allows third-party developers to create and deploy applications inexpensively.
- that is global and supports established internationalization technologies and practices.

2. To define a general-purpose application programming model:
   - that is rich and enables interactive applications on current and future wireless devices.
   - that is based on the Internet's WWW programming model including both browsing and scripting services.
   - that provides access to common mobile device functionality and services such as phonebooks, messaging services, and call control services.
   - that enables applications to be accessible to a wide range of devices.
   - that enables creation of applications that behave well on all WAP-compliant devices.
   - that allows developers to leverage specific functionality of specific devices.
3. To provide network operators the means to enhance and extend network services.
4. To enable multivendor interoperability.

Goals and Requirements

The following list summarizes the requirements of the wireless application environment.

- WAE must enable simple yet efficient, meaningful, and powerful application development and execution environments.
- WAE must provide a general framework. WAE cannot assume that a browser is the controlling agent in the device, nor can it assume that a browser is running at all times. Other applications may exist in the device, in which case WAE must not prevent such applications from coexisting or even integrating with a browser. In addition, those other applications should be able to access and leverage common WAE services on the device where appropriate.
- WAE must not dictate or assume any particular user interface. WAE implementations must be able to introduce new user interface models or use existing HCI models. Implementers must be able to present end users with a consistent and meaningful HCI suitable to the targeted device.

- WAE must be suited for a wide variety of limited-capability devices. WAE must have a small memory footprint and limited computational power requirement. WAE must be suitable for the current generation of wireless devices without jeopardizing its ability to evolve and support future generations of those devices.
- WAE must promote and incorporate efficient means to reduce the amount and frequency of over-the-air data exchanges with origin servers. WAE must provide the means to communicate device capabilities to origin servers, which would enable origin server-side optimizations and further minimize over-the-air resource consumption. In addition, WAE network services must be based on WAP's network protocol stack.
- WAE must support internationalization and localization using standard or well-accepted practices and methods.
- WAE must not compromise WAP's security model. WAE must include meaningful access control mechanisms that ensure secure processing of network accessed content.
- WAE must promote and enable interoperable implementation between various manufacturers and content or service providers.
- WAE must include extensions to allow the means for call control and messaging, as well as enabling a standard set of value-added call and feature control capabilities.
- WAE must allow network operators to introduce new operator-specific features to their implementations.

WAE Architecture Overview

user agents

The WAE architecture includes all elements of the WAP architecture related to application specification and execution. At this point, the WAE architecture is predominately focused on the client-side aspects of WAP's system architecture; namely, items relating to **user agents.** Specifically, the WAE architecture is defined primarily in terms of networking schemes, content formats, programming languages, and shared services. Interfaces are not standardized and are specific to a particular implementation. This approach allows WAE to be implemented in a variety of ways without compromising interoperability or portability. This approach has worked particularly well with a browser (a class of user agents) model such as that used in the Web. The Internet and the WWW are the inspiration and motivation behind significant parts of the WAE specification; and consequently, a similar approach is used within WAE.

THE WWW MODEL

The Internet's World Wide Web provides a flexible and powerful logical model. Applications present content to a client in a set of standard data formats that are browsed by client-side user agents known as web browsers (or simply browsers). Typically, a user agent sends requests for one or more named data objects (or content) to an origin server. An origin server responds with the requested data expressed in one of the standard formats known to the user agent (e.g., HTML).

The WWW standards include the following mechanisms necessary to build a general-purpose environment.

- All resources on the WWW are named with Internet-standard uniform resource locators (URLs).
- All classes of data on the WWW are given a specific type, allowing the user agent to correctly distinguish and present them appropriately. Furthermore, the WWW defines a variety of standard content formats supported by most browser user agents. These include HTML, JavaScript scripting language, and others such as bitmap image formats.
- The WWW also defines a set of standard networking protocols allowing any browser to communicate with any origin server. One commonly used protocol on the WWW today is HTTP.

The WWW infrastructure and model has allowed users to easily reach a large number of third-party content and applications. It has allowed authors to easily deliver content and services to a large community of clients using various user agents (e.g., Netscape Navigator and Microsoft Internet Explorer).

THE WAE MODEL

WAE adopts a model that closely follows the WWW model. All content is specified in formats that are similar to the standard Internet formats. Content is transported using standard protocols in the WWW domain and an optimized HTTP-like protocol in the wireless domain. WAE has borrowed from WWW standards, including authoring and publishing methods, when possible. The WAE architecture allows all content and services to be hosted on standard web origin servers that can incorporate proven technologies (e.g., CGI). All content is located using WWW standard URLs.

WAE enhances some of the WWW standards in ways that reflect the device and network characteristics. WAE extensions are added to support mobile network services such as call control and messaging. Careful attention goes to the memory and CPU processing constraints found in mobile terminals. Support for low bandwidth and high latency networks is included in the architecture as well.

WAE assumes the existence of gateway functionality responsible for encoding and decoding data transferred from and to the mobile client. The purpose of encoding content delivered to the client is to minimize the size of data sent to the client over the air as well as to minimize the computational energy required by the client to process that data. The gateway functionality can be added to origin servers or placed in dedicated gateways as illustrated earlier in Figure 4–4.

Following are the major elements of the WAE model.

- WAE user agents—client-side in-device software that provides specific functionality (e.g., display content) to the end user. User agents (such as browsers) are integrated into the WAP architecture. They interpret network content referenced by a URL. WAE includes user agents for the two primary standard contents: encoded WML and compiled WMLScript.
- Content generators—applications (or services) on origin servers (e.g., CGI scripts) that produce standard content formats in response to requests from user agents in the mobile terminal. WAE does not specify any standard content generators, but expects that there will be a great variety available running on typical HTTP origin servers commonly used in WWW today.
- Standard content encoding—a set of well-defined content encoding, allowing a WAE user agent (e.g., a browser) to conveniently navigate web content. Standard content encoding includes compressed encoding for WML, bytecode encoding for WMLScript, standard image formats, a multipart container format, and adopted business and calendar data formats.
- Wireless telephony applications (WTAs)—a collection of telephony-specific extensions for call and feature control mechanisms that provide authors (and ultimately end users) advanced mobile network services.

The resulting WAE architecture fits within a model that does the following:

- Leverages the Internet (i.e., the model takes advantage of standards, technology, and infrastructure developed for the Internet)
- Leverages thin-client architecture advantages (e.g., service deployment has significantly lower cost per device due to the device-independent nature of WAE and the centralized management of the services at the origin servers)
- Provides end-user advanced mobile network services through network operator–controlled telephony value-added services
- Provides the means for vendors to build differentiating user-friendly services that can take advantage of WWW and mobile network services
- Provides an open extensible framework for building wireless services

Typically, a user agent on the terminal initiates a request for content; however, not all content delivered to the terminal will result from a terminal-side request. For example, WTA includes mechanisms that allow origin servers to deliver generated content to the terminal without a terminal's request.

In some cases, what the origin server delivers to the device may depend on the characteristics of the device. The user agent characteristics are communicated to the server via standard capability negotiation mechanisms that allow applications on the origin server to determine characteristics of the mobile terminal device. WAE defines a set of user agent capabilities that will be exchanged using WSP mechanisms. These capabilities include such global device characteristics as WML version supported, WMLScript version supported, floating-point support, and image formats supported.

URL Naming

WAE architecture relies heavily on WWW's URL and HTML semantics. WAE assumes the existence of a generalized architecture for describing gateway behavior for different types of URLs, and support for connection to at least one WAP gateway.

In particular, the URL naming mechanisms used in WAE are motivated by the following scenarios.

- A secure service (e.g., banking or brokerage) where an end-to-end secure connection using WTLS is necessary, mandating a secure gateway controlled by the content provider
- A content provider who wants to provide a caching gateway that will cache the encoded content to improve performance
- A specialized service with a built-in server that will be accessible only to WAP devices and, therefore, wants to use WSP natively rather than incur the higher overhead of processing HT~sessions

WAE is based on the architecture used for WWW proxy servers. The situation in which a user agent (e.g., a browser) must connect through a proxy to reach an origin server (i.e., the server that contains the desired content) is similar to a wireless device accessing a server through a gateway.

Most connections between the browser and the gateway use WSP, regardless of the protocol of the destination server. The URL, used to distinguish the desired content, always specifies the protocol used by the destination server regardless of the protocol used by the browser to connect to the gateway. In other words, the URL refers only to the destination server's protocol and has no bearing on what protocols may be used in intervening connections.

In addition to performing protocol conversion by translating requests from WSP into other protocols and the responses back into WSP, the gateway also performs content conversion. This is analogous to HTML/HTTP proxies available on the Web today. For example, when an HTTP proxy receives an FTP or Gopher directory list, it converts the list into an HTML document that presents the information in a form acceptable to the browser. This conversion is analogous to the encoding of content destined to WAE user agents on mobile devices.

Currently, only one scheme is expected to be supported by WAE user agents: HTTP. The browser, in this case, communicates with the gateway using WSP. The gateway in turn would provide protocol conversion functions to connect to an HTTP origin server. As an example, a user with a WAP-compliant telephone requests content using a specific URL. The telephone browser connects to the operator-controlled gateway with WSP and sends a Get request with that URL. The gateway resolves the host address specified by the URL and creates an HTTP session to that host. The gateway performs a request for the content specified by the URL. The HTTP server at the contacted host processes the request and sends a reply (e.g., the requested content). The gateway receives the content, encodes it, and returns it to the browser.

Components of WAE

WAE is divided into two logical layers: user agents, which include such items as browsers, phonebooks, and message editors, and services and formats, which include common elements and formats accessible to user agents such as WML, WMLScript, image formats, and vCard and vCalendar formats.

WAE separates services from user agents and assumes an environment with multiple user agents. This logical view, however, does not imply or suggest an implementation. For example, WAE implementations may choose to combine all the services into a single user agent. Others may choose to distribute the services among several user agents. The resulting structure of a WAE implementation is determined by the design decisions of its implementers and should be guided by the specific constraints and objectives of the target environment.

WAE User Agents

The WML user agent is a fundamental user agent of the WAE; however, WAE is not limited to a WML user agent. WAE allows the integration of domain-specific user agents with varying architectures and environments. In particular, a WTA user agent has been specified as an extension to the WAE specification for the mobile telephony environments. The WTA extensions allow authors to access and interact with mobile telephone features (e.g., call control) and other applications assumed on the telephones such as phonebooks and calendar applications.

WAE SERVICES AND FORMATS

The WAE services and formats layer includes the bulk of technical contribution of the WAE effort. The following section provides an overview of the major components of WAE, including the wireless markup language, the wireless markup scripting language, WAE applications, and WAE-supported content formats.

Wireless Markup Language

WML is a tag-based document language; in particular, it is an application of a generalized markup language. WML shares a heritage with the WWW's HTML and hand-held device markup language (HDML). WML is specified as an XML document type. It is optimized for specifying presentation and user interaction on limited capability devices such as telephones and other wireless mobile terminals.

WML and its supporting environment were designed with certain small narrowband device constraints in mind, including small displays, limited user-input facilities, narrowband network connections, limited memory resources, smaller displays, limited user-input facilities, narrowband network connections, limited memory resources, and limited computational resources. Given the wide and varying range of terminals targeted by WAP, considerable effort was put into the proper distribution of presentation responsibility between the author and the browser implementation.

WML is specified in a way that allows presentation on a wide variety of devices, yet allows for vendors to incorporate their own user interfaces. For example, WML does not specify how implementations request input from a user. Instead, WML specifies the intent in an abstract manner. This allows WML to be implemented on a wide variety of input devices and mechanisms. Implementations may, for example, choose to solicit user input visually like many WWW user agents, or it may choose to use a voice-based interface. The user agent must decide how to best present all elements within a card depending on the device capabilities. For example, certain user agents on devices with larger displays may choose to present all the information in a single card at once. Others with smaller displays may break the content up across several units of displays.

WML provides these features:

- *Support for text and images.* WML provides the authors with a means to specify text and images to be presented to the user. This may include layout and presentation hints. As with other markup languages, WML requires the author to specify the presentation in general terms and gives the user agent freedom to determine exactly how the information is presented to the end user. WML provides a set of text markup elements including various *emphasis* elements (e.g., bold, italic, big.); various *line breaks* models (e.g., line wrapping, line wrapping suppression); and *tab columns* that support simple tabbing alignment.
- *Support for user input.* WML supports several elements to solicit user input. The elements can be combined into one or more cards. All requests for user input are made in abstract terms, allowing the user agent the freedom to optimize features for the particular device. WML includes

a small set of input controls. For example, WML includes a text entry control that supports text and password entry. Text entry fields can be masked, preventing the end user from entering incorrect character types. WML also supports client-side validation by allowing the author to invoke scripts at appropriate times to check the user's input. WML includes an option selection control that allows the author to present the user with a list of options that can set data, navigate among cards, or invoke scripts. WML supports both single and multiple option selections. WML also includes task invocation controls. When activated, these controls initiate a navigation or a history management task such as traversing a link to another card (or script) or popping the current card off of the history stack. The user agent is free to choose how to present these controls. The user may, for example, bind them to physical keys on the device, render button controls in a particular region of the screen (or inline within the text), or bind them to voice commands.

- *Navigation and history stack.* WML allows several navigation mechanisms using URLs. It also exposes a first-class history mechanism. Navigation includes HTML-style hyperlinks, intercard navigation elements, and history navigation elements.
- *International support.* WML's document character set is Unicode. This enables the presentation of most languages and dialects.
- *MMI independence.* WML's abstract specification of layout and presentation enables terminal and device vendors to control the MMI design for their particular products.
- *Narrowband optimization.* WML includes a variety of technologies to optimize communication on a narrowband device, including the ability to specify multiple user interactions (cards) in one network transfer (a deck). It also includes a variety of state management facilities that minimize the need for origin server requests. WML includes other mechanisms to help improve response time and minimize the amount of data exchanged over the air. For example, WML allows the author to parameterize (or pass variables to) a subsequent context. It supports variable substitution and provides out-of-band mechanisms for client-side variable passing without having to alter URLs. The out-of-band passing of variables without changing the way URLs appear attempts to improve client-side cache hits.
- *State and context management.* WML exposes a fiat context (i.e., a linear, nonnested context) to the author. Each WML input control can introduce variables. The state of the variables can be used to modify the contents of a parameterized card without having to communicate with the server. Furthermore, the lifetime of a variable state can last longer than a single deck and can be shared across multiple decks without having to use a server to save the intermediate state between deck invocations.

WMLScript

WMLScript is a lightweight procedural scripting language. It enhances the standard browsing and presentation facilities of WML with behavioral capabilities, supports more advanced UI behavior, adds intelligence to the client, provides a convenient mechanism to access the device and its peripherals, and reduces the need for roundtrips to the origin server.

WMLScript is loosely based on a subset of the JavaScript WWW scripting language. It is an extended subset of JavaScript and forms a standard means for adding procedural logic to WML decks. WMLScript refines JavaScript for the narrowband device, integrates it with WML, and provides hooks for integrating future services and in-device applications.

WMLScript provides the application programmer with the capability to check the validity of user input before it is sent to the content server, access device facilities and peripherals, and interact with the user without introducing roundtrips to the origin server (e.g., display an error message).

Key WMLScript features include:

- *JavaScript-based scripting language.* WMLScript starts with an industry-standard solution and adapts it to the narrowband environment. This makes WMLScript easy for a developer to learn and use.
- *Procedural logic.* WMLScript adds the power of procedural logic to WAE.
- *Event based.* WMLScript may be invoked in response to certain user or environmental events.
- *Compiled implementation.* WMLScript can be compiled down to a more space-efficient bytecode that is transported to the client.
- *Integrated into WAE.* WMLScript is fully integrated with the WML browser. This allows authors to construct their services using both technologies, using the most appropriate solution for the task at hand. WMLScript has access to the WML state model and can set and get WML variables. This enables a variety of functionality (e.g., validation of user input collected by a WML card).
- *International support.* WMLScript character set is Unicode. This enables the presentation of most languages and dialects.
- *Efficient extensible library support.* WMLScript can be used to expose and extend device functionality without changes to the device software.

One objective in designing the WMLScript language was to be close to the core JavaScript. In particular, WMLScript was based on the ECMA-262 Standard ECMAScript Language specification. The originating technologies for the ECMA standard include, most notably, JavaScript and JScript. WMLScript is not fully compliant with ECMAScript. The standard has been used only as the basis for defining WMLScript language. The resulting WMLScript is a weakly typed language. Variables in the language are not formally typed in that a variable's type may change throughout the life cycle of the variable depending on the data it contains. The following basic data types are supported: boolean, integer, floating point, string, and invalid. WMLScript attempts to automatically convert between the different types as needed. In addition, support for floating-point data types may vary depending on the capabilities of the target device.

URLs

WAE assumes a rich set of URL services that user agents can use; in particular, WAE relies heavily on HTTP and HTML URL semantics. In some cases, WAE components extend the URL semantics, such as in WML, where URL fragments have been extended to allow linking to particular WMLScript functions.

WAE Content Formats

WAE includes a set of agreed-upon content formats that facilitate interoperable data exchange. The method of exchange depends on the data and the targeted WAE user agents. The two most important formats defined in WAE are the encoded WML and the WMLScript bytecode formats. WAE defines WML and WMLScript encoding formats that make transmission of WML and WMLScript more efficient and that minimize the computational efforts needed on the client.

WAE also defines and adopts the following formats for data types.

- *Images.* WAE assumes visual environments that will support several image formats. The selection of formats was an attempt to meet several competing requirements, including support of multiple choices of pixel depth, support of colorspace tables, small encoding, very low CPU and RAM decoding and presentation demands, and availability of common tools and other developer support.
- *Multipart messages.* WAE leverages a multipart-encoding scheme optimized for exchanging multiple typed content over WSP. See the official WAP specification at http://www.wapforum.org for additional details.
- User agent-specific formats. WAE adopts two additional content formats specific to exchanging data among user agents suitable for both **client-server communication** and **peer-to-peer** communication: electronic business cards (vCard.2.1) and electronic calendaring and scheduling exchange format.

client-server communication

peer-to-peer

WML and WMLScript Exchanges

Figure 4–6 presents the different parts of the logical architecture assumed by a WML user agent. Origin servers provide application services to the end user. The service interaction between the end user and the origin server is packaged as standard WML decks and scripts. Services may rely on decks and scripts that are statically stored on the origin server, or they may rely on content produced dynamically by an application on the origin servers.

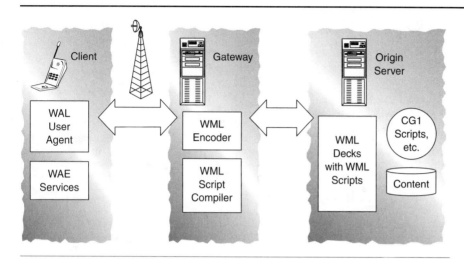

Figure 4–6
WML Logical
Architecture

Several stages are involved when origin servers and WML user agents exchange WML and WMLScript. In particular, a user wishing to access a particular service from an origin server submits a request to the origin server using a WML user agent. The user agent requests the service from the origin server on behalf of the user using some URL scheme operation (e.g., HTTP Get request method).

The origin server honoring the user's request replies by sending back a single deck. Presumably, this deck is initially in a textual format. On their way back to the client, textual decks are expected to pass through a gateway where they are converted into formats better suited for over-the-air transmission and limited device processing. In principle, once the gateway receives the deck from the origin server, the gateway does all the necessary conversions between the textual and binary formats. A WML encoder (or tokenizer) in the gateway converts each WML deck into its binary format. Encoded content is then sent to the client to be displayed and interpreted. Some optimization may be done at the gateway based on any negotiated features with the client.

The user agent may submit one or more additional requests (using some URL scheme) for WMLScript as the user agent encounters references to them in a WML deck. On its way back, a WMLScript compiler takes the script as input and compiles it into bytecode that is designed for low bandwidth and thin mobile clients. The compiled bytecode is then sent to the client for interpretation and execution.

The existence of a gateway is not mandatory, as illustrated in Figure 4–7. In particular, the location where the actual encoding and compilation is done is not of particular concern to WAE. It is conceivable that some origin servers will have built-in WML encoders and WMLScript compilers. It may also be

Figure 4–7
WML Logical
Architecture without a
Gateway

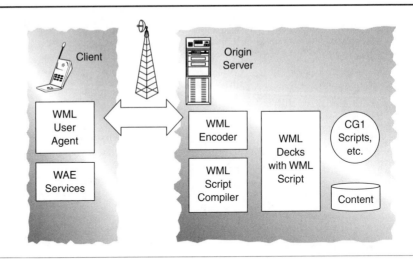

possible, in certain cases, to statically store (or cache) particular services in tokenized WML and WMLScript bytecode formats, thereby eliminating the need to perform any on-the-fly conversion of the deck.

INTERNATIONALIZATION

The WAE architecture is designed to support mobile terminals and network applications using a variety of languages and character sets. This work is collectively described as internationalization (referred to as I18N). It is a design goal of WAE to be fully global in its nature in that it supports any language.

WAE models a significant amount of its I18N architecture based on WWW and, in particular, on SGML and HT~technologies. For example, it is assumed that HTTP headers are used to specify the current character encoding and language of any content delivered to the user agent.

The WAE architecture makes the assumption regarding I18N that WAE user agents will have a current language and will accept content in a set of well-known character encoding sets. Origin server-side applications can emit content in one or more encoding sets and can accept input from the user agent in one or more encoding sets.

The IANA registry of character sets and languages is used to define the encoding and language characteristics. Content fetched by a WAE user agent via WSP will be described by two attributes: the character set (i.e., the encoding used in the content) and the language (i.e., the default language for the document). These attributes are encapsulated in the WSP/HTTP content-type and content-language headers.

WAE has adopted the Unicode 2.0 standard as the basis for all character data. Unicode contains the majority of the characters and symbols present in human languages and is widely supported in the internet community. In particular, most components of WAE contain I18N-specific support: WML contains additional support allowing the user agent and the origin server to negotiate the transmission encoding of user input sent from the user agent to the server; WMLScript defines string manipulation functionality to use Unicode collation order; and all content and data types are transmitted via protocols, which support the declaration of language and encoding, or they include such information in the data format itself.

SECURITY AND ACCESS CONTROL

WAE leverages WTLS when services require authenticated and/or secure exchanges. In addition, both WML and WMLScript include access control constructs that communicate to the client URL-based access restrictions. In particular, constructs allow the authors of WML decks and WMLScript to grant public access to the **content** (i.e., the deck or script can be referenced from other   **content** content) or to restrict access to the content to a set of "trusted" decks or scripts.

WTA ARCHITECTURE OVERVIEW

WTA is a collection of telephony-specific extensions for call and feature control mechanisms that make advanced mobile network services available to authors and end users (http://www.wapforum.org, 2001). WTA merges the features and services of data networks with the services of voice networks. It introduces mechanisms that ensure secure access to important resources within mobile devices. The WTA framework allows real-time processing of events that are important to the end user while browsing. Within the WTA framework, the client and server coordinate the set of rules that govern event handling via an event table. WTA origin servers can adjust the client's rules by pushing (or updating) a client's event table if required.

The wireless telephony application framework has four main goals.

1. To enable network operators to provide advanced telephony services that are well integrated and have consistent user interfaces
2. To enable network operators to create customized content to increase demands and accessibility for various services in their networks
3. To enable network operators to reach a wide range of devices by leveraging generic WAE features which allow the operator to create content independent of device-specific characteristics and environments
4. To enable third-party developers to create network-independent content that accesses basic features (i.e., nonprivileged)

Most of the WTA functionality is reserved for the network operators, as in-depth knowledge and access to the mobile network are needed to fully take advantage of the mobile network's features. Nevertheless, a limited set of basic WTA functions, such as initiating phone calls, is available to all WTA authors.

WTA FRAMEWORK COMPONENTS

The following sections describe the key components of the WTA framework.

WTA Libraries

WTA exposes its services to content authors as a set of libraries and interfaces. WTA functionality is divided across several libraries according to its sensitivity and application. WTA defines three classes of WTA services:

- *Common network services.* These WTA services are available independent of network type. They are common to all networks (e.g., answering an incoming call). Access to these services is restricted to content running within a WTA user agent
- *Network-specific services.* These WTA services target a specific type of network. They are extensions to common network services that expose unique and common features of a particular type of network (e.g., IS-136 includes a send flash service). Like common network services, network-specific services are restricted to content running in a WTA user agent.
- *Public services.* These WTA services are available to any anonymous or third-party content (e.g., initiate call setup). There are no access restrictions on such services. Any user agent is free to access public services.

Classifying and separating services enables secure and reliable execution of content. It limits functions available to authors and developers at large. Access to WTA services can be done directly from either the WML, using the WTA! URL scheme, or WMLScript functions by calling WTAI library functions.

WTA URL Scheme

WTA introduces a URL scheme that allows authors to invoke library services. The services may reside on the device or be delegated to a server. Using this scheme, authors can pass data to a service and receive data back from the service without having to leave the current browsing context.

WTA Event Handling

The WTA framework provides a variety of means for authors to deal with telephony-based events in a real-time and pseudo-real-time fashion. Fundamentally, telephony-based events can be sent to the WTA agent with any required event-specific parameters and content. This allows a network operator to deploy content (e.g., decks) with call control and network event-handling aspects. Clients can maintain event tables that describe how a user agent should deal with incoming events. This event table is coordinated with a WTA origin server controlled by the network operator.

For the most part, content sent with the event (or content already residing on the client) will be sufficient to handle most events; however, the framework does not prevent more advanced scenarios which require additional content to be retrieved from an origin server based on end-user demand. How a network operator chooses to handle events depends largely on the type of events, reliability and latency requirements, and quality desired.

WTA Network Security

The operator is assumed to have control over what resources are to be made accessible to any anonymous or third-party content in both the mobile network and the client. The integrity of both the mobile network and the client is enforced because of a restricted WTA content delivery. In particular, content with privileged WTA services can only be executed when it is delivered to the WTA user agent through a dedicated WTA port running WTLS protocols. This allows network operators to use standard network security elements to protect their networks. For example, origin servers that are delivering content can be identified by the operator as either trusted WTA content servers, which are under the control of the device's operator, or as untrusted third-party content servers, which may include any public origin server on the internet. Network operators can then use standard firewall technologies to regulate access to a mobile's ports. **Port** access can then be used to determine the credentials given to content, which determines its access privileges to WTA services in the network and for the client.

port

TELEPHONY-SPECIFIC EXCHANGES

WTA user agents, defined by WAE as telephony-specific extensions, use similar exchange constructs as a WML user agent. However, WTA user agents rely on additional and extended interactions needed to deliver meaningful telephony-based services.

The elements of the logical WTA network, presented in Figure 4–8, are content and content generators, firewalls (optional), and mobile switching framework.

The WTA user agent is connected to the mobile network using dedicated signaling connections. The WTA server (an origin server) communicates with the client using the WAP stack. The WTA server may be connected to the mobile network and is responsible for deploying content to its clients. In the case of call handling, for example, the mobile network sets up the call to the client,

Figure 4–8
WTA Logical Architecture

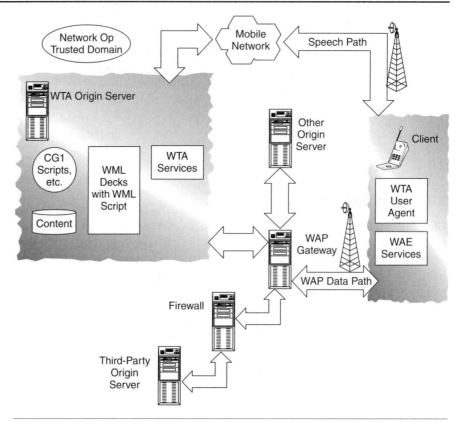

the server delivers the event-handling content, and the user agent invokes the event-handler content and manages the presentation of the call-handling service to the user.

The WTA user agent is a content interpreter that extends a typical WML user agent. It supports extended libraries and executes WML decks and WMLScript similar to a WML user agent; however, unlike a typical WML user agent, the WTA user agent has a rigid and real-time context management component. For example, the user agent drops outdated (or stale) events, does not place intermediate results on the history stack, and typically terminates after the event is handled.

WTA Origin Servers

The WTA origin server is assumed to be under the control of the network operator and is therefore to be regarded as a trusted content server. The operator's server is assumed to control, in varying degrees, the mobile network switch. The success of the WTA content (e.g., handling call control) is to some extent dependent on the operator's ability to access and control the features and characteristics of the mobile network.

The operator has information about latency, capacity, and reliability for the different bearers in the mobile network. Because the operator is able to provide the WTA services without needing to rely on the Internet, the operator can have more control over the behavior of the services than a third-party service provider, and can better optimize its services to achieve good, real-time characteristics.

Third-Party Origin Servers

Content from third-party providers does not handle any extensive set of WTA functions. Developing advanced WTA applications requires, in most cases, in-depth knowledge of the mobile network. Due to the limitations imposed by the operator as to which third party is granted access to the mobile network resources and WTA services, third-party content providers are limited to handling WAE content using the subscriber's standard WML user agent.

Mobile Network

The network operator controls the mobile network. The mobile network handles switching and call setup to the mobile subscribers (or terminals). The mobile network connects with the client using in-band or out-of-band signaling connections. The mobile network-to-client signaling is exposed to the content running in a user agent that is using WTA network events. Even though the mobile network-to-client messaging uses network-level system-specific signaling, at the content level signaling is converted to more generic and abstract WTA network events.

Although the mobile network is involved in the execution of the WTA network services, operations and services in the mobile network are not within the scope of the WAP effort. WTA services only make assumptions on the availability of basic network features such as call setup and call accept. Call control features in the mobile network are made available to the WTA user agent through the device's WTA interface.

Wireless Markup Language Specification

Wireless markup language is designed with the constraints of small narrowband devices in mind, including small display and limited user input facilities, narrowband network connection, and limited memory and computational resources.

WML includes five major functional areas.

1. *Text presentation and layout.* WML includes text and image support, including a variety of formatting and layout commands. For example, boldfaced text may be specified.
2. *Deck/card organizational metaphor.* All information in WML is organized into a collection of cards and decks. Cards specify one or more units of user interaction (e.g., a choice menu, a screen of text, or a text entry field). Logically, a user navigates through a series of WML cards, reviews the contents of each, enters requested information, makes choices, and moves on to another card.
3. *Card decks.* A WML deck is similar to an HTML page in that it is identified by a URL and is the unit of content **transmission.**
4. *Intercard navigation and linking.* WML includes support for explicitly managing the navigation between cards and decks. WML also includes provisions for event handling in the device, which may be used for navigational purposes or to execute scripts.
5. *String parameterization and state management.* All WML decks can be parameterized using a state model. Variables can be used in the place of strings and are substituted at runtime. This parameterization allows for more efficient use of network resources.

WML is designed to meet the constraints of a wide range of small narrowband devices. These devices are primarily characterized in the following ways.

- *Display size: smaller screen size and resolution.* A small mobile device such as a phone may have only a few lines of textual display, each line containing 8 to 12 characters.
- *Input devices: limited or special purpose.* A phone typically has a numeric keypad and a few additional function-specific keys. A more sophisticated device may have software-programmable buttons but may not have a mouse or other pointing device.
- *Computational resources: low-power CPU and small memory size.* These are often limited by power constraints.
- *Narrowband network connectivity: low bandwidth and high latency.* Devices with 300 bps to 10 Kbps network connections and 5- to 10-second roundtrip latency are not uncommon.

WWW is a network of information and devices. The areas of specification that ensure widespread interoperability are (1) a unified naming model—naming is implemented with URLs which provide a standard way to name any network resource; (2) standard protocols to transport information (e.g., HTTP); and (3) standard content types (e.g., HTML, WML).

WML is an extensible markup language and inherits the XML document character set. In SGML nomenclature, a document character set is that of all logical characters a document type may contain (e.g., the letter *T* and a fixed

transmission

integer identifying that letter). An SGML or XML document is simply a sequence of these integer tokens, which taken together form a document.

In a WML reference processing model, user agents must implement this processing model, or one that is indistinguishable from it. The user agent must correctly map the external character encoding of the document to Unicode before processing the document in any way. Any processing of entities also is done in the document character set.

WML is targeted at devices with limited hardware resources, including significant restrictions on memory size. It is important that the author have a clear expectation of device behavior in error situations, including those caused by lack of memory.

WMLScript Language Specification

WMLScript is part of the WAP application layer. It can be used to add client-side procedural logic. The language is based on ECMAScript, but it has been modified to better support low bandwidth communication and thin clients. WMLScript can be used with WML to provide intelligence to the clients, but it has also been designed so that it can be used as a stand-alone tool.

One difference between ECMAScript and WMLScript is that WMLScript has a defined bytecode and an interpreter reference architecture. This way, the narrowband communication channels available today can be optimally utilized and the memory requirements for the client kept to the minimum. Many of the advanced features of the ECMAScript language have been dropped to make the language smaller, easier to compile into bytecode, and easier to learn. For example, WMLScript is a procedural language and it supports locally installed standard libraries.

WMLScript is designed to provide general scripting capabilities to the WAP architecture. Specifically, WMLScript can be used to complement the wireless markup language, which is based on the extensible markup language. It is designed to be used to specify application content for narrowband devices such as cellular phones and pagers. This content can be represented with text, images, selection lists, and so forth. Simple formatting can be used to make the user interfaces more readable, as long as the client device used to display the content can support it. However, all this content is static and there is no way to extend the language without modifying WML itself.

The following capabilities are not supported by WML.

- Check the validity of user input (validity checks for the user input).
- Access facilities of the device; for example, while on a phone, allow the programmer to make phone calls, send messages, add phone numbers to the address book, and access the SIM card.
- Generate messages and dialogues locally, thus reducing the need for expensive roundtrip to show alerts, error messages, and confirmations.
- Allow extensions to the device software and configure a device after it has been deployed.

WMLScript was designed to overcome these limitations and to provide programmable functionality which can be used over narrowband communication links in clients with limited capabilities.

Many of the services utilized with thin mobile clients can be implemented with WML. Scripting enhances the standard browsing and presentation facilities of WML with behavioral capabilities. They can be used to support more advanced user interface (UI) functions, add intelligence to the client, provide access to the device and its peripheral functionality, and reduce the amount of bandwidth needed to send data between the server and the client.

WIRELESS TELEPHONY APPLICATION INTERFACE SPECIFICATION

The WAP WTAI features provide the means to create telephony applications, using a WTA user agent with the appropriate WTAI function libraries. A typical example is to set up a mobile-originated call using the WTAI functions accessible from either a WML deck/card or WMLScript.

WTA events

The application model for WTA is based on a WTA user agent, executing WML and WMLScript. The WTA user agent uses the WTAI function libraries to make function calls related to network services. The WTA user agent is able to receive **WTA events** from the mobile network and push content, such as WML decks and WTA events, from the WTA server. WTA events and WTAI functions make it possible to interact and handle resources (e.g., call control) in the mobile network.

Wireless Telephony Application Specification

The WTA user agent is an extension of the WAE user agent using WTA interface. WTA is intended for use in specifying wireless telephony applications that interface with local and network telephony infrastructure.

WTA interfaces are of three types. The first two types, network common and network specific, are reserved for the network operators, because the mobile network operators control and maintain the services for users on their mobile network. The third type, public, is a limited set of WTA functions, such as initiating a mobile phone call, available to content from any content developer.

Wireless telephony applications have four main goals.

1. *Advanced end-user services.* WTA makes it possible for network operators to provide advanced services with a consistent interface toward the end users.
2. *Increased utilization of network.* Network operators can utilize WTA to increase the use of the network services.
3. *Interoperability.* WAP applications written using the WTA interface can execute on a variety of telephony devices.

4. *Network-independent applications.* WTA content developers write telephony applications that span various networks using different protocols.

WTA enables content written in WML and WMLScript to utilize telephony features in the device and the mobile network. The WTA server, the mobile network, and the WTA client can be thought of as a single application with parts in the client and the WTA server. The WTA server acts as the principal content generator. The WTA server may be connected to the mobile network where it could have the means to control the mobile network services. Content may be customized by the WTA server and downloaded to the client.

WTA extends the basic WAE application model in three ways: It provides a means for a WTA server to push content to the device; it provides a means for mobile network events to trigger the rendering of content in the device; and it provides telephony functions on the device that can be accessed from WML or WMLScript.

The WSP push feature is used by the WTA server to push down content to the WTA client. Although content can be essentially anything, two fundamentally different types of content formats are standard WAE content formats (e.g., WML, WMLScript, or WBMP) and a specific WTA content format called WTA event. The WTA framework allows a flexible implementation for the client which allows it to support pushed standard content, WTA event, or both. The WTA server must decide what features to use based on the user agent characteristics and the profile specified for the specific WTA user agent.

Two modes available in the WTA model are **server-centric** mode and client-centric mode. The fundamental differences between pushing a WTA event or content are illustrated in Figures 4–9 and 4–10. The coordinating network element has been used for naming the particular model. The two important network elements in this case are the WTA server and the WTA client. In both models, the WTA server acts as the content generator toward the WTA client.

server-centric

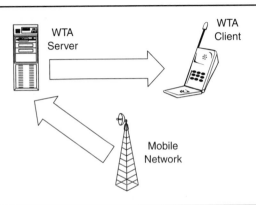

Figure 4–9
Server-centric WTA
Mode

Figure 4–10
Client-centric WTA Mode

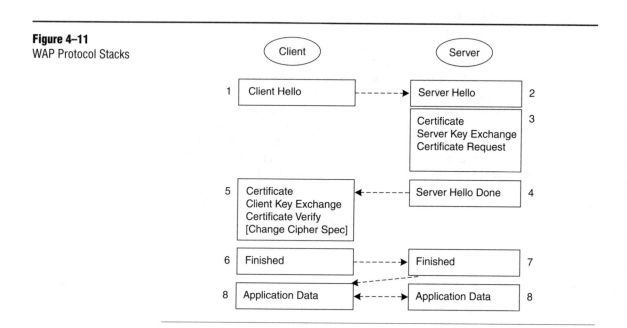

Figure 4–11
WAP Protocol Stacks

Server-centric mode (Figure 4–11) handles events, such as incoming calls, that occur in the mobile network and provides the WTA client with content that is used for handling the specific task. The WTA server deploys content customized for the occurring event.

Client-centric mode (Figure 4–12) handles events, such as incoming calls, received from the mobile network. Event bindings in the WTA client can be

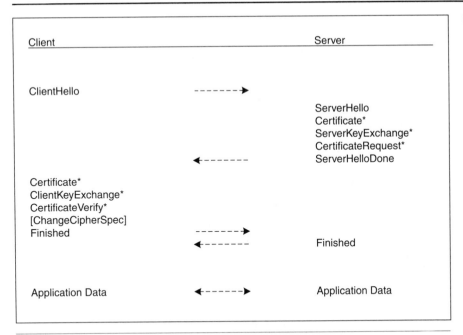

Figure 4–12
Message Flow—Full Handshake

used to associate the event to executable content which is used for handling the specific task.

By combining the server- and client-centric models in the same architecture, the operator can take advantage of the features that best fit the particular type of mobile network and applications.

SUPPORTED CONTENT FORMATS

This section describes the specific types of content formats to be supported and the required behavior of the WTA client depending on the mode (client centric or server centric) when receiving the content.

Following are the fundamental methods for content transfer.

- Mobile network event involves the representation of the network-specific signaling that can be handled. The WTA user agent encounters the network event in an abstract form of the WTA event (abstract means no physical WTA event content is needed).
- Acknowledged content requires an acknowledgment to the sending party; an example is the event table.
- Unacknowledged content does not require an acknowledge; an example is a WML deck.

WTA URIs and URLs

Wireless devices come in many flavors with regard to available bearers and types of telephony features. Many standards, such as GSM, support device access to telephony features of the mobile network. URLs form a unifying naming model for how to identify features independently of the internal structure in the device and the mobile network.

In WTA, URLs are used when specifying the identity of local resources such as logical indicators, and when specifying telephony features such as the setup of a mobile-originated call.

WTA Session Management

A WTA session is used by the WTA user agent to facilitate interaction with the WTA server and the mobile network. Setting up a WTA session also includes the setup procedure for the event table when the WTA client supports the client-centric mode. The WTA server must then use a reliable connection for the download of the event table.

WIRELESS SESSION PROTOCOL SPECIFICATION

The session layer protocol family in the WAP architecture is called the Wireless Session Protocol (WSP). WSP provides the upper level application layer of WAP with a consistent interface for two session services. The first is a connection mode service that operates above a transaction layer protocol WTP, and the second is a **connectionless session service** that operates above a secure or nonsecure datagram transport service.

connectionless session service

WSPs currently offer services most suited for browsing applications (WSP/B). WSP/B provides HTTP 1.1 functionality and incorporates new features such as long-lived sessions, a common facility for data push, capability negotiation, and session suspend/resume. The protocols in the WSP family are optimized for low-bandwidth bearer networks with relatively long latency.

WSP Architectural Overview

Wireless Session Protocol is a session level protocol family for remote operations between a client and proxy or server.

Reference Model

A model of layering the protocols in WAP is illustrated in Figure 4–11. WAP protocols and their functions are layered in a style resembling that of the OSI reference model. Layer management entities handle protocol initialization, configuration, and error conditions (such as loss of connectivity due to the mobile station roaming out of coverage) that are not handled by the protocol itself.

WSP is designed to function on the transaction and datagram services. Security is assumed to be an optional layer above the transport layer. The security layer preserves the transport service interfaces. The transaction, session, or application management entities are assumed to provide the additional support that is required to establish security contexts and secure connections. This support is not provided by the WSPs directly. In this regard, the security layer is modular. WSP itself does not require a security layer; however, applications that use WSP may require it.

WSP/B Features

WSP provides a means for organized exchange of content between cooperating client-server applications. Specifically, it provides the applications a means to do the following:

- Establish a reliable session from client to server and release that session in an orderly manner.
- Agree on a common level of protocol functionality using **capability negotiation.**
- Exchange content between client and server using compact encoding.
- Suspend and resume the session.

capability negotiation

The currently defined services and protocols (WSP/B) are most suited for browsing-type applications. WSP/B actually defines two protocols: One provides connection-mode session services over a transaction service, and another provides nonconfirmed, connectionless services over a datagram transport service. The connectionless service is most suitable when applications do not need reliable delivery of data and do not care about confirmation. It can be used without actually having to establish a session.

In addition to the general features, WSP/B offers more:

- HTTP/1.1 functionality
- Extensible request-reply methods
- Composite objects
- Content-type negotiation
- Exchange of client and server session headers
- Interruption of transactions in process
- Push of content from server to client in an unsynchronized manner
- Support for multiple, simultaneous asynchronous transactions

Basic Functionality

The core of the WSP/B design is a binary form of HTTP. Consequently, the requests sent to a server and responses going to a client may include both headers (meta-information) and data. All the methods defined by HTTP/1.1 are supported. In addition, capability negotiation can be used to agree on a set of extended request methods so that full compatibility to HTTP/1.1 applications can be retained.

WSP/B provides typed data transfer for the application layer. The HTTP/1.1 content headers are used to define content type, character set encoding, languages, and so forth in an extensible manner. However, compact binary encodings are defined for the well-known headers to reduce protocol overhead. WSP/B also specifies a compact composite data format that provides content headers for each component within the composite data object. This is a semantically equivalent binary form of the MIME "multipart/mixed" format used by HTTP/1.1.

WSP/B itself does not interpret the header information in requests and replies. As part of the session creation process, request and reply headers that remain constant over the life of the session can be exchanged between service users in the client and the server. These may include acceptable content types, character sets, languages, device capabilities, and other static parameters. WSP/B will pass through client and server session headers as well as request and response headers without additions or removals.

The life cycle of a WSP/B session is not tied to the underlying transport. A session can be suspended while the session is idle to free up network resources or save battery power. A lightweight session reestablishment protocol allows the session to be resumed without the overhead of full-blown session establishment. A session may be resumed over a different bearer network.

Extended Functionality

WSP/B allows extended capabilities to be negotiated between the peers, providing for high performance, feature-full implementation, and simple, basic, and small implementations. WSP/B provides an optional mechanism for attaching header information (meta-data) to the acknowledgment of a transaction. This allows the client application to communicate specific information about the completed transaction back to the server.

WSP/B optionally supports asynchronous requests so that a client can submit multiple requests to the server simultaneously. This improves utilization of airtime in that multiple requests and replies can be coalesced into fewer messages, and improves latency as the results of each request can be sent to the client when they become available.

WSP/B partitions the space of well-known header field names into header code pages. Each code page can define only a fairly limited number of encodings for well-known field names, which permits them to be represented more compactly. Running out of identities for well-known field names on a certain code page is still not a problem because WSP/B specifies a mechanism for shifting from one header code page to another.

WSP Elements of Layer-to-Layer Communication

The session layer in WAP provides both connection-mode and connectionless services. They are defined using an abstract description technique based on service primitives. Some of the terms and concepts used to describe the communication mechanisms are borrowed from the OSI reference model.

WIRELESS TRANSACTION PROTOCOL SPECIFICATION

A transaction protocol is defined to provide the services necessary for interactive "browsing" (request/response) applications. During a browsing session, the client requests information from a server, which may be fixed or mobile, and the server responds with the information. The request/response duo is referred to as a "transaction" in this document. The objective of the protocol is to reliably deliver the transaction while balancing the amount of reliability required for the application with the cost of delivering the reliability.

WTP runs on top of a datagram service and optionally on a security service. WTP has been defined as a lightweight transaction-oriented protocol that is suitable for implementation in "thin" clients (mobile stations) and that operates efficiently over wireless datagram networks. Benefits of using WTP include improved reliability over datagram services. WTP relieves the upper layer from retransmissions and acknowledgments, which are necessary if datagram services are used. Another benefit is improved efficiency over connection-oriented services. WTP has no explicit connection setup or teardown phases. WTP also is message oriented and designed for services oriented toward transactions such as "browsing."

Protocol Overview

The following list summarizes the features of WTP.

- Three classes of transaction service are class 0—unreliable invoke message with no result message; class 1—reliable invoke message with no result message; and class 2—reliable invoke message with exactly one reliable result message.
- Reliability is achieved through the use of unique transaction identifiers, acknowledgments, duplicate removal, and retransmissions.
- There are no explicit connection setup or teardown phases, because explicit connection open and/or closed imposes excessive overhead on the communication link.
- User-to-user reliability allows the WTP user to confirm every received message. This feature is optional.
- The last acknowledgment of the transaction may contain out-of-band information related to the transaction (e.g., performance measurements). This feature is optional.
- Concatenation may be used, when applicable, to convey multiple protocol data units in one service data unit of the datagram transport.

- Message orientation means the basic unit of interchange is an entire message and not a stream of bytes.
- The protocol provides mechanisms to minimize the number of transactions being replayed as the result of duplicate packets.
- Abort of outstanding transactions, includes flushing of unsent data both in client and server. The abort can be triggered by the user canceling a requested service.
- For reliable invoke messages, both success and failure are reported. If an invoke cannot be handled by the **responder,** an abort message will be returned to the **initiator** instead of the result.
- The protocol allows for asynchronous transactions. The responder sends back the result as the data become available.

responder

initiator

WIRELESS TRANSPORT LAYER SECURITY SPECIFICATION

The security layer protocol in the WAP architecture is called the wireless transport layer security (WTLS). The WTLS layer operates above the transport protocol layer. The WTLS layer is modular and its use depends on the required security level of the given application. WTLS provides the upper level layer of WAP with a secure transport service interface that preserves the transport service interface below it. In addition, WTLS provides an interface for managing (e.g., creating and terminating) secure connections.

The primary goal of the WTLS layer is to provide privacy, data integrity, and authentication between two communicating applications. WTLS provides functionality similar to TLS 1.0 and incorporates new features such as datagram support, optimized handshake, and dynamic key refreshing. The WTLS protocol is optimized for low bandwidth bearer networks with relatively long latency.

WTLS Architectural Overview

WTLS is designed to function on connection-oriented and/or datagram transport protocols. Security is assumed to be an optional layer above the transport layer. The security layer preserves the transport service interfaces. The session or application management entities are assumed to provide additional support required to manage (e.g., initiate and terminate) **secure connections.**

secure connections

HANDSHAKE PROTOCOL OVERVIEW

The cryptographic parameters of the secure session are produced by the WTLS Handshake Protocol, which operates on top of the WTLS record layer. When a WTLS client and server first start communicating, they agree on a protocol version, select cryptographic algorithms, optionally authenticate each other, and use public-key encryption techniques to generate a shared secret.

The WTLS Handshake Protocol involves the following steps.

1. Exchange hello messages to agree on algorithms, exchange random values.
2. Exchange the necessary cryptographic parameters to allow the client and server to agree on a premaster secret.
3. Exchange certificates and cryptographic information to allow the client and server to authenticate themselves.
4. Generate a master secret from the premaster secret and exchanged random values.
5. Provide security parameters to the record layer.
6. Allow the client and server to verify that their peer has calculated the same security parameters and that the **handshake** occurred without tampering by an attacker.

handshake

These goals are achieved by the **handshake protocol,** which can be summarized as follows: The client sends a client hello message to which the server must respond with a server hello message, or else a fatal error will occur and the secure connection will fail. The client hello and server hello are used to establish security enhancement capabilities between client and server. The client hello and server hello establish the following attributes: protocol version, key exchange suite, cipher suite, compression method, key refresh, and sequence number mode. Additionally, two random values are generated and exchanged: ClientHello.random and ServerHello.random.

handshake protocol

The actual key exchange uses up to four messages: the server certificate, the server key exchange, the client certificate, and the client key exchange. New key exchange methods can be created by specifying a format for these messages and defining the use of the messages to allow the client and server to agree upon a shared secret. This secret should be quite long. For wireless environments, 20 bytes can be considered suitable.

Following the hello messages, the server will send its certificate, if it is to be authenticated. Additionally, a server key exchange message may be sent, if it is required (e.g., the server does not have a certificate, or if its certificate is for signing only). The server may request a certificate from the client (or get the certificate from some certificate distribution service), if that is appropriate to the key exchange suite selected. Now the server will send the server hello done message, indicating that the hello-message phase of the handshake is complete. (The previous handshake messages are combined in one lower layer message.) The server will then wait for a client response. If the server has sent a certificate request message, the client must send the certificate message.

The client key exchange message is now sent if the client certificate does not contain enough data for key exchange or if it is not sent at all. The content of that message will depend on the public key algorithm selected between the client hello and the server hello. If the client is to be authenticated using a certificate with a signing capability (e.g., RSA), a digitally signed certificate verify message is sent to explicitly verify the certificate.

At this point, a change CipherSpec message is sent by the client, and the client copies the pending CipherSpec into the current CipherSpec. The client then immediately sends the finished message under the new algorithms, keys, and secrets. From now on, the CipherSpec indicator is set to 1 in the messages. When the server receives the change CipherSpec message it also copies the pending CipherSpec into the current CipherSpec. In response, the server will send its own finished message under the new CipherSpec. At **full handshake** this point, the handshake is complete (**full handshake**) and the client and server may begin to exchange application layer data (see Figure 4–12).

session resume When the client and server decide to resume a previous secure session (**session resume**) instead of negotiating new security parameters, the message flow is as follows: The client sends a ClientHello using the session ID of the secure session to be resumed. The server then checks its secure session cache for a match. If a match is found, and the server is willing to reestablish the secure connection under the specified secure session, it will send a ServerHello with the same session ID value. At this point, the server must send a ChangeCipherSpec message and proceed directly to the finished message to which the client should respond with its own finished message. Once the **abbreviated handshake** reestablishment is complete (**abbreviated handshake**), the client and server may begin to exchange application layer data (see Figure 4–13). If a session ID match is not found, the server generates a new session ID and the TLS client and server perform a full handshake.

Note that many simultaneous secure connections can be instantiated under one secure session. Each secure connection established from the same secure session shares some parameters with the others (e.g., master secret).

The shared secret handshake means that the new secure session is based on a shared secret already implanted in both ends (e.g., physically). In this case, the shared secret KeyExchangeSuite is requested by the client. The message flow is similar to the abbreviated handshake in Figure 4–13.

Figure 4–13
Message Flow—
Abbreviated Handshake

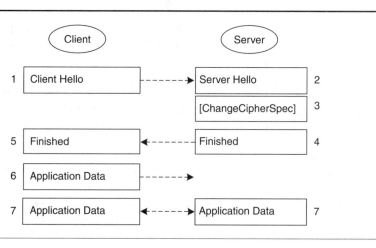

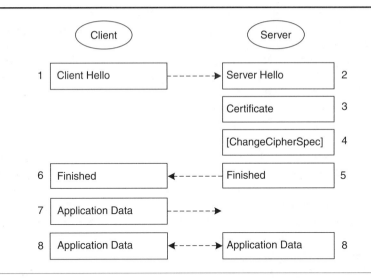

Figure 4–14
Message Flow—
Optimized Handshake

Another variation is that the server, after receiving the ClientHello, can retrieve the client's certificate using a certificate distribution service or its own sources (**optimized handshake**). In a Diffie-Hellman-type key exchange method, assuming the Diffie-Hellman parameters are provided in the certificates, the server can calculate the premaster secret and master secret at this point. In this case, the server sends its certificate, a ChangeCipherSpec, and a Finished message. See Figure 4–14 for more information.

optimized handshake

Handshake Reliability over Datagrams

In the datagram environment, handshake messages may be lost, out of order, or duplicated. To make the handshake reliable over datagrams, WTLS requires that the handshake messages going in the same direction must be concatenated in a single transport **service data unit (SDU)** for transmission, that the client retransmits the handshake messages if necessary, and that the server must appropriately respond to the retransmitted messages from the client.

service data unit (SDU)

The handshake may consist of multiple messages to be delivered in one direction before any responses are required from the other end. Those messages must be concatenated into a single-transport SDU for transmission or retransmission to guarantee that all the messages in the same SDU arrive in order. For instance, ServerHello, ChangeCipherSpec, and Finished messages can be sent in a single-transport SDU for the abbreviated handshake. The maximum size of SDU for the underlying transport service layer must be sufficient to contain all those messages.

For the full handshake, the client must retransmit ClientHello and Finished messages if the expected response messages are not received from the server for a predefined timeout period. Note that the whole transport SDU that contains the Finished message must be retransmitted. After the number of retransmissions exceeds the maximum predefined retransmission counter, the client terminates the handshake. Those predefined timeout and counter values may be obtained from the WTP stack through the management entity if the WTP stack is present above the WTLS stack.

For the optimized and abbreviated handshakes, like the full handshake, the client retransmits ClientHello, if necessary. In addition, the client must also append the Finished message with the Application Data message until an Application Data message from the server is received and decrypted successfully, or a duplicated Finished Received alert (warning) is received from the server. However, the first Finished message can be either sent alone or appended with the application data message, if any.

service primitive

For the full handshake, the server must retransmit the transport SDU that contains the ServerHello message upon receiving a duplicated ClientHello message. However, if the ClientHello is new, the server must start a new handshake and SEC-Create.ind **service primitive** must be generated. The server must also retransmit the transport SDU that contains the Finished message upon receiving a duplicated Finished message from the client.

For the optimized and abbreviated handshakes, the server behaves the same as that in the full handshake for handling the duplicated or new ClientHello messages. In addition, the server must ignore a duplicated Finished message and keep the committed secure connection intact. If the server has no application data to send to the client, it should send a duplicated Finished Received alert (warning).

Handshake Protocol

The WTLS Handshake Protocol is one of the defined higher level clients of the WTLS Record Protocol. This protocol is used to negotiate the secure attributes of a secure session. Handshake messages are supplied to the WTLS record layer, where they are encapsulated within one or more WTLSPlaintext structures, which are processed and transmitted as specified by the current active connection state.

WIRELESS DATAGRAM PROTOCOL SPECIFICATION

The transport layer protocol in the WAP architecture consists of the Wireless Transaction Protocol and the Wireless Datagram Protocol. The WDP layer operates above the data-capable bearer services supported by the various network types. As a general datagram service, WDP offers a consistent service to the upper layer protocols (security, transaction, and session layers) of WAP and communicates transparently over one of the available bearer services.

The protocols in the WAP family are designed for use over narrowband bearers in wireless telecommunications networks. Because the WDPs provide a common interface to the upper layer protocols (security, transaction, and session layers), they are able to function independently of the underlying wireless network. This is accomplished by adapting the transport layer to specific features of the underlying bearer.

WDP Architectural Overview

The WDP operates above the data-capable bearer services supported by multiple network types. WDP offers a consistent service to the upper layer protocols (security, transaction, and session) of WAP and communicates transparently over one of the available bearer services.

Reference Model

The model of protocol architecture for the WDP is given in Figure 4–15.

Services offered by WDP include application addressing by port numbers, optional segmentation and reassembly, and optional error detection. The services allow for applications to operate transparently over different available bearer services. The model of protocol architecture for the Wireless Transport Protocol is given in Figure 4–16.

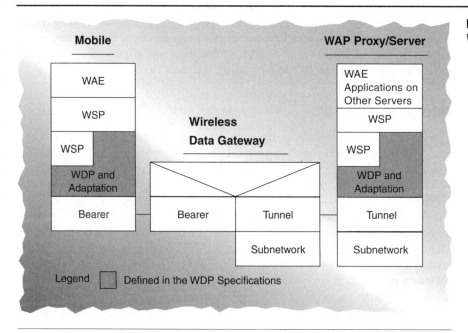

Figure 4–15
WDP Architecture

Figure 4–16
WTP Architecture

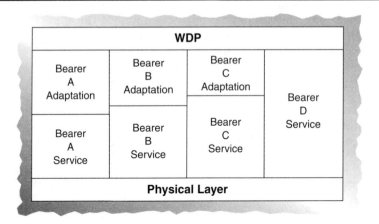

WDP offers a consistent service at the transport service access point to the upper layer protocol of WAP. This consistency of service allows for applications to operate transparently over different available bearer services. The varying heights of each of the bearer services shown in Figure 4–16 illustrate the difference in functions provided by the bearers and thus the difference in WDP necessary to operate over those bearers to maintain the same service offering at the transport service access point.

WDP can be mapped onto different bearers, with different characteristics. To optimize the protocol with respect to memory usage and radio transmission efficiency, the protocol performance over each bearer may vary. However, the WDP service and service primitives will remain the same, providing a consistent interface to the higher layers.

Description of the WDP

The WDP layer operates above the data-capable bearer services supported by the various network types. As a general datagram service, WDP offers a consistent service to the upper layer protocols (security, transaction, and session) of WAP and communicates transparently over one of the available bearer services.

WDP supports several simultaneous communication instances from a higher layer over a single underlying WDP bearer service. The port number identifies the higher layer entity above WDP. This may be another protocol layer such as the Wireless Transaction Protocol or the Wireless Session Protocol or an application such as electronic mail. By reusing the elements of the underlying bearers, WDP can be implemented to support multiple bearers and yet be optimized for efficient operation within the limited resources of a mobile device.

At the mobile, the WDP consists of the common WDP elements shown by the layer labeled WDP, as shown in Figure 4–17. The adaptation layer of the WDP maps the WDP functions directly onto a specific bearer. The adaptation layer is different for each bearer and deals with the specific capabilities and characteristics of that bearer service. The bearer layer is a service such as GSM **SMS, USSD, IS-136 R-Data,** or CDMA Packet Data. At the gateway, the adaptation layer terminates and passes the WDP packets on to a WAP proxy/server via a tunneling protocol, which is the interface between the gateway that supports the bearer service and the WAP proxy/server. For example, if the bearer were GSM SMS, the gateway would be a GSM SMSC and would support a specific protocol (the tunneling protocol) to interface the SMSC to other servers. The subnetwork is any common networking technology that can be used to connect two communicating devices; examples are wide area networks based on TCP/IP or X.25, or LANs operating TCP/IP over Ethernet. The WAP proxy/server may offer application content or may act as a gateway between the wireless WTP suites and the wired Internet.

SMS, USSD, IS-136 R-Data

WDP Management Entity

The WDP management entity is used as an interface between the WDP layer and the environment of the device. The WDP management entity provides information to the WDP layer about changes in the device's environment that may impact the correct operation of the WDP.

The WDP is designed around an assumption that the operating environment is capable of transmitting and receiving data. For example, this assumption includes the following basic capabilities that must be provided by the mobile.

- The mobile is within a coverage area applicable to the bearer service being invoked.
- The mobile has sufficient power and the power is on.
- Sufficient resources (processing and memory) within the mobile are available to WDP.
- The WDP is correctly configured.
- The user is willing to receive/transmit data.

The WDP management entity would monitor the state of the preceding services and capabilities of the mobile's environment and would notify the WDP layer if one or more of the assumed services were not available. For example, if the mobile roamed out of coverage for a bearer service, the bearer management entity should report to the WDP management entity that transmission/reception over that bearer is no longer possible. In turn, the WDP management entity would indicate to the WDP layer to close all active connections over that bearer. Other examples such as low battery power would be handled in a similar way by the WDP management entity.

In addition to monitoring the state of the mobile environment, the WDP management entity may be used as the interface to the user for setting various configuration parameters used by WDP, such as device address. It could also be used to implement functions available to the user, such as a "drop-all-data connections" feature. In general, the WDP management entity will deal with all issues related to initialization, configuration, dynamic reconfiguration, and resources as they pertain to the WDP layer.

Because the WDP management entity must interact with various components of a device that are manufacturer specific, the design and implementation of the WDP management entity is considered outside the scope of the WDP specification and is an implementation issue.

WIRELESS CONTROL MESSAGE PROTOCOL SPECIFICATION

WCMP contains control messages that resemble the Internet Control Message Protocol (ICMP) messages. WCMP can also be used for diagnostics and informational purposes.

WCMP Architectural Overview

Figure 4–17 shows a general model of the WAP architecture and how WCMP fits into that architecture.

Figure 4–17
WCMP in the WAP
Architecture

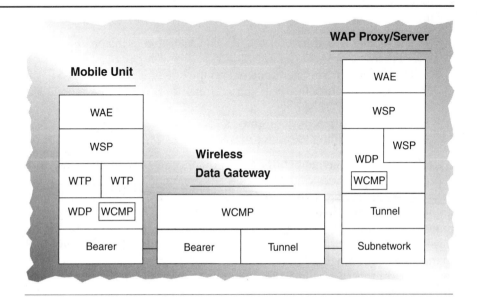

The transport layer protocol in the WAP architecture is the Wireless Datagram Protocol. The WDP operates above the data-capable bearer services supported by multiple network types. WDP offers a consistent but unreliable service to the upper level protocols of WAP and communicates transparently over one of the available bearer services.

WCMP is used by WDP nodes and wireless data gateways to report errors encountered in processing datagrams. WCMP can also be used for informational and diagnostic purposes.

WCMP Description

The WCMP is used in environments that do not provide an IP bearer. WCMP is used by WDP nodes and wireless data gateways to report errors encountered in processing datagrams. WCMP messages are usually generated by the WDP layer, the management entity, or a higher layer protocol. WCMP can also be used for informational and diagnostic purposes. The WCMP provides an efficient error-handling mechanism for WDP, resulting in improved performance for WAP protocols and applications.

WCMP in IP Networks

In IP-based networks, the functionality of the WCMP is implemented by using the ICMP.

THE FUTURE

Probably the oldest promise in the history of mobile phones is the ability to access important data via a wireless connection. For nearly 10 years, the promise has been greeted by snickers and outright laughter, but now it finally appears to be garnering some credibility due to WAP.

It should be made clear that WAP is not a replacement for Internet standards. It is based on many of the standards currently running on the Web. WAP simply handles them in a more efficient way.

Internet standards such as HTTP and TCP are not designed for long latencies, intermittent coverage, and the limited bandwidth of wireless networks. HTTP sends commands in bulky text format, and security standards require heavy client-server communication that makes them nearly useless over wireless links.

WAP has the answers, according to nearly all major carriers, handset vendors, and software developers in the wireless market. The protocol uses binary transmission for greater compression of data, and is optimized for long latency and low bandwidth. It uses wireless markup language, a derivative of XML, for WAP content, which makes optimum use of small screens, including support for two-line text displays or full graphic screens.

WAP appears to have a great deal of potential, with the WAP Forum already working on end-to-end security, smart card interfaces, connection-oriented transport protocols, persistent storage, billing interfaces, and push technology.

Two examples from opposite sides of the world show the importance of WAP. These examples show that WAP, in less than 5 years, has captured the eye of both industry and users alike. The first example comes from Japan. The number of cellular telephone subscribers in Japan with WAP-enabled telephones has hit 1 million, according to a recent survey (www.cnet.com, 2002). WAP-based services began in Japan in April 1999 when DDI Cellular and Nippon launched the service on a new CDMA network. The Tu-Ka Cellular Group, which was acquired by DDI in 2001, has also launched a similar service on its personal digital communications (PDC) network, a Japanese cellular standard. There are several hundred WAP sites in Japan offering everything from business news and weather to train timetables, horoscopes, and chat services, some of which are available on a low-priced monthly subscription.

The second example of WAP involves Europe. In July 2000, Scandinavia's largest bank announced a rollout of wireless banking services to all of its customers. Svenska Handelsbanken said the Nordic region, where 60 percent of the population carries mobile phones, is a perfect testing ground for wireless banking services, which are expected to widely surpass Internet banking in popularity. Handelsbanken, as the Stockholm-based bank is commonly called, currently has signed up more than 150,000 customers for its Internet banking services. The wireless services are based on the WAP specification for mobile data communications and further Handelsbanken's stated aim to enable its customers to do their banking anytime and anywhere. In the initial phase of the trial, which began in October 1999, Handelsbanken provided its financial advisers and selected customers with WAP-enabled telephones with which they could access general and personalized stock market information. The handsets were provided by Finland's Nokia Corp. and Sweden's L.H. Ericsson Telephone Co.

In July 2000, Handelsbanken began offering all of its customers a full range of wireless banking services, from access to account and stock market information to money transfers, purchases, stock sales, or bill payments. Handelsbanken's technology partner in rolling out the wireless initiative is IBM Corp., which is providing the application development tools and specialized software for translating the bank's data into a format based on extensible markup language that can be received via a WAP-enabled mobile phone. Nokia, meanwhile, is providing the server through which the information is sent onto the wireless network, as well as handsets for accessing the services.

WAP services and applications, as well as usage, are expected to have a slower growth rate in the United States. This is typical for many standards-based products as the marketplace is very different here than in some countries overseas where governmental entities provide "guidance" on adherence to standards-based approaches. However, it is expected that WAP-enabled devices and the associated WAP infrastructure will be mostly in place by 2004-2005, at the latest.

It is clear that WAP has come a long way. Yet, it still has a way to go before it is widely accepted in the United States. For example, this comment from Jeff Dumont, a Cellular One agent, with a large Georgia-based Cellular One reseller, sums up the current situation nicely: "So far, only 1 in maybe 200 or 300 ask for WAP-enabled phones. The providers charge extra fees for this access, around 99 cents a minute and up, which is affecting sales. And, people just don't seem to want to do any advanced stuff from their cellular phones; they want to talk. Sure, a few check the weather and such, but who wants to spend *extra* to do that? Maybe that will change, but for now, that's what I hear. And, Cingular, a major cellular U.S. telephony provider, currently charges an extra $6.99 per month for simple weather and sports score lookups on WAP-enabled phones."

■ SUMMARY

In this chapter, we discuss the Wireless Application Protocol, specifically its role and importance in the wireless Internet community. We examine the various components of WAP and how they fit together to create a stable yet flexible architecture. The WAP Forum is the controlling body responsible for developing, enhancing, and maintaining the WAP specification. We learn the objectives of this organization and the requirements met by the WAP.

In the next chapter, we address another emerging wireless technology, Bluetooth.

REVIEW QUESTIONS

1. The organizational body controlling the WAP specifications is called the _____.
   a. WSP Group
   b. WAP Forum
   c. WDP Interface Consortium
   d. none of the above

2. This controlling group was founded in _____.
   a. 1990
   b. 1999
   c. 1997
   d. none of the above

3. A goal of this controlling group is _____.

   a. to redesign the browsers for Internet use
   b. to create a wireless protocol that works across all wireless technologies
   c. to create a new standards-based specification for wireless technologies
   d. none of the above

4. In WAP, when discussing factors relative to the hand-held market, one must consider that _____.
   a. these devices have less-powerful CPUs
   b. these devices have less memory
   c. these devices have much smaller displays
   d. all of the above apply

5. One of the most difficult aspects of wireless technologies is typically _____.
   a. latency
   b. power spikes
   c. 5-4-3 rule
   d. none of the above

6. The WAP programming model is similar to the _____ programming model.
   a. WWW
   b. OSI
   c. electromagnetic spectrum
   d. none of the above

7. WML is similar to _____.
   a. HTTP
   b. HTML
   c. FTP
   d. none of the above

8. WSP provides an efficient interface for _____.
   a. WML
   b. the browser
   c. session services
   d. none of the above

9. WTP provides a _____ protocol tailored for "thin" clients.
   a. transaction-oriented
   b. transport
   c. session layer
   d. none of the above

10. WTLS stands for _____.
   a. Wireless transfer telephony layering service
   b. wireless telephony layer service
   c. wireless transport layer security
   d. none of the above

11. WAP stands for _____.
   a. Wireless Appropriate Protocol
   b. Wireless Application Protocol
   c. Wavelength Application Protocol
   d. none of the above

12. WDP is the _____ layer in the WAP architecture.
   a. session
   b. data link

c. transport
d. none of the above

13. The upper layers function independently of the underlying wireless network due to _____.
   a. adaptation of the transport layer
   b. configuration of underlying bearer (i.e., network) services
   c. virtual layering processes
   d. none of the above

14. WAE stands for _____.
   a. wow! another entity to remember
   b. wireless appropriate environment
   c. wireless application environment
   d. none of the above

15. WTLS provides the features of _____.
   a. data integrity, privacy, and authentication
   b. data security, denial-of-service functionality, and C2 protection
   c. data security, integrity control, and performance enhancements
   d. none of the above

16. WAE assumes the presence of a _____ responsible for encoding/decoding.
   a. proxy
   b. security layer
   c. gateway
   d. none of the above

17. The major components of the WAE model include _____.
   a. user agents, content generators, standard content encoding, and WTA
   b. user agents, mail transfer agents, and standard content encoding
   c. hand-held device display screens, users, and proxy functionality
   d. none of the above

18. True/False: The WAE architecture model leverages the Internet.

19. WMLScript is similar to _____.
   a. HTTP
   b. HTML

c. JavaScript

d. none of the above

20. True/False: WAE relies heavily upon URL and HTML semantics.

21. True/False: WTA allows real-time and near real-time processing of events important to the user.

22. True/False: The WAE model leverages thin client technologies.

23. True/False: WMLScript provides procedural logic capability.

24. True/False: WTP offers three classes of service.

25. WCMP is similar to _____ in the Internet domain.

a. ICMP

b. DNS

c. RARP

d. none of the above

26. Wireless Application Protocol is a product of the industry group called _____.

a. WAP Coalition

b. CCITT

c. WAP Forum

d. IEEE

27. A WAP phone is similar to a _____.

a. miniature browser

b. cellphone with voice recognition

c. Internet Explorer

d. Bluetooth

28. The WAP programming model provides guidance for _____.

a. designers

b. programmers

c. users

d. none of the above

29. Wireless telephony operates in an environment of _____.

a. less powerful CPUs

b. small hard disk space

c. higher bandwidth

d. significant power consumption

30. WAP is based upon a _____ architecture.

a. limited

b. comprehensive and thorough

c. standardized

d. WWW

HANDS-ON EXERCISES

1. Using your browser and your favorite search engine, locate sites that explain WAP. Create a table listing at least five sites, including their similarities and differences. For example, do they all cover WAE, WDP, and the like, or are they limited in their coverage?

2. Using your browser, go to www.wapforum.org. Provide a three-page research paper summarizing your results after categorizing the data available on this site.

3. Using your browser, go to several sites and look for recent articles on WAP. Examples are www.zdnet.com or www.pcmag.com. What do these articles say about the future of WAP?

4. Research and compare the WMLScripting language to JavaScript. Produce a table listing five similarities and five differences between the two.

5. Research and compare the WMLScripting language to JavaScript. How do these differ from EC-MAScript, another standards-based scripting language used primarily in Europe?

6. Produce a five-page research paper summarizing the various components of WAP (WDP, WTA, WML, etc.). Which of these are most likely to incur dramatic revisions over time due to market and technical factors, and why?

5 Bluetooth

OBJECTIVES

After reading this chapter and completing the exercises, you will be able to:

- Discuss the importance of the new Bluetooth standard
- Describe devices that are appropriate for this standard
- Describe the components and operation of Bluetooth
- Describe how Bluetooth relates to WAP

In the previous chapter, we discussed the Wireless Application Protocol, or WAP. WAP is an emerging standard in the cellular arena. Within the more generic personal devices (laptops, wireless computer devices, and so on) is another emerging standard, namely Bluetooth.

BLUETOOTH MAKES LIFE EASIER

Bluetooth will unlock the potential in the thousands of devices and systems people encounter every day.

In the Office . . .

You arrive at the office and put down your briefcase. While in your office, your personal digital assistant (PDA) automatically synchronizes with your desktop PC and transfers files, e-mails, and schedule information.

While in a meeting, you access your PDA to send your presentation to the electronic whiteboard. You record meeting minutes on your PDA and wirelessly transfer these to the attendees before they leave the meeting.

You are the factory supervisor for Widgets, Inc. As you walk through the factory, you are able to check the status of every piece of test equipment you encounter because you can instantly download a user interface for every machine. You request product defect rates and piece part failures at selected workstations.

In the Home . . .

Upon arriving at your home, the door automatically unlocks for you, the entryway lights come on, and the heat is adjusted to your preset preferences.

An alarm notifies you that your toddler has just left the house.

Your PDA morphs from business to personal use as you enter your home. An electronic bulletin board in the home automatically adds your scheduled activities to the family calendar, and alerts you of any conflicts.

You have a home security system composed of Bluetooth technology devices. You have just upgraded the system and added devices. Because they all use Bluetooth technology, they automatically reconfigure and recognize each other.

On the Road . . .

You arrive at the airport. A long line is formed for ticketing and seat assignment. You avoid the line, using your PDA to present an electronic ticket and automatically select your seat. The airline's online system checks identification via the "ID-tag" feature built into your PDA and confirms your reserved seat.

You enter the airport waiting lounge, equipped with Bluetooth technology Internet ports. Via the ports, you and other guests use Bluetooth technology laptops, PDAs, and other devices to access your office or home-based servers via the airline server. Using voice-over IP, you also make "free" Internet voice calls courtesy of your airline.

You get on the Rent-A-Car bus. Your reservation is automatically transferred to the Rent-A-Car database, and you are dropped off at your car. You get in the Bluetooth technology rental car. Your hotel reservations are automatically queried from your PDA and the GPS system offers you directions to your hotel.

You arrive at the hotel. As you enter, you are automatically checked in and your room number and electronic key are transferred to your PDA. As you approach the room, the door automatically opens.

In the Car . . .
As you enter a national park, a map of the park appears on your display. You can view the schedule of activities for the park and your own personal electronic tour guide is downloaded to your vehicle.

As you approach your vehicle, the door unlocks automatically, the radio tunes in your favorite station, and the seat adjusts to your preferred settings.

As you enter your vehicle, you are reminded of the items on your daily calendar and the results of a recent diagnostic test of your vehicle.

You receive a new message en route, which is verbally transmitted to you via the vehicle's speakerphone.

In Social Settings . . .
Anxious to see the first-run movie, you arrive at the theater to find a long line at the ticket counter. Using your Bluetooth technology PDA to wirelessly confirm and pay for your tickets, you avoid the long line, enter the theater, and take your preferred seat.

At the racetrack, your PDA is used to download information on selected horses and jockies, to perform statistical analysis using historical information, to place bets, to request slow-motion replays, and to order food and beverage.

You are attending an industry trade show. You have preloaded your preferences for product information into your PDA. As you walk through the exhibits, your PDA detects other Bluetooth PDAs and exchanges preference information. Bluetooth facilitates the exchange of information, and enables you to meet others with common interests.

As you enter an upscale bar, you are handed a Bluetooth technology device. This device allows you to send messages and communicate with others in the bar, to order and pay for food and beverage, and to participate in games such as Trivia and Clue.

Source: www.bluetooth.com, www.123wapinfo.com

Monument honoring Bluetooth in Lund, Sweden
Source: www.bluetooth.com

PURPOSE AND BEGINNINGS

This chapter discusses the Bluetooth standard. The heart of the Bluetooth brand identity is the name, which comes from a Danish king, Harald Blatand, who unified Denmark and Norway in 908 C.E. Blatand translates to "blue tooth," which was chosen as the code name for the project that developed a wireless communication standard for unifying Denmark and Norway. Ericsson created the Bluetooth wireless standard in Lund, Sweden, in 1994.

DESIGN AND PRINCIPLES OF OPERATION

Bluetooth is a wireless standard. Bluetooth devices communicate via radio modules or transceivers which are encoded onto microprocessor chips. In addition, the Bluetooth standard specifies software, called a link manager, that enables Bluetooth devices to communicate. It supports up to seven simultaneous links within a radius of 10 meters.

Bluetooth

A Bluetooth system consists of a radio unit, a link control unit, and a support unit for link management and **host terminal interface (HTI)** functions, as shown in Figure 5–1. HTI is the interface between **Bluetooth host** and **Bluetooth unit.** Bluetooth Host is a computing device, peripheral, cellular telephone, and/or an access point to PSTN.

host terminal interface (HTI)

Bluetooth host

Bluetooth unit

In February 1998, Ericsson, Nokia, Toshiba, IBM, and Intel came together to form a special interest group (SIG). The group contained the right mix of business areas: two market leaders in mobile telephony, two market leaders in laptop computing, and a market leader in core digital-signal-processor technology. Total membership is now more than 1,800 companies (www.bluetooth.com 2000).

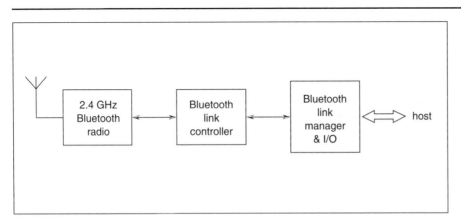

Figure 5–1
Functional Components

Since the beginnings of wireless communications, there have been many differing standards. It was apparent there needed to be a single, unifying approach with powerful players providing support. Since the organization of the SIG, the Bluetooth wireless technology has grown to become a unifying force. For example, the Cahners In-Stat Group estimates that by 2005, the Bluetooth wireless technology will be a built-in feature in more than 670 million products (www.cnet.com 2001).

Bluetooth provides an inexpensive standard for linking a wide range of small devices. In short, Bluetooth technology allows for the replacement of many cables that connect one device to another with one universal short-range radio link. For example, using a laptop and cellular phone equipped with Bluetooth technology—essentially a microprocessor-controlled radio link—would eliminate having to use a cable to transfer data between the two devices. Working within a 33-foot range, Bluetooth also allows two devices to exchange information. Examples include the following:

- Home Web terminals
- Computer mice
- Optical character recognition (OCR) wands
- Cameras
- Camcorders
- Computers, particularly laptops
- Personal digital assistants (PDAs)
- Cellular phones via WAP

The Bluetooth wireless technology allows users to make almost instant connections between various communications devices, such as mobile phones and desktop and notebook computers. The Bluetooth specification also provides efficient transmission and ensures safety from interference and security of data.

The Bluetooth radio is built into a small microchip and operates in a globally available frequency band ensuring communication compatibility worldwide. The Bluetooth specification has two power levels defined: a lower power level that covers the shorter personal area within a room or small office, and a higher power level that can cover a medium range, such as within a home or larger office. Both software controls and identity coding built into each microchip ensure that only those units preset by their owners can communicate.

The Bluetooth wireless technology supports both point-to-point and point-to-multipoint connections. With the current specification, up to seven subordinate devices can be set to communicate with a predominant radio in one device. Several of these small coverage areas, or **piconets,** can be established and linked in ad hoc **scatternets** to allow communication among continually flexible configurations. Scatternets are groupings of piconets, aligned in a pattern to assist in communications along a specific direction. All devices in the same piconet have priority synchronization, but other devices can be set to enter at any time. The topology can best be described as a flexible, multiple piconet structure (www.bluetooth.com 2000).

piconets

scatternets

Bluetooth operates in the unlicensed 2.4 GHz industrial, scientific, medical (ISM) frequency band, and can connect a maximum of eight devices on a shared 1 Mbps radio channel over a range of up to 10 meters. The radio protocol copes with a certain amount of interference, so transmission can coexist with Bluetooth-based devices used by other people nearby, or with other radio technologies using the same frequency band, such as 802.11 wireless local area network (LAN) systems.

The founders of the SIG set as one of its chief goals that Bluetooth microchips would occupy only the minimum of space on a circuit board and consume very little power. These are obviously crucial factors in the protocol's target market, a market comprising the replacement of the cumbersome, proprietary, and expensive cables and PC card devices used to connect laptops to mobile phones.

Another important factor for the SIG was target cost. Specifically, the protocol should add no more than $5 to a Bluetooth-enabled device. More than this would deter the advancement of the standard and lead to more de facto standards and proprietary solutions—an undesirable event. This low additional cost would also mean that Bluetooth could be added to any consumer electronic device from refrigerators to electronic keyboards, and those devices would remain competitive in price.

Finally, the Bluetooth SIG wanted a standard that would provide compatibility between the radios used in the system and define the quality of the system. We will discuss these aspects in some depth in this chapter.

Frequency Bands

As mentioned, the Bluetooth system operates in the 2.4 GHz ISM band. The frequency of this band can vary depending on geographical location. In most countries, the range is 2,400 MHz to 2,483.5 MHz.

Channel spacing, which is 1 MHz, is used between channels to minimize interference from neighboring channels. The channel spacings cannot, or should not, be used by any transmitter. To comply with out-of-band regulations in each country, a guard band is used at the lower and upper band edges. The guard band is a larger allocation of frequency than channel spacing. Table 5–1 shows these guard bands.

Geography	Lower Guard Band	Upper Guard Band
USA	2 MHz	3.5 MHz
Europe (except Spain and France)	2 MHz	3.5 MHz
Spain	4 MHz	26 MHz
France	7.5 MHz	7.5 MHz
Japan	2 MHz	2 MHz

Table 5–1
Guard Bands

TRANSMITTER CHARACTERISTICS

The power level at the antenna connector of the equipment is a crucial issue in order for the Bluetooth network to operate effectively. Out-of-range power levels can prevent interoperability. The equipment is classified into three power classes, as shown in Table 5–2. A power control is required for power class 1 equipment. The power control is used for limiting the transmitted power.

SPURIOUS EMISSIONS

Spurious emissions are those not from the original source. The spurious emission, in band and out of band, is measured with a frequency hopping transmitter hopping on a single frequency. This means that the synthesizer changes frequency between the receive slot and transmit slot, but always returns to the same transmit frequency. Table 5–3 provides more detailed information.

Receiver Characteristics

Receiving devices must adhere to careful sensitivity requirements to achieve interoperability. This is typically called the actual sensitivity level and is defined as the input level for which a raw bit error rate (BER) of 0.1 percent is

Table 5–2
Bluetooth Equipment
Power Classes

Power Class	Maximum Output Power (Pmax)	Nominal Output Power	Minimum Output Power	Power Control
1	100 mW (20 dBm)	N/A	1 mW (0 dBm)	Pmin<+4 dBm to Pmax Optional: Pmin to Pmax
2	2.5 mW (4 dBm)	1 mW (0 dBm)	0.25 mW (−6 dBm)	Optional: Pmin to Pmax
3	1 mW (0 dBm)	N/A	N/A	Optional: Pmin to Pmax

Table 5–3
Frequency Band
Requirements and
Spurious Transmissions

Geography	Regulatory Range	RF Channels
USA, Europe and most other countries	2.400–2.4835 GHz	f=2402+k MHz, k=0, . . . , 78
Spain	2.445–2.475 GHz	f=2449+k MHz, k=0, . . . , 22
France	2.4465–2.4835 GHz	f=2454+k MHz, k=0, . . . , 22

met. The requirement for a Bluetooth receiver is an actual sensitivity level of 70 dBm or better. The receiver must achieve the 70 dBm sensitivity level with any Bluetooth transmitter. Compliance standards are set according to Bluetooth technical specifications. All Bluetooth-compliant vendors must agree to this standard.

BASEBAND CHARACTERISTICS

Baseband consists of the physical layer characteristics defined in the Bluetooth standard according to the Bluetooth specification. In OSI terminology, the baseband is the physical layer of the Bluetooth standard. It manages physical channels and links apart from other services such as error correction, hop selection, and Bluetooth security. The baseband layer lies on top of the Bluetooth radio layer in the Bluetooth stack. The baseband protocol is implemented as a link controller, which works with the link manager for carrying out link level routines such as link connection and power control. The baseband also manages asynchronous and synchronous links, handles packets, and does paging and inquiry to access and inquire Bluetooth devices in the area. The baseband transceiver applies a time-division duplex (TDD) scheme (alternate transmit and receive). Therefore, apart from different hopping frequencies (frequency divisions), the time is also slotted. Keep in mind that Bluetooth is a short-range radio link intended to replace the cable(s) connecting portable and/or fixed electronic devices. Key features provided via the standard created by the Bluetooth SIG are robustness, low complexity, low power, and low cost.

baseband

As mentioned, Bluetooth operates in the unlicensed ISM band at 2.4 GHz. A frequency hop transceiver is applied to combat interference and fading. A slotted channel is applied with a nominal slot length of 625 microseconds (ms). For full duplex transmission, a TDD scheme is used. On the channel, information is exchanged through **packets.** Each packet is transmitted on a different hop frequency. A packet nominally covers a single slot, but can be extended to cover up to five slots.

packets

As an example of its flexibility, the Bluetooth protocol uses a combination of circuit and packet switching. Slots can be reserved for synchronous packets. Bluetooth can support an asynchronous data channel, up to three simultaneous synchronous voice channels, or a channel which simultaneously supports asynchronous data and synchronous voice. Each voice channel supports a 64 Kbps synchronous (voice) channel in each direction. The asynchronous channel can support maximal 723.2 kb/s asymmetric (and still up to 57.6 kb/s in the return direction), or 433.9 kb/s symmetric. This range is truly amazing and contributes to its continued popularity and widespread global applications.

As another example of its flexibility, the Bluetooth system provides a point-to-point connection (only two Bluetooth units involved), or a point-to-multipoint connection, as shown in Figures 5–2 and 5–3.

Figure 5–2
Single Point-to-point
Connection

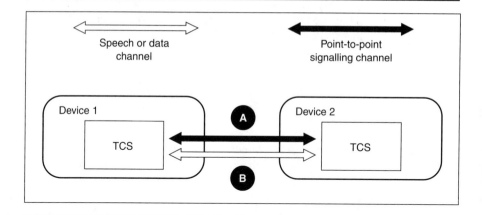

Figure 5–3
Multiple Point-to-point
Connections:
(a) Piconets with a
Single Slave Operation;
(b) Multislave Operation;
and (c) Scatternet
Operation

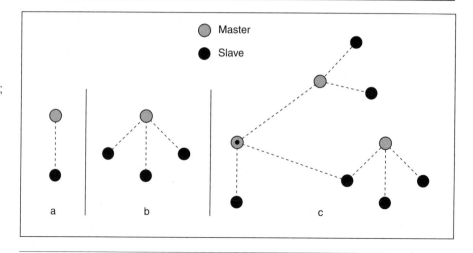

In the point-to-multipoint connection, the channel is shared among several Bluetooth units. Two or more units sharing the same channel form a piconet (units sharing a common channel). One Bluetooth unit acts as the master of the piconet, whereas the other unit(s) acts as a slave(s). Up to seven slaves can be active in the piconet. In addition, many more slaves can remain locked to the master in a so-called parked state. Note that these parked slaves cannot be active on the channel, but remain synchronized to the master. Both for active and parked slaves, the channel access is controlled by the master. Multiple piconets with overlapping coverage areas form a scatternet. **Coverage areas** are where two Bluetooth units can exchange messages. Each piconet can have only one master; however, slaves can participate in differ-

coverage areas

ent piconets on a time-division multiplex basis. In addition, a master in one piconet can be a slave in another piconet. The piconets shall not be time or frequency synchronized. Each piconet has its own hopping channel.

PHYSICAL CHANNEL

The **physical channel** is represented by a pseudo-random hopping sequence, hopping through the seventy-nine or twenty-three RF channels. The hopping sequence is unique for the piconet and is determined by the Bluetooth device address of the master; phase in the hopping sequence is determined by the Bluetooth clock of the master.

physical channel

The channel is divided into **time slots,** where each slot corresponds to an RF hop frequency. Consecutive hops correspond to different RF hop frequencies. All Bluetooth units participating in the piconet are time and hop synchronized to the channel. The channel is divided into time slots, each 625 ms in length. The time slots are numbered according to the Bluetooth clock of the piconet master. In the time slots, master and slave can transmit packets.

time slots

A scheme is used where master and slave alternatively transmit, as shown in Figure 5–4. The master shall start its transmission in even-numbered time slots only, and the slave shall start its transmission in odd-numbered time slots only. The packet start shall be aligned with the slot start. Packets transmitted by the master or the slave may extend to five time slots.

The RF hop frequency remains fixed for the duration of the packet. For a single packet, the RF hop frequency to be used is derived from the current Bluetooth clock value. For a multislot packet, the RF hop frequency to be used for the entire packet is derived from the Bluetooth clock value in the first slot

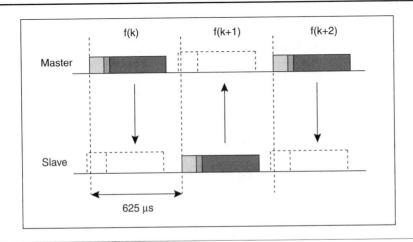

Figure 5–4
Baseband TDD Scheme

of the packet. The RF hop frequency in the first slot after a multislot packet shall use the frequency as determined by the current Bluetooth clock value. If a packet occupies more than one time slot, the hop frequency applied shall be the hop frequency as applied in the time slot where the packet transmission was started.

PHYSICAL LINKS

Between master and slave(s), different types of links can be established. Two link types are synchronous connection-oriented (SCO) link and asynchronous connectionless (ACL) link.

The SCO link is a point-to-point link between a master and a single slave in the piconet. The master maintains the SCO link by using reserved slots at regular intervals. The ACL link is a point-to-multipoint link between the master and all slaves participating on the piconet. In the slots not reserved for the SCO link(s), the master can establish an ACL link on a per-slot basis to any slave, including the slave(s) already engaged in an SCO link.

SCO Link

The SCO link is a symmetric, point-to-point link between the master and a specific slave. The SCO link reserves slots and can therefore be considered as a circuit-switched connection between the master and the slave. The SCO link typically supports time-bounded information such as voice. The master can support up to three SCO links to the same slave or to different slaves. A slave can support up to three SCO links from the same master, or two SCO links if they originate from different masters. SCO packets are never retransmitted.

The master will send SCO packets at regular intervals to the slave in the reserved master-to-slave slots. The SCO slave is always allowed to respond with an SCO packet in the following slave-to-master slot unless a different slave was addressed in the previous master-to-slave slot. If the SCO slave fails to decode the slave address in the packet header, it is still allowed to return an SCO packet in the reserved SCO slot.

The SCO link is established by the master sending an SCO setup message.

ACL Link

In the slots not reserved for SCO links, the master can exchange packets with any slave on a per-slot basis. The ACL link provides a packet-switched connection between the master and all active slaves participating in the piconet.

asynchronous

isochronous

Both **asynchronous** and **isochronous** (simultaneous) services are supported. Between a master and a slave, only one ACL link can exist. For most ACL packets, packet retransmission is applied to ensure data integrity.

A slave is permitted to return an ACL packet in the slave-to-master slot only if it has been addressed in the preceding master-to-slave slot. If the slave fails to decode the slave address in the packet header, it is not allowed to transmit.

ACL packets not addressed to a specific slave are considered as broadcast packets and are read by every slave. If there is no data to be sent on the ACL link and no polling is required, no transmission shall take place.

GENERAL FORMAT

The baseband controller interprets the first bit arriving from a higher software layer as the first bit to be sent over the air. Furthermore, data fields are generated internally at baseband level, such as the packet header fields and payload header.

The data on the piconet channel is conveyed in packets. The general packet format is shown in Figure 5–5. Note that each packet consists of the access code, the header, and the payload. The number of bits per entity also is indicated.

The access code and header are of fixed size: 72 bits and 54 bits, respectively. The payload can range from 0 bit to a maximum of 2,745 bits. Different packet types have been defined. Packets may consist of the (shortened) access code only, of the access code and header, or of the access code and header and payload. Each will be briefly discussed next.

Access Code

Each packet starts with an access code. If a packet header follows, the access code is 72 bits long, otherwise the length of access code is 68 bits. This access code is used for synchronization, DC offset compensation, and identification. The access code identifies all packets exchanged on the channel of the piconet. All packets sent in the same piconet are preceded by the same channel access code. The access code also is used in **paging** (are you there? messages to other Bluetooth units) and inquiry procedures. In this case, the access code itself is used as a signaling message and neither a header nor a payload is present. A Bluetooth unit transmits **inquiry** messages to discover what other Bluetooth units are active in the coverage area.

paging

inquiry

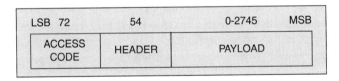

Figure 5–5
Standard Packet Format

Packet Header

The header contains link control (LC) information and consists of six fields.

- AM ADDR: 3-bit active member address
- TYPE: 4-bit type code
- FLOW: 1-bit flow control
- ARQN: 1-bit acknowledge indication
- SEQN: 1-bit sequence number
- HEC: 8-bit header error check

This makes the total header, including the HEC, to be 18 bits.

Packet Types

The packets used on the piconet are related to the physical links in which they are used. Until now, two physical links are defined: the SCO link and the ACL link. For each of these links, twelve different packet types can be defined. Four control packets will be common to all link types; their type code is unique irrespective of the link type. To indicate the different packets on a link, the 4-bit type code is used.

Payload Format

The previous packet overview considered several payload formats. In the payload, two fields are distinguished: the (synchronous) voice field and the (asynchronous) data field. The ACL packets have only the data field and the SCO packets have only the voice field, with the exception of the DV packets which have both.

Voice Field

The voice field has a fixed length. For the HV packets, the voice field length is 240 bits; for the DV packet, the voice field length is 80 bits. No payload header is present.

Data Field

The data field consists of three segments: a payload header, a payload body, and possibly a CRC code (the AUX1 packet does not carry a CRC code).

Payload Header. Only data fields have a payload header. The payload header is 1 or 2 bytes long. Packets in segments one and two have a 1-byte payload header; packets in segments three and four have a 2-byte payload header. The payload header specifies the logical channel (2-bit L CH indication), controls the flow on the logical channels (1-bit flow indication), and has a payload length indicator (5 bits and 9 bits for 1-byte and 2-byte payload header, respectively). **Logical channels** are different types of channels on a physical link.

logical channels

Payload Body. The payload body includes the user host information and determines the effective user throughput. The length of the payload body is indicated in the length field of the payload header.

CRC Code Generation. The 16-bit cyclic redundancy check code in the payload is generated by the CRC-CCITT polynomial 210041 (octal representation).

TRANSMIT/RECEIVE TIMING

The Bluetooth transceiver applies a TDD scheme, which means that it alternately transmits and receives in a synchronous manner. The mode of the Bluetooth unit determines the exact timing of the TDD scheme. In the normal connection mode, the master transmission shall always start at even-numbered time slots (master CLK1 = 0) and the slave transmission shall always start at odd-numbered time slots (master CLK1 = 1). Due to packet types that cover more than a single slot, master transmission may continue in odd-numbered slots and slave transmission may continue in even-numbered slots.

CHANNEL CONTROL

This section describes how the channel of a piconet is established and how units can be added to and released from the piconet. Several states of operation of the Bluetooth units are defined to support these functions.

Master-Slave Definition

The channel in the piconet is characterized entirely by the master of the piconet. The Bluetooth device address of the master determines the frequency hopping sequence and the channel access code; the system clock of the master determines the phase in the hopping sequence and sets the timing. In addition, the master controls the traffic on the channel by a polling scheme. By definition, the **master** is represented by the Bluetooth unit that initiates the connection (to one or more **slave** units). Note that the names *master* and *slave* refer only to the protocol on the channel: The Bluetooth units themselves are identical; that is, any unit can become a master of a piconet. Once a piconet has been established, master-slave roles can be exchanged.

master

slave

The Clock

Every Bluetooth unit has an internal system clock that determines the timing and hopping of the transceiver. It must be consistently accurate or interoperability may be prevented. This free-running native clock is never adjusted and is always on. For synchronization with other units, only offsets are used that, added to the native clock, provide temporary Bluetooth clocks which are mutually synchronized.

The Bluetooth clock provides the central aspect of the Bluetooth transceiver. Its resolution is at least half the transmit (TX) or receiver (RX) slot length, or 312.5 ms.

The timing and the frequency hopping on the channel of a piconet are determined by the Bluetooth clock of the master. When the piconet is established, the master clock is communicated to the slaves. Each slave adds an offset to its native clock to be synchronized to the master clock. Because the clocks are free running, the offsets require regular updates.

BLUETOOTH SECURITY

The Bluetooth technology provides peer-to-peer communications over short distances. To provide usage protection and information confidentiality, the system provides security measures at both the application layer and the link layer. This means that in each Bluetooth unit, the authentication and encryption routines are implemented in the same way. Four different entities are used for maintaining security at the link layer: a public address, which is unique for each user; two secret keys; and a random number, which is different for each new transaction.

The four entities and their sizes as used in Bluetooth are summarized in Table 5–4.

The Bluetooth device address is the 48-bit IEEE address which is unique for each Bluetooth unit. The Bluetooth addresses are publicly known, and can be obtained via MMI interactions or automatically via an inquiry routine by a Bluetooth unit.

The secret keys are derived during initialization and are never disclosed. Normally, the encryption key is derived from the authentication key during the authentication process. For the authentication algorithm, the size of the key used is always 128 bits. For the encryption algorithm, the key size may vary between 1 and 16 octets (8 to 128 bits).

Table 5–4
Bluetooth Security Entities

Entity	Size
BD_ADDR	48 bits
Private user key, authentication	128 bits
Private user key, encryption configurable length (byte-wise)	8-128 bits
RAND	128 bits

The size of the encryption key is configurable for two reasons. The first reason deals with the various requirements imposed on cryptographic algorithms in different countries due to export regulations and official attitudes toward privacy in general. The second reason is the need to facilitate a future upgrade path for the security without the need of a costly redesign of the algorithms and encryption hardware; increasing the effective key size is the simplest way to combat increased computing power at the opponent side. Currently, an encryption key size of 64 bits gives satisfying protection for most applications.

The encryption key is entirely different from the authentication key (even though the latter is used when creating the former). Each time encryption is activated, a new encryption key is generated. Thus, the lifetime of the encryption key does not necessarily correspond to the lifetime of the authentication key.

LINK MANAGER PROTOCOL

The link manager (LM) carries out link setup, authentication, link configuration, and other protocols. It discovers other remote LMs and communicates with them via the Link Manager Protocol (LMP). To perform its service provider role, the LM uses the services of the underlying link controller (LC).

The LMP essentially consists of several protocol data units (PDUs), which are sent from one device to another, determined by the AM ADDR in the packet header. LM PDUs are always sent as single-slot packets and the payload header is therefore 1 byte.

LMP messages are specialized control messages used for link setup, security, and control. They are transferred in the payload. Figure 5–6 shows where LMP fits into the architecture.

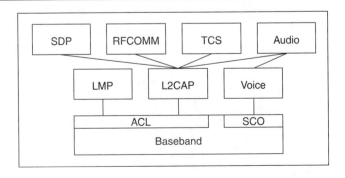

Figure 5–6
Link Manager Protocol in
Bluetooth Architecture

Link manager messages have higher priority than user data. Thus, if the link manager needs to send a message, it shall not be delayed by the L2CAP traffic, although it can be delayed by many retransmissions of individual baseband packets.

LOGICAL LINK CONTROL AND ADAPTATION LAYER PROTOCOL

The Logical Link Control and Adaptation Layer Protocol (L2CAP) is layered over the Baseband Protocol and resides in the data link layer. L2CAP provides connection-oriented and connectionless data services to upper layer protocols with protocol multiplexing capability, segmentation and reassembly (SAR) operation, and group abstractions. L2CAP permits higher level protocols and applications to transmit and receive L2CAP data packets up to 64 kilobytes in length.

The two link types supported for the baseband layer are SCO links and ACL links. SCO links support real-time voice traffic using reserved bandwidth. ACL links support best-effort traffic. The L2CAP specification is defined for only ACL links and no support for SCO links is planned.

L2CAP is one of two link level protocols running over the baseband, as shown in Figure 5–7. L2CAP is responsible for higher level protocol multiplexing, MTU abstraction, group management, and conveying quality of service (QoS) information to the link level.

Protocol multiplexing is supported by defining channels. Each channel is bound to a single protocol in a many-to-one fashion. Multiple channels can be bound to the same protocol, but a channel cannot be bound to multiple protocols. Each L2CAP packet received on a channel is directed to the appro-

Figure 5–7
L2CAP in Bluetooth
Layers

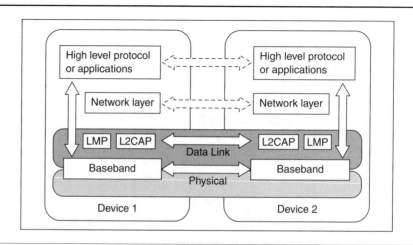

priate higher level protocol. L2CAP abstracts the variable-size packets used by the Baseband Protocol. It supports large packet sizes up to 64 kilobytes using a low overhead SAR mechanism.

Group management provides the abstraction of a group of units, allowing more efficient mapping between groups and members of the Bluetooth piconet. Group communication is connectionless and unreliable. When composed of only a pair of units, groups provide a connectionless channel alternative to L2CAP's connection-oriented channel.

L2CAP conveys QoS information across channels and provides some admission control to prevent additional channels from violating existing QoS contracts.

L2CAP provides connection-oriented and connectionless data services to upper layer protocols with protocol multiplexing capability, SAR operation, and group abstractions. L2CAP permits higher level protocols and applications to transmit and receive L2CAP data packets up to 64 kilobytes in length.

The two link types are SCO and ACL links. SCO links support real-time voice traffic using reserved bandwidth. ACL links support best-effort traffic. The L2CAP specification is defined for only ACL links and no support for SCO links is planned. For ACL links, use of the AUX1 packet on the ACL link is prohibited. This packet type supports no data integrity checks (no CRCs). Because L2CAP depends on integrity checks in the baseband to protect the transmitted information, AUX1 packets must never be used to transport L2CAP packets.

L2CAP Functional Requirements

As mentioned, the functional requirements for L2CAP include protocol multiplexing, SAR, and group management. Figure 5–7 illustrates how L2CAP fits into the Bluetooth protocol stack. L2CAP lies above the Baseband Protocol and interfaces with other communication protocols such as the Bluetooth Service Discovery Protocol (SDP), RFCOMM, and Telephony Control (TCS). Voice-quality channels for audio and telephony applications are usually run over baseband SCO links. Packetized audio data, such as IP telephony, may be sent using communication protocols running over L2CAP.

Following are the essential protocol requirements for L2CAP.

- *Simplicity and low overhead.* Implementations of L2CAP are applicable for devices with limited computational resources.
- *Low power consumption.* L2CAP does not consume excessive power because that would significantly sacrifice power efficiency achieved by the Bluetooth radio.
- *Minimum memory requirements.* Memory requirements for protocol implementation are kept to a minimum.
- *Diverse device set.* The Bluetooth Protocol complexity is acceptable to personal computers, PDAs, digital cellular phones, wireless headsets, joysticks, and other wireless devices supported by Bluetooth.
- *Bandwidth efficiency.* The protocol is designed to achieve reasonably high bandwidth efficiency.

- *Protocol multiplexing.* L2CAP supports protocol multiplexing, because the Baseband Protocol does not support any type field identifying the higher layer protocol being multiplexed above it. L2CAP must be able to distinguish between the upper layer protocols, such as the SDP.
- *RFCOMM.* The RFCOMM protocol provides emulation of serial ports over the L2CAP protocol. RFCOMM is a simple transport protocol, with additional provision for emulating the 9 circuits of RS-232 (EIA-TIA-232-E) serial ports and supports up to 60 simultaneous connections between two Bluetooth devices.
- *Telephony Control.* The Telephony Control Specification (TCS) CS contains the following functionality: signalling for the establishment and release of speech and data calls between Bluetooth devices, signaling to ease the handling of groups of Bluetooth devices, and provisions to exchange signaling not related to an ongoing call.
- *Segmentation and reassembly.* Compared with other wired physical media, the data packets defined by the Baseband Protocol are limited in size. Exporting a maximum transmission unit (MTU) associated with the largest baseband payload (341 bytes) limits the efficient use of bandwidth for higher layer protocols that are designed to use larger packets. Large L2CAP packets must be segmented into multiple smaller baseband packets prior to their transmission over the air. Similarly, multiple received baseband packets may be reassembled into a single larger L2CAP packet following a simple integrity check. The SAR functionality is absolutely necessary to support protocols using packets larger than those supported by the baseband.
- *Quality of service.* The L2CAP connection establishment process allows the exchange of information regarding the QoS expected between two Bluetooth units. Each L2CAP implementation must monitor the resources used by the protocol and ensure that QoS contracts are honored.
- *Groups.* Many protocols include the concept of a group of addresses. Baseband Protocol supports the concept of a piconet, a group of devices synchronously hopping together using the same clock. The L2CAP group abstraction permits implementations to efficiently map protocol groups on to piconets.

GENERAL OPERATION

L2CAP is based around the concept of channels. Each of the end points of an L2CAP channel is referred to by a channel identifier.

Operation Between Devices

Figure 5–8 illustrates the use of channel identifier numbers (CIDs) in a communication between corresponding peer L2CAP entities in separate devices. The connection-oriented data channels represent a connection between two

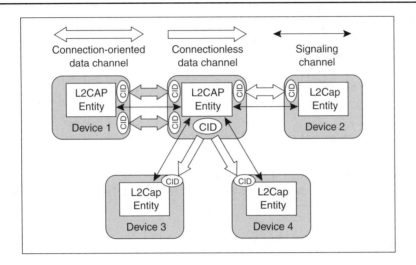

Figure 5–8
Communications
Between Devices

devices, where a CID identifies each end point of the channel. The connection-tionless channels restrict data flow to a single direction. These channels are used to support a channel group, where the CID on the source represents one or more remote devices. There are also a number of CIDs reserved for special purposes. The signaling channel is one example of a reserved channel. This channel is used to create and establish connection-oriented data channels and to negotiate changes in the characteristics of these channels. Support for a signaling channel within an L2CAP entity is mandatory. Another CID is reserved for all incoming connectionless data traffic.

Operation Between Layers

L2CAP implementations transfer data between higher layer protocols and the lower layer protocol. Each implementation supports a set of signaling commands for use between L2CAP implementations. L2CAP implementations also are prepared to accept certain types of events from lower layers and generate events to upper layers. How these events are passed between layers is an implementation-dependent process.

Segmentation and Reassembly

SAR operations are used to improve efficiency by supporting a MTU size larger than the largest baseband packet. This reduces overhead by spreading the network and transport packets used by higher layer protocols over several baseband packets. All L2CAP packets may be segmented for transfer over baseband packets. The protocol performs no SAR operations, but the packet format supports adaptation to smaller physical frame sizes. An

L2CAP implementation exposes the outgoing (i.e., the remote host's receiving) MTU and segments higher layer packets into chunks that can be passed to the link manager via the host controller interface (HCI), whenever one exists. On the receiving side, an L2CAP implementation receives chunks from the HCI and reassembles those chunks into L2CAP packets using information provided through the HCI and from the packet header.

DATA PACKET FORMAT

L2CAP is packet based, but follows a communications model based on channels. A channel represents a data flow between L2CAP entities in remote devices. Channels may be connection oriented or connectionless.

Connection-Oriented Channel

Figure 5–9 illustrates the format of the L2CAP packet (also referred to as the L2CAP PDU) within a connection-oriented channel. The fields shown are as follows:

- Length: 2 octets (16 bits). Length indicates the size of information payload in bytes, excluding the length of the L2CAP header. The length of an information payload can be up to 65,535 bytes. The length field serves as a simple integrity check of the reassembled L2CAP packet on the receiving end.
- Channel ID: 2 octets. The channel ID identifies the destination channel end point of the packet. The scope of the channel ID is relative to the packet-receiving device.
- Information: 0 to 65,535 octets. This contains the payload received from the upper layer protocol (outgoing packet), or delivered to the upper layer protocol (incoming packet).

Connectionless Data Channel

In addition to connection-oriented channels, L2CAP also exports the concept of a group-oriented channel. Data sent to the group channel is sent to all

---

Figure 5–9
L2CAP Packet Format

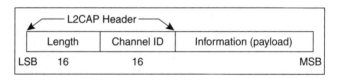

members of the group in a best-effort manner. Groups have no QoS associated with them. Group channels are unreliable; L2CAP makes no guarantee that data sent to the group successfully reaches all members of the group. If reliable group transmission is required, it must be implemented at a higher layer.

Transmissions to a group must be nonexclusively sent to all members of that group. The local device cannot be a member of the group, and higher layer protocols are expected to loop back any data traffic being sent to the local device. Nonexclusive implies that nongroup members may receive group transmissions, and higher level (or link level) encryption can be used to support private communication.

SERVICE DISCOVERY PROTOCOL

The Service Discovery Protocol (SDP) provides a means for applications to discover which services are available and to determine the characteristics of those available services (e.g., **service discovery**).

service discovery

A specific SDP is needed in the Bluetooth environment, as the set of services that are available changes dynamically based on the RF proximity of devices in motion. This is qualitatively different from service discovery in traditional network-based environments. The SDP defined in the Bluetooth specification is intended to address the unique characteristics of the Bluetooth environment.

Capabilities

The following capabilities have been identified as requirements for version 1.0 of the SDP.

- SDP provides the ability for clients to search for needed services based on specific attributes of those services.
- SDP permits services to be discovered based on the class of service.
- SDP enables browsing of services without a prior knowledge of the specific characteristics of those services.
- SDP provides the means for the discovery of new services that become available when devices enter RF proximity with a client device as well as when a new service is made available on a device that is in RF proximity with the client device.
- SDP provides a mechanism for determining when a service becomes unavailable when devices leave RF proximity with a client device as well as when a service is made unavailable on a device that is in RF proximity with the client device.
- SDP provides for services, classes of services, and attributes of services to be uniquely identified.

- SDP allows a client on one device to discover a service on another device without consulting a third device.
- SDP is suitable for use on devices of limited complexity.
- SDP provides a mechanism to incrementally discover information about the services provided by a device. This is intended to minimize the quantity of data that must be exchanged in order to determine that a particular service is not needed by a client.
- SDP supports the caching of service discovery information by intermediary agents to improve the speed or efficiency of the discovery process.
- SDP is transport independent.
- SDP functions while using L2CAP as its transport protocol.
- SDP permits the discovery and use of services that provide access to other service discovery protocols.
- SDP shall support the creation and definition of new services without requiring registration with a central authority.

Operation

The service discovery mechanism provides the means for client applications to discover the existence of services provided by server applications as well as the attributes of those services. The attributes of a service include the type or class of service offered and the mechanism or protocol information needed to utilize the service. Figure 5–10 shows this interaction.

SDP involves communication between an SDP server and an SDP client. The server maintains a list of service records that describe the characteristics

Figure 5–10
SDP Interaction

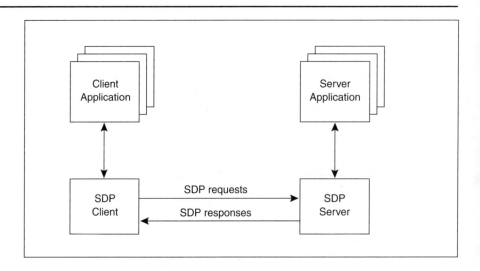

of services associated with the server. Each service record contains information about a single service. A client may retrieve information from a service record maintained by the SDP server by issuing an SDP request.

If the client, or an application associated with the client, decides to use a service, it must open a separate connection to the service provider to utilize the service. SDP provides a mechanism for discovering services and their attributes (including associated service access protocols), but it does not provide a mechanism for utilizing those services (such as delivering the service access protocols).

There is a maximum of one SDP server per Bluetooth device. (If a Bluetooth device acts only as a client, it needs no SDP server.) A single Bluetooth device may function both as an SDP server and as an SDP client. If multiple applications on a device provide services, an SDP server may act on behalf of those service providers to handle requests for information about the services that they provide. Similarly, multiple client applications may utilize an SDP client to query servers on behalf of the client applications.

The set of SDP servers that are available to an SDP client can change dynamically based on the RF proximity of the servers to the client. When a server becomes available, a potential client must be notified by a means other than SDP so that the client can use SDP to query the server about its services. Similarly, when a server leaves proximity or becomes unavailable for any reason, there is no explicit notification via the SDP. However, the client may use SDP to poll the server and may infer that the server is not available if it no longer responds to requests.

Protocol Description

SDP is a simple protocol with minimal requirements on the underlying transport. It can function over a reliable packet transport (or even unreliable, if the client implements timeouts and repeats requests as necessary). SDP uses a request/response model, in which each transaction consists of one request protocol data unit (PDU) and one response PDU; however, the requests may potentially be pipelined and responses may potentially be returned out of order. In the specific case when SDP utilizes the Bluetooth L2CAP transport protocol, multiple SDP PDUs may be sent in a single L2CAP packet, but only one L2CAP packet per connection to a given SDP server may be outstanding at a given instant. Limiting SDP to sending one unacknowledged packet provides a simple form of flow control.

RFCOMM

The RFCOMM protocol provides emulation of serial ports over the L2CAP protocol. RFCOMM is a simple transport protocol, with additional provisions for emulating the nine circuits of RS-232 (EIATIA-232-E) serial ports.

The RFCOMM protocol supports up to sixty simultaneous connections between two BT devices. The number of connections that can be used simultaneously in a BT device is implementation specific.

Device Types

For the purposes of RFCOMM, a complete communication path involves two applications running on different devices (the communication end points) with a communication segment between them.

RFCOMM is intended to cover applications that make use of the serial ports of the devices in which they reside. In the simple configuration, the communication segment is a BT link from one device to another (direct connect), as shown in Figure 5–11. Where the communication segment is another network, BT is used for the path between the device and a network connection device such as a modem. RFCOMM is only concerned with the connection between the devices in the direct connect case, or between the device and a modem in the network case.

RFCOMM can support other configurations, such as modules that communicate via BT on one side and provide a wired interface on the other side, as shown in Figure 5–12. These devices are not actually modems but offer a similar service.

Basically two device types exist that RFCOMM must accommodate. Type 1 devices are communication end points such as computers and printers. Type 2 devices are those that are part of the communication segment (e.g., modems). Though RFCOMM makes no distinction between these two

Figure 5–11
RFCOMM Direct Connect

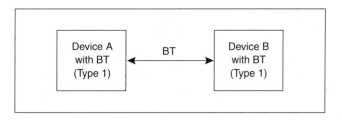

Figure 5–12
RFCOMM and Legacy
Devices

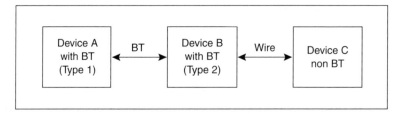

device types in the protocol, accommodating both impacts the RFCOMM protocol.

The information transferred between two RFCOMM entities has been defined to support both type 1 and type 2 devices. Some information is only needed by type 2 devices, whereas other information is intended to be used by both. In the protocol, there is no distinction between type 1 and type 2. It is therefore up to the RFCOMM implementers to determine if the information passed in the RFCOMM protocol is of use to the implementation. Because the device is not aware of the type of the other device in the communication path, each must pass on all available information specified by the protocol.

RFCOMM Service Overview

RFCOMM emulates RS-232 (EIATIA-232-E) serial ports. The emulation includes transfer of the state of the nondata circuits. RFCOMM has a built-in scheme for null modem emulation.

In the event that a baud rate is set for a particular port through the RFCOMM service interface, it will not affect the actual data throughput in RFCOMM; that is, RFCOMM does not incur artificial rate limitation or pacing. If either device is a type 2 device (relays data onto other media), however, or if data pacing is done above the RFCOMM service interface in either or both ends, then actual throughput will on an average reflect the baud rate setting.

RFCOMM supports emulation of multiple serial ports between two devices and also emulation of serial ports between multiple devices.

Service Interface Description

RFCOMM is intended to define a protocol that can be used to emulate serial ports. In most systems, RFCOMM will be part of a port driver, which includes a serial port emulation entity.

Service Definition Model

Figure 5–13 shows a model of how RFCOMM fits into a typical system. The figure represents the RFCOMM reference model.

TELEPHONY CONTROL PROTOCOL SPECIFICATION

The Telephony Control Protocol Specification (TCS) does not discriminate between user and network side, but merely between outgoing side (the party originating the call) and incoming side (the party terminating the call). Figure 5–14 shows TCS in the Bluetooth layers.

Figure 5–13
RFCOMM Service
Definition Model

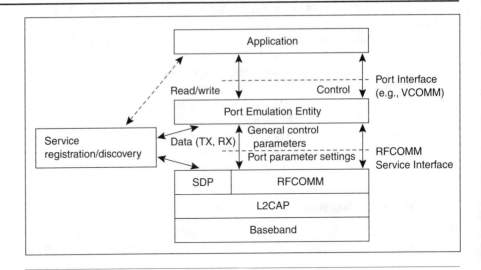

Figure 5–14
TCS in Bluetooth

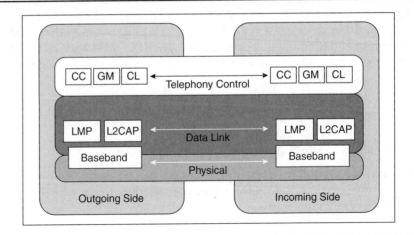

Operation Between Layers

TCS implementations should follow the general architecture shown in Figure 5–15. The internal structure of TCS binary contains the functional entities of call control (CC), group management (GM), and connection less (CL) as described earlier, complemented with the protocol discrimination which, based on the TCS internal protocol discriminator, routes traffic to the appropriate functional entity.

To handle more calls at once, multiple instances of TCS binary may exist simultaneously. Discrimination between the multiple instances can be based on the L2CAP channel identifier.

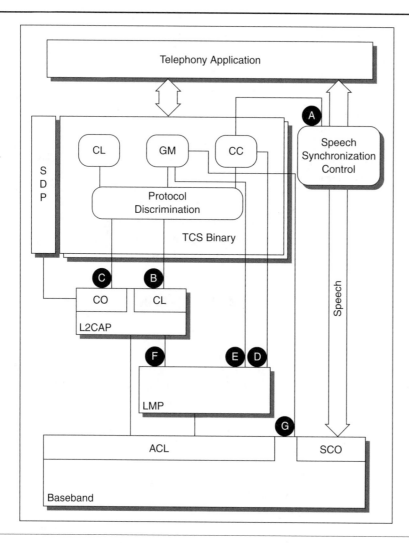

Figure 5–15
TCS Architecture

TCS binary interfaces with a number of other (Bluetooth) entities to provide its (telephone) services to the application. The interfaces are identified in Figure 5–15, and information is exchanged across these interfaces for the following purposes.

- The call control entity provides information to the speech synchronization control about when to connect (disconnect) the speech paths. This information is based on the call control messages (e.g., reception of Connect Acknowledge or Disconnect).

- To send a Setup message using point-to-multipoint signaling, it is delivered on this interface to L2CAP for transmission on the connectionless channel. Conversely, L2CAP uses this interface to inform TCS of a Setup message received on the connectionless channel. The connectionless L2CAP channel maps on to the piconet broadcast.
- When a TCS message needs to be sent using point-to-point signaling, it is delivered on this interface to L2CAP for transmission on a connection-oriented channel. During L2CAP channel establishment, specific QoS to be used for the connection will be indicated, in particular the usage of low power modes (L2CAP will inform LMP about this—interface F).
- The call control entity controls the LMP directly, for the purpose of establishing and releasing SCO links.
- The group management entity controls the LMP and LC/baseband directly during initialization procedures to control (for example) the inquiry, paging, and airing—interface E and G.

INTEROPERABILITY REQUIREMENTS FOR BLUETOOTH AS A WAP BEARER

This section is intended for Bluetooth implementers who wish to take advantage of the dynamic, ad hoc characteristics of the Bluetooth environment in providing access to value-added services using the WAP environment and protocols. Bluetooth provides the physical medium and link control for communications between WAP client and server. Also described is how PPP may be used to achieve this communication.

WAP in the Bluetooth Piconet

In many ways, Bluetooth can be used like other wireless networks with regard to WAP. Bluetooth can be used to provide a bearer for transporting data between the WAP client and its adjacent WAP server. Bluetooth's ad hoc nature also provides capabilities that are exploited uniquely by the WAP protocols.

WAP Server Communications

The traditional form of WAP communications involves a client device that communicates with a server/proxy device using the WAP protocols. In this case, the Bluetooth medium is expected to provide a bearer service as specified by the WAP architecture.

Initiation by the Client Device
When a WAP client is actively listening for available Bluetooth devices, it can discover the presence of a WAP server using Bluetooth's SDP. In Figure 5–16,

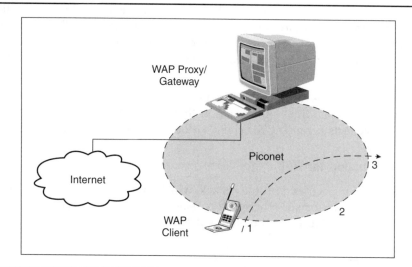

Figure 5–16
WAP Server/Proxy in
Bluetooth Piconet

stage 1, the WAP client device is moving into range of the WAP proxy/ gateway's piconet. When the client detects the presence of the WAP proxy/ gateway, it can automatically, or at the client's request, connect to the server.

The client must also be able to determine the specific nature of the WAP proxy/gateway that it has detected. The Bluetooth SDP is utilized to learn the following information about the server.

- Server name—a user-readable descriptive name for the server
- Server home page document name—the optional home page URL for the server
- Server/proxy capability—indicates if the device is a WAP content server, a proxy, or both. If the device is a proxy, then it must be able to resolve URLs that are not local to the server/proxy device

In Figure 5–16, stage 2, the device is communicating with the WAP proxy/gateway.

Termination by the Client Device

Termination by the client device is also an event carried out by the SDP. In Figure 5–16, stage 3, the device is exiting the piconet. When the device detects that communication has been lost with the WAP proxy/gateway, it may optionally decide to resume communications using the information obtained at discovery.

Initiation by the Server Device

An alternative method of initiating communications between a client and server is for the server to periodically check for available client devices. When the server device discovers a client that indicates it has WAP client capability, the server may optionally connect and push data to the client. The client device has the option of ignoring pushed data at the end user's discretion. In addition, through the Bluetooth SDP, the server can determine the following information about the client.

- Client name—a friendly format name that describes the client device
- Client capabilities—information that allows the server to determine basic facts regarding the client's Bluetooth-specific capabilities

Network Support for WAP

The following specifies a protocol stack, which may be used below the WAP components. Support for other protocol stack configurations is optional, and must be indicated through the Bluetooth SDP.

PPP/RFCOMM

Devices that support Bluetooth as a bearer for WAP services using PPP provide protocol stack support, as shown in Figure 5–17. The baseband, LMP, and L2CAP are the OSI layer 1 and 2 Bluetooth protocols. RFCOMM is the Bluetooth adaptation of GSM TS 07.10. SDP is the Bluetooth protocol. PPP is the IETF point-to-point protocol. WAP is the protocol stack and application environment.

Figure 5–17
Protocol Support for
WAP

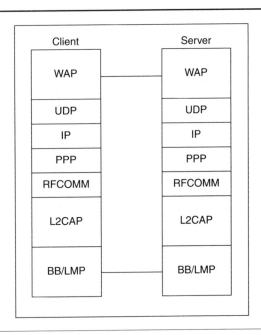

So, where can you find Bluetooth-enabled devices? Well, there are still not many available in the marketplace. You can see a full list of all the available Bluetooth-enabled devices at www.bluetooth.com. An example is the newly announced Sony DSC-FX77 camera. It will include a built-in Bluetooth transmitter that will allow users to transmit images from as far away as 30 feet to other Bluetooth-equipped devices, such as lap tops and cell phones. It is expected to be available in November, 2002 (www.cnet.com 2002.)

ALTERNATIVES TO BLUETOOTH

Although Bluetooth is the hottest standard for wireless networking, it is not the solution for all settings. Bluetooth is slow; its personal area network transfers 1 megabyte each second. Compare this with ubiquitous Ethernet cable found in most offices or a wireless LAN standard, such as 802.11b, where data travel as fast as 11 Mbps. Because of its speed limit, Bluetooth limits itself to passing only small packets of data.

One alternative to Bluetooth, 802.11b, is probably better known by Apple iBook owners. 802.11b and HomeRF are Bluetooth networking rivals. Although Intel says Bluetooth compliments the alternatives, great time is being spent ensuring products are compatible.

Problems still being ironed out with Bluetooth include:

- Cost—To make Bluetooth products affordable, the cost of chips must be near $5. They are currently hovering at $20.
- Security—One analyst compared the situation with posting your social security number in a chat room.
- Battery drain—This is a particular problem with mobile devices, making them unreliable.

THE OUTLOOK

The world is going wireless. The Gartner Group, a technology forecasting firm, estimates Bluetooth will play a vital role in uniting the 70 percent of new cell phones and 40 percent of new PDAs accessing the Web by 2004. Estimates for Bluetooth-enabled devices in the marketplace vary widely. It is encouraging that the number of Bluetooth-enabled devices is growing. See http://qualweb.bluetooth.org for a list of these devices.

RECENT EVENTS IN THE BLUETOOTH WORLD

Bluetooth has experienced delays in implementation, and so Bluetooth-enabled devices are still scarce. There are reasons for this delay. First, cost continues to be a major issue. In December 2000, the prices for Bluetooth chips were around the $15 to $20 range, well beyond the $5 cost envisioned by the Bluetooth SIG. "You can't put a $20 Bluetooth chip in these products" and make money, says Jan Edbrooke, marketing director at Logitech, which will not release its own Bluetooth-enabled keyboards and mice until chip prices fall. In fact, analysts predict the standard's first commercial successes are more likely in Europe and Asia, where mobile phones are more popular.

Security and privacy issues are also problems. The original Bluetooth specification has security provisions, but they are up to vendors to implement. Bluetooth devices could be easy prey for virus-mongers without a security system, who could undermine its promise of convenience. The Bluetooth SIG is diligently working on this issue.

Manufacturers of devices that use the Bluetooth local wireless networking standard admit that the first products were suitable only for evaluation purposes, following research by IT Week Labs that found problems with power consumption, usability, and price. This admission is the first indication from manufacturers that Bluetooth products may not be ready for general deployment.

Version 1.1, a more complete Bluetooth standard, was not ratified until early 2002. Products based on version 1.1 did not appear until several months after that. Clearly some of the advantages of Bluetooth products, such as low power consumption and low price, are not being delivered in the first crop of products because they use first-generation Bluetooth chipsets.

Bluetooth also may have suffered from the unreasonable expectations. Similar technologies (e.g., GSM) took 5 years from concept to product, with no hype until 6 months prior to launch. This is in contrast to Bluetooth, which received significant hype before the standard was even specified. Even now, it is only part way into what should be a 5-year cycle.

Recently, Bluetooth took at big upturn, when Japanese electronics giant, Toshiba, said it would commercialize large-scale integration (LSI) chips that would enhance the performance of devices using Bluetooth technology.

The Radiocommunications Agency has also reported that Bluetooth could be hampered by interference from wireless LANs, which reduces connectivity to 2 meters.

A product now on the market is Mitac Cat PDA, a hand-held computer powered by the Linux operating system. It also will go into mass production with a view to selling on the Taiwanese consumer market. This product comes with the Linpus Linux 0.7, a multilingual version of the Linux operating system, and features Bluetooth connectivity. Mitac also lists the product between $150 and $199.

Larger manufacturers also are part of the action. Hewlett-Packard unveiled a host of mobile and Bluetooth offerings at the CeBIT IT show in March 2001, including new PDAs, 1 GHz notebooks, and Bluetooth adapters for its printers.

Microsoft's website says it this way: "Support for Bluetooth wireless technology is not in the first release of Windows XP, because there is not a sufficient array of production-quality devices that conform to the Bluetooth specification for Microsoft to test. However Microsoft is actively developing support for Bluetooth technology and will ship this support in a future release" (http://www.microsoft.com/hwdev/tech/network/blvetooth/default/asp 2000). In addition, an HP adapter will add Bluetooth capability to its current Deskjet and LaserJet printer ranges. A product to link notebooks and the Deskjet 350C via Bluetooth is in the working stages, as are third-party products to link the Jornada to other Bluetooth devices.

However, this deployment is not without problems. For example, Microsoft did not add support for Bluetooth to its Windows XP system. This situation is especially important, because the firm is one of the lead members of the trade group that is developing the Bluetooth standard. This lack of support from Microsoft is a major blow to Bluetooth, and it will likely slow the technology's once strong momentum. Without Microsoft, Bluetooth adoption becomes slightly more onerous for hardware manufacturers and software developers as the software giant will not deliver a family of device drivers or other software to simplify how the technology gets incorporated.

Although Bluetooth and 802.11b complement each other in several ways, the two standards also compete because consumers will likely use them for the same function. With Bluetooth, data from a notebook would be sent by an internal Bluetooth radio chip to a cell phone. Under 802.11b, data are transmitted directly to an 802.11b receiver, which would be wired into an Ethernet network. To work, 802.11b requires that the user be inside or near a structure containing a transmitter.

Although some thought this could be an impediment, airports, hotels, and office buildings are rapidly installing 802.11b transmitters. Notebook manufacturers are also opting to install 802.11b transmitters over Bluetooth transmitters. The 802.11b standard defines transfer rates of 11 Mbps in the 2.4 GHz radio band.

Another factor in its favor is that 802.11b depends on the same networking protocols and standards as traditional networking, so adaptation should be fairly easy.

Nevertheless, Bluetooth's demise is not evident, as shown by the $35 million that British Bluetooth network supplier Red-M received via venture-capital funding. This funding boost is designed to give Red-M an early boost in the Bluetooth market.

Despite recent pronouncements that Bluetooth is mired in nearly impossible technical and production problems, Red-M and other companies are resolutely optimistic the wireless networking system is simply going through early growing pains. Simon Gawne, Red-M's vice president of marketing and business development, points out that Cambridge Silicon Radio recently shipped its one millionth Bluetooth chip and that Ericsson, Palm, and other device makers are planning to sell Bluetooth-integrated products soon.

"This Christmas, any new mobile phone at the high end will come with Bluetooth," he said. "At that point, it shifts from a niche to the general consumer market."

For more information and updates on all topics discussed in this section, go to www.zdnet.co.uk/news/2000 and www.news.cnet.com/news/0-1006-200-5520038.html.

■ SUMMARY

Bluetooth is the subject of this chapter; it is an emerging wireless connectivity standard for a host of small devices from PDAs to laptops. It has a leading role in the wireless industry, because of its technical characteristics and recent developments.

In the next chapter we delve into the well-known world of cellular technology. After all, how many of us do not believe that a cell phone is a necessity?

REVIEW QUESTIONS

1. The Bluetooth transceiver operates in the _____ band.
   a. 2.4 GHz ISM
   b. 900 MHz
   c. 1.2 MHz ISM

2. The actual sensitivity level is _____.
   a. the limited error factor
   b. input level with a limited raw error rate
   c. error rate restriction based on the transmitter

3. SNR is the _____.
   a. signal-to-error negativity response
   b. signal-to-interference ratio
   c. none of the above

4. True/False The Bluetooth Protocol uses a combination of circuit and packet switching.

5. True/False Bluetooth can support both data and voice channels.

6. True/False Three or more units sharing the same channel is known as a piconet.

7. True/False Each piconet has its own hopping channel.

8. True/False Bluetooth utilizes time slots.

9. The two link types supported by Bluetooth are:
   a. SCO and SCO-A
   b. ACL and ACL-S
   c. SCO and ACL

10. TDD is used by Bluetooth as a _____.
    a. frequency multiplexing algorithm
    b. time displacement/slotting algorithm
    c. none of the above

11. Multiple piconets with overlapping coverage are called _____.
    a. multinets
    b. scatternets
    c. m-piconets

12. The SCO is a _____ link.
    a. point-to-point
    b. point-to-multipoint
    c. multipoint-to-multipoint

13. True/False Only data fields have a payload header.

14. True/False A master unit initiates the connection.

15. To establish new connections, the _____ and _____ procedures are utilized.

a. Initiate/Begin

b. Inquiry/Paging

c. Begin/Connect

16. True/False Multiple piconets can cover the same geographical area.

17. True/False The Standby state is the default state in a Bluetooth unit.

18. True/False The ACL link provides a packet-switched connection.

19. True/False Bluetooth is intended typically for communications over a limited geographical area.

20. True/False L2CAP provides connection-oriented and connectionless data services.

21. True/False Memory requirements are not really a factor in the Bluetooth arena due to its unique operation.

22. True/False Segmentation and reassembly operations are utilized to improve efficiency.

23. True/False L2CAP provides a group-oriented channel.

24. True/False The SDP service permits services to be discovered based on the class of service.

25. True/False SDP is a protocol with minimal requirements.

26. True/False RFCOMM provides emulation of parallel ports.

27. Bluetooth is an emerging _____ standard.

a. close proximity

b. LAN

c. WAN

d. none of the above

HANDS-ON EXERCISES

1. Using your browser and a search engine, conduct a search for sites referencing Bluetooth. Are these sites primarily in the United States or Europe?

2. Using your browser and a search engine, conduct research and discover how Bluetooth began. Why was there a need for this standard?

3. Conduct research, via the Net and library resource (if any), and compare Bluetooth and WAP. List at least ten characteristics, from a network perspective, in which they are similar.

4. Using an image construction package such as Adobe Photoshop, construct a protocol hierarchy for Bluetooth.

5. Based on step 4, construct a WAP hierarchy next to your Bluetooth hierarchy. Are their similarities and differences? Are they distinctly different or are there overlaps?

6. Using an image construction package such as Adobe Photoshop, construct a drawing showing connectivity between at least three different Bluetooth devices. Hint: Under what circumstances would you utilize Bluetooth in the office, for example?

7. Research any similarities and dissimilarities between WAP and Bluetooth as far as how they have emerged onto the technology scene.

6 Cellular Telephony

OBJECTIVES

After reading this chapter and completing the exercises, you will be able to:

- Discuss the impact and prevalence of cellular telephony
- Discuss the history of cellular telephony
- Describe the various components and operation of cellular telephony
- Discuss recent developments in cellular telephony
- Discuss the immediate future of the cellular industry

GSM ASSOCIATION AND WAP FORUM™ FORMALIZE COOPERATIVE ALLIANCE TO BENEFIT WIRELESS INDUSTRY

London, UK and Mountain View, Calif. USA—The GSM Association, the world's leading wireless industry body, and the WAP Forum™, the international industry association that develops and fosters the growth of the Wireless Application Protocol (WAP), today announced a cooperative alliance to formalize and strengthen their working relationship.

Set in force for the next three years, this alliance is designed to renew and strengthen the organizations' shared objective of promoting international standardisation through contribution to the establishment of global standards in the wireless telecommunications industry. In addition to common aims and objectives, the WAP Forum and GSM Association share several key member companies and often collaborate on projects, such as the GSM Association's m-Services, which is based on WAP 2.0.

"The GSM Association and the WAP Forum have agreed to work together to improve and evolve the wireless Internet," says Rob Conway, Chief Executive Officer of the GSM Association. "This is good for the consumer, for our members and for the global wireless industry. Both organizations recognize the need to collaborate closely to ensure a smooth, consistent and co-ordinated development path in the future."

The two organizations plan to collaborate on work areas such as: identification of network operator and service provider requirements, device, protocol and application requirements; roaming and billing; development of WAP specifications; and interoperability. The WAP Forum and the GSM Association intend to use this open standards industry agreement to ensure that quality services develop more quickly.

"We are enthusiastic about the opportunity to work more seamlessly with our colleagues at the GSM Association to continue to work toward our common goals and ensure interoperability of the wireless Internet," says Robert L. (Bob) Brown, CEO of the WAP Forum. "We are continuously deepening relationships and building strategic bridges with key industry groups and standardization organisations to jointly drive mobile specific Internet application technologies across all relevant industry segments."

This agreement is one of many important relationships for both organizations. Both the WAP Forum and the GSM Association will continue to work with other bodies to promote the interests of its member companies.

About the WAP Forum™

The WAP Forum is the industry association that is responsible for developing and fostering the growth of the Wireless Application Protocol (WAP), the open global de facto standard that allows mobile users of wireless hand-held devices to securely access and interact with Internet-based content, applications, and services. The WAP Forum is comprised of hundreds of members, representing 99 percent of the handsets sold worldwide and more than 450 million global subscribers. Members include worldwide device manufactures, carriers, infrastructure providers, software developers, and other wireless solution providers. For more information about the WAP Forum, including a listing of its members, visit www.wapforum.org.

About the GSM Association
The GSM Association is the world's leading wireless industry representative body, consisting of more than 574 second and third generation wireless network operators and key manufacturers and suppliers to the wireless industry. Membership of the Association spans 171 countries of the world.

The GSM Association is responsible for the deployment and evolution of the GSM Family of technologies (GSM, GPRS, EDGE, and 3GSM) for digital wireless communications. The Association's members provide digital GSM wireless services to more than 612 million customers (end October 2001). The GSM Platform accounts for approximately 70 percent of the total digital cellular wireless market.

Source: Press release. Retrieved November 8, 2001, www.gsmworld.com/ news/press.

HISTORY OF CELLULAR TELEPHONY

Cellular telephony has had a long and evolutionary history. The concept actually began immediatly after World War I. At that time, mobile radio technology was still in its infant stages. For example, issues such as simply designing and then constructing a radio system that could receive radio broadcasts in a moving vehicle were deemed insurmountable. Experiments were conducted to determine how far radio broadcasts would be adequately received in vehicles going through various terrain and the antenna height required at both transmitting and receiving sites. For example, one early suggestion was that automobiles should have a 20-foot antenna to adequately receive radio broadcasts. In all cases, these problems were finally solved after much experimentation.

Predecessors

amplitude modulation (AM)

The predecessors to cellular telephony were broadcast radio and mobile telephony systems. Early experiments in radio broadcasting were based on conventional **amplitude modulation (AM)** radio,[1] because that technology

---

[1] Amplitude modulation (AM) is a way of placing data on to an alternating-current (AC) carrier waveform. A problem exists, however, in that the carrier consumes 75 to 80 percent of the power. This makes AM an inefficient mode of transmission.

had been a stable and functional technology since 1908. Lee de Forest is credited with making several early radio broadcasts, including a program of phonograph records from the Eiffel Tower in Paris in 1908; the voice of Enrico Caruso in a performance of *Cavalleria Rusticana* at the Metropolitan Opera House in January 1910; self-styled "radio concerts" several nights a week in 1916 from an experimental broadcasting station at High Bridge, New York; and news bulletins of the Woodrow Wilson–Charles Evans Hughes presidential election in November 1916.

Size

The size issue was critical in the development of cellular telephony. The first practical solution to this problem came in 1920 when the Detroit Police Department conducted some truly pioneering experiments with broadcast radio messages to receivers in police cars in the greater Detroit area. The major problem, however, was that transmission was only one way, which meant that the officers had to stop at a wireline telephone station to communicate with the sender. If this sounds familiar, it should: This is similar to today's paging systems.

In the early 1930s, mobile transmitters were developed, and the first two-way mobile system was implemented by the Bayonne (New Jersey) Police Department. Two problems arose, however. First, operating a powerful transmitter in close proximity to a very sensitive receiver limited the system to half duplex, or "push-to-talk" transmission.[2] Second, the mobile units were quite large. The receiving radio equipment consumed the entire trunk of a typical police vehicle, forcing the officer to store anything else under the front seats.

Experimentation continued until a full-duplex system was invented in the early 1930s. In fact, by 1934, there were 194 municipal police radio systems and 58 state police radio stations serving more than 5,000 radio-equipped police cars. The size of the equipment also was reduced to that of a small suitcase.

FM

In 1935, **frequency modulation (FM)** was invented by Edwin H. Armstrong. This was important, because FM transmissions are preferable to AM. The absence of background noise and the lack of distortion in FM circuits made radio transmissions stand out against a velvety silence with a presence that was new in auditory experiences. Why is FM radically better than AM? Instead of varying the strength (amplitude) of the radio signal, FM varies the frequency of the signal. It seems that Armstrong had found a way to conquer that ever-present danger to communications, namely static, otherwise known as random amplitude noise from natural and man-made sources. In addition, the range was at least tripled compared with AM, giving an unparalleled crispness suitable for broadcasting fine music. Also critical for the operation of

frequency modulation (FM)

---

[2] This is where you push a button to talk and release it to then listen.

mobile radio, FM needed much less power. This meant that smaller, less powerful vehicular transmitters and sensitive receivers could be designed and subsequently implemented. FM exhibited other desirable properties, notably the "capture effect": an FM receiver tends to "lock in" on the stronger of two competing signals and to reject the unwanted signal almost completely (unlike AM where two competing signals will often be superimposed upon one another). Finally, FM proved to be much more resistant to the peculiar propagation problems of mobile radio transmission, especially the problem of "fast fading" or flutter, which virtually destroyed mobile AM signals.

During World War II, many advancements made in radio technology supported the war effort. FM was in fact one of the areas of crucial technological superiority of the United States over its enemies in the war. During this period, the main contribution was the creation of a commercial FM manufacturing capability. Size, reliability, cost, and performance were dramatically improved through successive product generations throughout the war.

The period after World War II witnessed a rapid expansion of true mobile telephone service into the commercial arena. Technical improvements were oriented toward two goals: to improve radio wave spectrum utilization and to increase carrying capacity.

During the 1940s, requests for radio wave spectrum for two-way mobile radio increased dramatically. In 1945, the FCC recommended mobile radio spectrum allocations for a wide range of private sector uses including police and fire departments; forestry services; electric, gas, and water utilities; and transportation services, including railroads, buses, streetcars, trucks, and taxis. In 1949, the FCC officially recognized mobile radio as a new class of service. This coincided with an increase in the number of mobile radio users. The number of mobile users exploded from a few thousand in 1940 to 86,000 by 1948, 695,000 by 1958, and almost 1.4 million by 1963 (Calhoun 1988).

The Public Telephone System and Mobile Users

The interconnection of mobile users to the public telephone landline network—to allow telephone calls from fixed stations to mobile users—was first introduced in 1946 and within a year was operating in more than twenty American cities. These systems were based on FM transmission and utilized one powerful transmitter, typically at the top of a tall tower on the highest structure or topographical feature available. This provided coverage of a 50-mile radius or more from the base, enough to encompass most metropolitan **base station**   areas. Although the first systems involved an operator at the **base station** who manually patched in the radio call to the wireline network, the achievement of automatic, direct-dial service was prevalent by 1950.

Demand Issues

Demand grew quickly and stayed ahead of available capacity in many of the large urban markets. The mobile telephone use was quite different from the use of dispatch-type mobile systems. In a dispatch mode of operation, each communication normally lasted only a few seconds. The amount of traffic or circuit usage generated by each mobile unit in a dispatch system is small. Thus, each dispatch channel can support a large number of mobile units, up to 100 or more, without serious interference. In a mobile telephone service, however, the normal phone call may run for several minutes. The amount of circuit usage per customer may be 10 times or more for a dispatch circuit. This reduces drastically the number of mobile telephone users who can share a single channel.

Another problem at this time was that mobile telephone subscribers were normally registered far in excess of the effective traffic capacity of the early systems. Loading of 50, 100, or more subscribers per channel was common, resulting in insufficient service. Signal blocking probabilities (the likelihood that a subscriber would be unable to obtain a circuit within a specified short period of time, such as 10 seconds) rose to as high as 65 percent. The actual usefulness of the mobile telephone decreased as users found that such blocking often completely prevented them from getting a circuit during the peak periods. Obviously, customers were dissatisfied and so usage decreased. It was clear that something had to be done or mobile radio would not be a significant presence in the telecommunications arena. It was also equally clear that more spectrum was needed to alleviate this problem.

This state of mobile telephony pointed to a factor not previously considered. FM transmission suffered a serious disadvantage: It required much more bandwidth than AM. Each radio channel had to be wider, and the guardbands separating each channel from adjacent channels had to be wider. Therefore, fewer such channels could be carved out of a given portion of the radio spectrum.

An important technological development during this period was the invention and application of automatic trunking radio systems. In the earliest radio systems, both transmitter and receiver were designed to operate on a specific fixed frequency. Each radio channel was dedicated to a specific user, or to a group of users who shared it like a party line. In a trunked radio system, by contrast, a group of channels are made available to an entire group of users, as shown in Figure 6–1. If channel A is in use, the next caller will be assigned to channel B, and so on. Each user may actually be assigned any one of the group of channels for a given call. No channel is dedicated solely to a single user; instead, all channels are available to all users.

The advantage of this method is that the total traffic (measured in call-seconds) that can be borne by a trunked radio system can be significantly higher (under certain conditions) than the traffic carried by an equal number of non-trunked, dedicated channels. For example, a single mobile channel can support only about two or three mobile telephone users (with a 10 percent chance of

Figure 6–1
Trunked Radio System

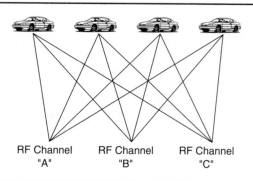

RF Channel   RF Channel   RF Channel
    "A"   "B"   "C"

blocking and an average of 150 call-seconds per customer per hour). Twenty such channels, not trunked, could support only about 50 customers, evenly assigned to different frequencies. If, however, those twenty channels are trunked together such that any user can access any idle channel, then more than 420 customers can be supported with the same grade of service and the same calling behavior. This is referred to as trunking efficiency (Figure 6–1). The only negative point is that trunking involves more sophisticated equipment because the unit must be able to efficiently tune into a variety of frequencies.

The idea of cellular telephony began to appear in Bell Corporation's system proposals to the FCC during the late 1940s. Technologists began to realize that the problem of spectrum congestion might be alleviated by restructuring the coverage areas of mobile radio systems. The traditional approach to mobile radio viewed the problem in terms similar to radio or television broadcasting. It involved setting up a high-power transmitter on top of the highest point in the area and blasting out the signal to the horizon (as much as 40 or 50 miles away). The result was fair-to-adequate coverage over a large area. It also meant, however, that the few available channels were locked up over a large area by a small number of calls. For example, in New York City in the 1970s, the Bell mobile system could support only twelve simultaneous mobile conversations. The thirteenth caller was blocked (Calhoun 1988).

Mobile Radio versus Cellular

The cellular telephony model is radically different. Cellular uses many low-power transmitters each specifically designed to serve only a small area, perhaps only a couple of miles across. Instead of covering an area like New York City with a single transmitter, the city would be blanketed with small coverage areas, called cells. See Figure 6–2 for an example. A problem quickly arose. To produce even a relatively efficient cellular system, it was clear that additional spectrum would be needed.[3] The handful of mobile telephone channels was far too small.

---

[3] See www.fcc.gov.

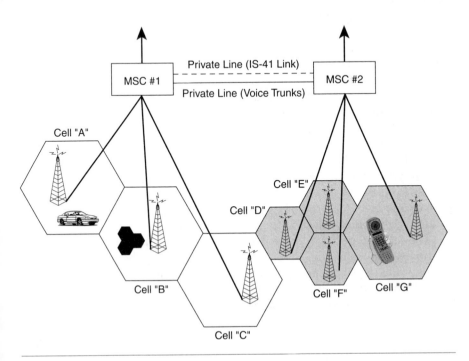

To/From: Public Switched
Telephone Network or
Other Networks

To/From: Public Switched
Telephone Network or
Other Networks

Figure 6–2
Typical Cell Layout

The first proposal for a large-capacity mobile telephone system was put forward by Bell engineers to the FCC in 1947, only a year after the introduction of mobile telephone service. The proposal was not acted upon by the FCC at that time, due to a series of political and procedural issues. Two years later, Bell prepared another proposal. Then, the FCC was in the process of creating a new and much expanded set of television allocations, in what would become known as the UHF band, encompassing more than 400 MHz of spectrum located between 470 MHz and 890 MHZ. Bell argued that mobile services were also entitled to additional spectrum and should not be squeezed out by the broadcasters; but, television was becoming popular, so the FCC deferred action on Bell's proposal.

Over the next two decades, however, issues came to the forefront that changed the FCC's lack of interest. First, mobile technology was improving and demand for mobile services grew robustly. Second, research continued to make cellular networking a more economically feasible endeavor. In 1970 it was assumed by almost everyone that the new mobile telephone service would be operated as an extension of AT&T's wireline telephone monopoly. The 1970 FCC decision specifically allocated the new mobile telephone spectrum to the wireline telephone companies only, which at that time meant AT&T for about 85 percent of the American population and for almost all the major cities. AT&T had developed the cellular idea originally, which it now elaborated into a series of technical proposals that became known as the Advanced Mobile Phone System (AMPS) architecture, which was strongly shaped by the assumption of monopoly service. The AMPS approach certainly had the look of a conventional utilitylike telephone network. AMPS would be heavily dependent upon a skeleton of switching and call-processing technology linking the dozens or hundreds of cells into an integrated metropolitan network under centralized control. In 1970, only AT&T had the expertise to build and operate such a system.

At this time, FM became the technology of choice for two reasons. First, it was not a proprietary technology. Any competent manufacturer could produce FM cellular radios. Second, the FCC firmly believed that competition and volume manufacturing should be encouraged by choosing a radio-link standard that was, in effect, in the public domain.

The cellular concept represented a very different approach to structuring a radio telephone network. It was an idea that held out the promise of virtually unlimited system capacity, breaking through the barriers that had restricted the growth of mobile telephony, and it did so without any fundamental technological leap forward. It simply worked smarter with the same resources. Indeed, cellular architecture was a system-level concept, essentially independent of radio technology. It appealed to mobile system engineers, because it kept them on relatively familiar hardware ground. It appealed to businesspeople and entrepreneurs, because it seemed to open the path to a really large market. Through the application of the cellular idea, mobile communication could become another first-class growth industry—like television, radio, or the telephone.

DESIGN AND PRINCIPLES OF CELLULAR OPERATION

Cellular telephony places transceivers in numerous small cells within an area. A central processor, or switch, controls user access to the transceivers, permitting the user to move among cells and maintain service. This provides the frequency reuse capability that gives cellular the potential for far greater capacity than early two-way mobile systems.

mobile switching center (MSC)

mobile telephone switching office (MTSO)

The central processor, or switch, is also referred to as the **mobile switching center (MSC)** or the **mobile telephone switching office (MTSO)**. It provides the switching and radio control functions. Links extend from the MSC to the cell sites and also provide connections to the **Public Switching Tele-**

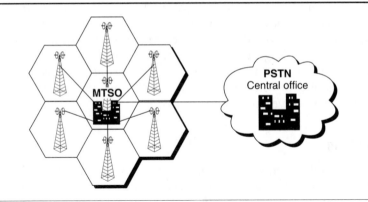

Figure 6–3
MTSOs and the PSTN

phone Network (PSTN). PSTN is the installed U.S. telephone network, originally AT&T as shown in Figure 6–3. The PSTN is the wired telephone network we use every day.

    Previously in Figure 6–2, the cell arrangements appear in a hexagonal pattern. Cellular systems are engineered using this hexagonal arrangement to determine the frequency reuse pattern. This method is used to ensure enough separation between any two cells using the same frequencies to prevent interference. Typically, the system may be designed using a four-cell, seven-cell, or twelve-cell arrangement to permit frequency reuse. If the world were flat and there were no obstructions, the radio waves could be confined to these hexagonal cells. In reality, however, the radio signals can at best only approximate the hexagonal shape. Special measures are needed for areas that are difficult to serve with normal antennas, such as tunnels. Still the use of hexagonal cells is a useful engineering tool.

Public Switching Telephone Network (PSTN)

The Cell Architecture

Early cellular systems used cell sites that typically covered a radius of 10 miles or more. As traffic increased, or better coverage was needed in certain areas, the size of the cell sites decreased. Presently, the average cell site has a radius of about 3 to 5 miles, but this is also diminishing. The cells can range in size from a radius of less than 1 mile to 25 miles or more, with the range being determined by the transmitted signal power and height of the antenna.

    As traffic grows in a cellular system, additional channels and cells are added until the entire available spectrum is in service in each cell. Metropolitan areas may use the entire spectrum, but rural service areas (RSAs) rarely need a full complement of channels.

cell site

The number of transceivers in a given cell depends on the traffic that the cell is expected to handle. The maximum number of radios in any given **cell site,** however, depends on the reuse pattern in the system design. This can be determined by dividing the total number of channels by the reuse pattern. For example, if the system is designed for a reuse pattern of N = 4, and the maximum number of voice channel pairs is 396, then the maximum number of channels in any given cell cannot exceed 99.

Cell Splitting

To minimize interference, a certain distance must be maintained between cells using the same frequencies; however, this distance can be reduced without disturbing the cell (i.e., frequency) reuse pattern. As the size of the cells is reduced, the same frequencies can be used in more cells, which in turn means more subscribers can be accommodated on the system. Particularly in congested areas, the cellular operator often splits an existing cell into two or more smaller cells (see Figure 6–4). New transceivers are placed and the power of the transmitters reduced to confine the signals to the newly created cells. For example, a cell that originally had a radius of 8 miles could be split into four cells with each new cell having a 2-mile radius. For the existing ana-

Figure 6–4
Dividing up a Macrocell
into Microcells

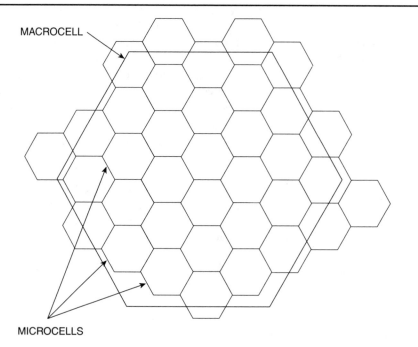

MACROCELL

MICROCELLS

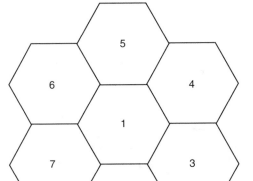

Figure 6–5
Cell Reuse

Each cell uses a different frequency.

log systems, cell splitting is an effective way to increase system capacity although some practical limitations are reached. Suitable locations for cell sites become more difficult and the processing load on the switch rapidly increases because handoffs are more frequent.

By reducing the coverage areas and creating a large number of small cells, it became possible (in theory) to reuse the same frequencies in different cells (see Figure 6–5). To understand how this changes the total picture, imagine that all the available frequencies could be reused in every cell. If this could be done, then instead of 12 simultaneous telephone circuits for the entire city there would be 12 circuits for every cell. If there were 100 cells (each about 10 miles across), there would be 1,200 circuits for the city, instead of only 12. See Figure 6–6 for a second view.

Early calculations indicated that, because of interference between mobile units operating on the same channel in adjacent cells, the same frequencies could not be used in every cell. It would be necessary to skip several cells before reusing the same frequencies. But the basic idea of reuse appeared to be valid. The system engineer could, in effect, create more than one usable mobile telephone circuit from the same channel, reused in different parts of the city.

The effects of interference also were not related to absolute distance between cells, but to the ratio of the distance between cells and the radius of the cells. The cell radius was determined by the transmitter power. It was within the power of the cellular firm to decide how many circuits would be created through reuse. If, for example, a grid of 10-mile-radius cells allowed reuse of

Figure 6–6

Cell Reuse—Another View

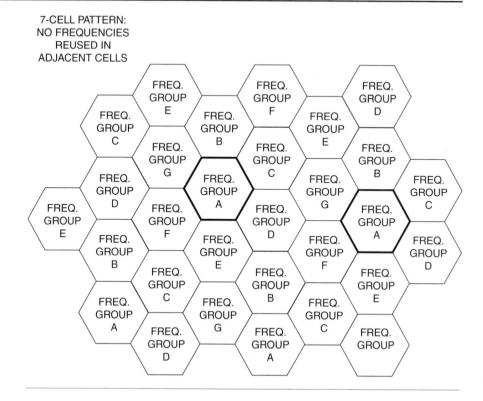

7-CELL PATTERN: NO FREQUENCIES REUSED IN ADJACENT CELLS

the frequencies in cell A at a distance of 30 miles, then a grid of 5-mile-radius cells would allow reuse at 15 miles, and 1-mile-radius cells would allow a hundred times as many circuits as a system based on 10-mile-radius cells. Speculation began about the effect of using cells only a few hundred feet in radius, a few city blocks in area. It was discovered that (on paper) they could create systems with thousands of cells and huge numbers of circuits per megahertz per square mile, capable of carrying the traffic of hundreds of thousands—or even millions—of paying customers in a metropolitan area.

Of course, it would have been enormously expensive to build hundred-cell systems from the beginning. It appeared, however, that large-radius cells could evolve gracefully into small-radius cells over time through a technique called cell splitting. When the traffic reached the point in a particular cell such that the existing allocation of channels in that cell could no longer support a good grade of service, that cell would be subdivided into a number of smaller cells—with even lower transmitter powers—fitting within the area of the former cell. The reuse pattern of radio frequencies could be repeated on the new, smaller scale and the total capacity multiplied for that area by a factor equal to the number of new cells. When, in time, the smaller cells were saturated, still smaller cells could be created. Even without going to ultrasmall cells only a few blocks across (later called **microcells**), the cellular architects conserva- **microcells** tively forecast at least three rounds of cell splitting (see Figure 6–4).

Cell splitting seemed to offer many advantages. It allowed the financial investment to be spread out as the system grew. New cells would only be added as the number of revenue-generating customers increased, providing the cash to support the continued investment. Moreover, it was thought that new cells could be created without scrapping the existing investment in the large-radius cell-site equipment; those transmitters would simply be powered down to fit within the new scale. Cell splitting could be applied in a geographically selective manner; the expense of smaller cells would only be necessary in the high-density traffic centers. In outlying areas, larger radius cells would suffice to carry the lower traffic densities in more suburban regions. Finally, cell splitting was flexible in the dimensions of time and space. In the minds of cellular architects, it was the ideal surgical technique for boosting capacity precisely where and when it was needed.

Cell Handoffs

The problem with small cells was that not all mobile calls could be completed within the boundaries of a single cell. A car moving at freeway speeds might zip through half a dozen very small cells in a single conversation. To deal with this situation, technologists developed the handoff technology. The cellular system would be endowed with its own system-level switching and control capability—a higher layer in the mobile network, operating above the individual cells. Through continuous measurements of signal strength received from the individual cell sites, the cellular system would be able to sense when a mobile user with a call in progress was passing from one cell to another, and then to switch the call from the first cell to the second cell "on the fly," without dropping or disrupting the call in progress. This technology required new techniques for determining into which of several possible new cells the mobile unit had strayed, as well as methods for tearing down and reestablishing the call in a very rapid manner, which posed some modest challenges to the cellular architects. The concept of handoff, however, seemed entirely feasible—small cells would work. Figure 6–7 shows a representation of cell handoff.

The essential principles of cellular architecture remain as follows:

1. Low power transmitters and small coverage zones or cells
2. Frequency reuse
3. Cell splitting to increase capacity
4. Handoff and central control

Certainly it was an ambitious architecture, but in principle handoff seemed no more challenging than other contemporary telecommunication developments; however, there was an obstacle. To realize a workable cellular system,

Figure 6–7
Cell Handoff

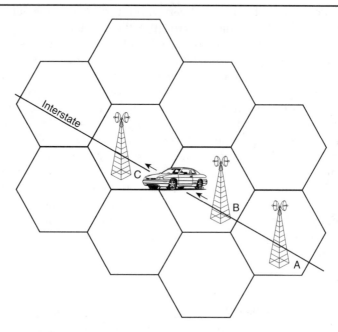

As a driver crosses from cell "B" to cell "C," the cell tower
of cell "B" hands off the call to the tower of cell "C."

additional spectrum would certainly be needed. The handful of mobile telephone channels available in conventional IMTS allocations was far too small to permit the realization of the cellular dream. The cellular idea might be able to "manufacture spectrum," but it needed a critical mass to begin. Alas, obtaining the initial allocation would not be easy.

Temporary Cell Sites

Because cell sites are time consuming and expensive to build, occasionally temporary cell sites can be used rather than building new ones. In these situations, cellular carriers may use **cell sites on wheels** (or **COWs**) or **cells on light truck sites** (or **COLTs**). COWs are fully equipped cell sites, including rather large antenna towers that are mounted on heavy-duty trucks or tractor trailers. The truck parks in the desired spot and begins service. COWs are normally used to temporarily repair a coverage deficiency or as a temporary replacement for a site that has failed due to equipment or a natural disaster. COLTS are temporary cells on small trucks that are useful for providing capacity at singular events such as a concert or a sporting event.

cell sites on wheels

cells on light truck sites

Cellular Power Levels

By design, cellular systems operate at a much lower signal power level than traditional two-way mobile systems. Base stations have a maximum power limitation of 100 watts effective radiated power (ERP), except for those located in RSAs, which can transmit as high as 500 watts ERP.

Mobile units designed for use in cars are capable of transmitting at 3 watts ERP, whereas portable hand-held units have a maximum power of 600 milliwatts, or 6/10 watt. The original bag-type cell phones issued only 10 years ago operated at this maximum 3 watts, whereas the newer trimline phones typically operate at 6/10 watt. This capacity is usually fine except in fringe areas where the full 3 watt maximum might allow connectivity whereas the 6/10 watt would not.

The power levels discussed earlier, though, represent maximum levels. The actual levels for cellular phones are adjusted automatically by instructions from the base stations. Cellular phones can transmit using power levels as low as 7 milliwatts. Base station levels are set according to system design parameters but are often much lower than the maximum 100 watts.

Performance Issues

Capacity of any wireless system is greatly affected by the number of voice channels that are available to serve subscribers. The number of subscribers that can use any given voice channel is a function of the traffic characteristics of the subscribers. These characteristics are the number of calls generated as well as the length of each call. There are traffic tables available that predict the number of channels required for a given traffic load, measured in a quantity called erlangs. One erlang is equal to the continuous occupancy of one channel. Presently, each cellular subscriber generates about 0.03 erlangs, or 30 millierlangs, of traffic.

Cellular systems are engineered to provide a B.01 grade of service, which is to say that one of every 100 call attempts may be blocked. Using this grade of service and a usage rate of 30 millierlangs per subscriber, a system can typically accommodate twenty to thirty subscribers for each voice channel in a set. For an analog system, only one subscriber can use a voice channel at any one time, but multiple subscribers can share a single channel because they do not all use the channel at the same time.

Cell Site to MSC Links

Several methods may be used to connect cell sites of any type to an MSC. The links may use two-wire or four-wire analog circuits, digital, fiber, or microwave facilities for these connections. The choice often depends on the type of microcell used, the facilities available in a given area, and cost. However, due to the ever-decreasing cost of fiber, and the near error-free nature and high-speed capability of fiber, fiber is increasingly becoming the media of choice.

Two-wire analog facilities often use copper wires in microcells having small capacities, such as stand-alone applications. These systems do not need to be connected to the MSC or switch, but do require a normal dial line connection to the local regional carrier's (LRC's) switch for access to the PSTN. LRCs are local telephone companies allowed to provide telephone service within a specified geographical area.

Four-wire analog circuits are not as common as they were when cellular first was introduced in the United States, because the LRCs have modernized their facilities in the last decade so that digital capabilities are more widely available. In some instances, still, when digital is not available or the voice channel requirements are very small, a four-wire analog link is used.

Digital links are quite common between cell sites and the MSC or switching center. These links may be several miles in length or contained within a building. Generally, a line at the digital signal (DS-1) rate (1.544 Mbps) is chosen for these applications, which requires some care in facility selection for proper performance. For intrabuilding applications, 22-gauge shielded wiring is the preferred choice to reduce the possibility of interference. Non-shielded wire in other sizes will work, but it is more susceptible to interference. Digital links at the DS-3 rate (45 Mbps) also are used for configurations requiring large numbers of voice channels.

Fiber links are a common way of connecting microcells to a donor cell. Because of its wide bandwidth capabilities, the fiber facility can transport the entire cellular spectrum if necessary. A number of these links use an analog signal that modulates the entire RF signal with an optical signal for transmission. Other equipment is available that digitizes the signal before it enters the fiber facility. This equipment is a little more tolerant of signal degradation than the analog variety. Both types typically can use fiber facilities that are more than 40 km in length without requiring an optical repeater.

macrocells   Microwave lengths from the MSC to cell site locations are quite common in cellular systems for normal **macrocells.** Microcells, because of their size and location, are generally more difficult to serve with the 2 GHz or 6 GHz microwave lengths normally employed in these situations. As an alternative, some companies have used 23 GHz or 38 GHz links to connect microcells. Because of the high frequency, a very small antenna may be used for the microwave; however, the range between the donor cell and microcell is limited and subject to fading.

Not all cellular carriers have a choice as to which form of transport will be used for these links. For example, some cellular carriers are part of a large holding company that also owns an LRC. These cellular carriers may be forced to use leased lines from the LRC, although microwave might be a less expensive alternative.

Mobile Switching Center

Only digital switches are now used for cellular service. This has been the case since the introduction of cellular service in the United States, although a few switches were initially based on an analog, electronic switch design. Digital switching, however, does not necessarily mean digital radio transmission in the cellular system. It simply refers to the manner in which calls are processed by the switch, or MSC, as it is known.

Like many landline switches, the MSCs are stored program controlled (SPC) devices whose functions are controlled by software instructions. The switches are usually full-availability, nonblocking switches, meaning any input port can be connected to any output port.

In addition to completing connections from the radio portion of the system to the landline network, or other wireless networks, the MSC usually also performs the mobility management function in an AMPS-type cellular network. Because of this feature, it has been estimated that an MSC takes 4 to 10 times the processing of a landline switch for an equivalent number of subscribers.

In most digital switches, the number of ports is configured in multiples of twenty-four to accommodate DS-l-rate input or output channel. Not all of the channels may be activated in any single DS-1. Some of the ports are connected to the cell sites by the MSC-cell site links. Other ports are connected to the PSTN or other networks, while still other ports provide links between MSCs themselves.

Repeaters

The general category of product used to extend a macrocell's coverage is called a **repeater.** Repeaters come in many shapes and sizes, but they all perform one basic function: They extend the wireless range of a macrocell. In that vein, they communicate directly with the macrocell either via copper, fiber optics, or a wireless link. Figure 6–8 shows the layout of a system using a macrocell and a repeater to reach automobiles within a tunnel.

repeater

Functionally, there is a significant difference between using a repeater to extend capacity and breaking down macrocells into microcells to increase capacity. Microcells add capacity because each microcell communicates directly with the MSC. Repeaters, because they communicate with the macrocell itself, actually take away capacity from the macrocell. Every person using the repeater's capacity inside the tunnel in Figure 6–8 means that one less person outside the tunnel can use the macrocell's capacity.

One of the fastest growing uses of repeaters is for in-building applications. In this situation, an antenna is placed on the roof of the building to transmit and receive mobile calls. The signal is then routed from the rooftop antenna, through the building, to a small repeater on every floor. The signals

Figure 6–8
Macrocells and
Repeaters

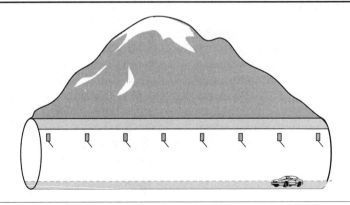

from the repeater are transmitted and received through an antenna no bigger than a smoke alarm. With in-building repeaters, you can begin a cellular phone call in your car, continue it while you enter the building—even in the elevator—and finish it after you arrive at your desk.

CELLULAR TELEPHONY OPERATIONS

The basic mobile telephone system service consists of the cell base station at the center of the cell and the MSC. The incoming mobile signals are captured at the base station and then forwarded to the MSC, which services some three to five cells. Each cell base station cell controller sends the call out to the mobile subscriber on one hand. It also acts as a receiver from the MSC as well as forwards the mobile call to the nearest MSC. The MSC performs the switching operations for all the cells it serves and is the connecting link to the broader world of PSTN communications. The MSC is thus the primary interface between the mobile network and the conventional telephone system, but the base station serves as the secondary interface in translating the call into the radio frequency (RF) signals for mobile transmissions. The MSC typically hands off a mobile call directly into the PSTN.

When a call is made to a mobile subscriber, the base station cell controller initially sends out a coded signal notifying the subscriber of the incoming call. Assuming the subscriber is within the sending area, the mobile telephone will automatically call the closest MSC to establish the connection. In short, the subscriber always "initiates" the call, regardless if the subscriber is calling from a mobile or remote location or is being called.

The communication from the mobile telephone is sent by a radio transmitter that has a range of 3 to 8 miles, depending on the density of traffic within the cell. The mobile telephone operates off the vehicle battery or generator, or if away from this power source, from a small battery within the handheld unit. These batteries hold sufficient charge for about 2 hours of service, but some thin silhouette, lightweight units have a capacity of only 30 minutes. The new lithium ion batteries will serve to either make cellular telephones lighter in weight or extend their usable power without frequent recharges.

The mobile telephone is technically a full-duplex transceiver, which means it is a radio device that is capable of transmission on one frequency and receiving on another. Half-duplex radio transceivers still used in dispatcher and other fleet services can require that the same frequency be used for transmission and reception, and thereby use a push-to-talk system. The most important frequency band allocated to analog mobile telephone services is the VHF band from 890 to 902 MHz, which is also shared with radio location services.

The transmission quality is often impaired by a variety of factors, including problems with effective handoff of the signal between cells; multipath, shadowing, and/or physical interference (see Chapter 3 for more information on these); narrowband channels; and congestion of channels in traffic-saturated cells. These problems help to explain why and how digital technology came about. The addition of more cells can likely serve to improve future performance in every area.

An important service feature in mobile communications is that of roaming—the ability to switch a mobile radio from one system to another and to have calls rerouted wherever you might travel. This feature allows you to change service from your local cellular system, even interconnection along interstate highways, to plug into other cellular systems or to link into satellite systems if available. The key is in advising the mobile routing system that you are not at your normal cellular address, but in fact are roaming to other systems. Ultimately, there will likely be a universal telephone number. Each subscriber will have a single unique number for use in all telephone systems. A worldwide intelligent signaling system will automatically forward calls. Today, the routing system is not "intelligent," and thus the user must indicate change of venue to the cellular call routing system to have calls properly forwarded.

Cellular radio telephone service is not only convenient for the consumer, but also flexible and increasingly multipurpose. Today cellular radio service is much more than a voice service; once the basic cellular telephone service is installed, many additional options are available. One can install a fax machine or a modem for data and operate it much like an ordinary PSTN line. The fax- or computer-based modem can be installed by landline or by a radio link via a wireless modem that connects to the mobile line. This allows a user in a remote location to connect directly to computer terminals around the country. Thus, one can connect either to send or receive faxes, e-mail, or obtain other data services.

With a wireless modem one can either be in the car or have access via a portable terminal within a reasonable proximity to the car's "wireless bus." In short, data, voice, or fax are available via mobile cellular services. Although voice provides the overwhelming source of cellular revenues today, most cellular operators project that a significant portion of revenues will be derived from data and fax services by 2005. The original assumption was to achieve this by 1999 but cost issues have delayed the deployment of advanced data networks. The newer cellular systems using CDPD techniques, being developed to interweave data services via packet radio techniques, will likely become a popular service in coming years.

There are two ways to operate cellular data services. The first is to connect a wireless modem to a cellular telephone and simply transmit the data over an open voice link. The difficulty with this approach is that it is difficult to maintain the link without losing the signal due to cell switching or other interference on the narrow 3 kHz channel allotted for voice transmission. The second option involves digital cellular packet data service. This service sets up a special data channel out of spare capacity that becomes available when setting up and breaking down regular cellular service. It is optimized for data services, especially short, bursty messages. This service, which can be more expensive than regular data service over a cellular phone, provides a higher grade of service with less downtime and fewer repeat transmissions.

ANALOG CELLULAR TELEPHONES

Channels

The AMPS system uses 832 full-duplex (two-way communications at the same time) channels, each consisting of a pair of simplex channels. There are 832 simplex transmission channels from 824 to 849 MHz, and 832 simplex receive channels from 869 to 894 MHz. Each of these simplex channels is 30 kHz wide. Thus, AMPS uses frequency division multiplexing access (FDMA) to separate the channels. See Table 6–1 for a list of analog technologies.

In the 800 MHz radio frequency band, radio waves are about 40 cm long and travel in straight lines. They are absorbed by trees and plants and bounce off the ground and buildings. It is possible that a signal sent by a mobile telephone will reach the base station by the direct path, but also slightly later after bouncing off the ground or a building. This may lead to an echo effect or signal distortion. Sometimes, it is even possible to hear a conversation that has bounced several times.

In the United States, the 832 channels in each city are allocated by the FCC. Of these, half are assigned to the local telephone company, the wireline carrier, or the B-side carrier. The other half are assigned to a new entrant in the cellular business, the A-side carrier. The idea is to encourage several competing cellular suppliers to promote competition and lower prices.

The distinction between a telephone company and a cellular phone company, however, is now blurred, because most telephone companies have a cellular partner. It frequently occurs that a company is an A-side carrier in some markets and a B-side carrier in others. Additional mixing occurs be-

Table 6–1
Analog Cellular
Technologies

AMPS	Advanced Mobile Phone System. Developed by Bell Labs in the 1970s and first used commercially in the United States in 1983. It operates in the 800 MHz band and is currently the world's largest cellular standard.
C-450	Installed in South Africa during the 1980s. Uses 450 MHz band. Much like C-Netz. Now known as Motorphone and run by Vodacom SA.
C-Netz	Older cellular technology found mainly in Germany and Austria. Uses 450 MHz.
Comvik	Launched in Sweden in August 1981 by the Comvik Network.
N-AMPS	Narrowband Advanced Mobile Phone System. Developed by Motorola as an interim technology between analog and digital. It has some 3 times greater capacity than AMPS and operates in the 800 MHz range.
NMT450	Nordic Mobile Telephones/450. Developed specially by Ericsson and Nokia to service the rugged terrain that characterizes the Nordic countries. Range 25 km. Operates at 450 MHz. Uses FDD FDMA.
NMT900	Nordic Mobile Telephones/900. The 900 MHz upgrade to NMT450 developed by the Nordic countries to accommodate higher capacities and hand-held portables. Range 25 km. Uses FDD FDMA technology.
NMT-F	French version of NMT900.
NTT	Nippon Telegraph and Telephone. The old Japanese analog standard. A high-capacity version is called HICAP.
RC2000	Radiocom2000. French system launched November 1985.
TACS	Total Access Communications System. Developed by Motorola and is similar to AMPS. It was first used in the United Kingdom in 1985, although in Japan it is called JTAC. It operates in the 900 MHz frequency range.

cause a provider carrier may sell or trade any or all of its 416 full-duplex channel licenses. The 832 channels are divided into four categories:

1. Control (base to mobile) to manage the system
2. Paging (base to mobile) to alert mobile users to calls
3. Access (bidirectional) for call setup and channel assignment
4. Data (bidirectional) for voice, fax, or data

Twenty-one channels are reserved for control, and these are coded into a PROM in each telephone. Because the same frequencies cannot be reused in nearby cells, the actual number of voice channels available per cell is much smaller than 832, typically about 50.

Call Management

Each mobile telephone in AMPS has a 32-bit serial number and a 10-digit telephone number in its **Programmable Read-Only Memory (PROM)** which is a chip in a device used to instruct the device on how to operate. The telephone number is represented as a three-digit area code, in 10 bits, and a seven-digit subscriber number, in 24 bits. When a phone is switched on, it scans a preprogrammed list of twenty-one control channels to find the most powerful

Programmable Read-Only Memory (PROM)

signal. Mobile phones are preset to scan for A-side only, B-side only, A-side preferred, or B-side preferred, depending on which service(s) the customer has chosen. From the control channel, it learns the numbers of the paging and access channels.

The phone then broadcasts its 32-bit serial number and 34-bit telephone number. Like the control information in AMPS, this packet is sent in digital form, multiple times, and with an error-correcting code, even though the voice channels themselves are analog.

When the base station hears the announcement, it tells the MSC, which records the existence of its new customer and also informs the customer's home MSC of its current location. During normal operation, the mobile telephone reregisters about once every 15 minutes.

To make a call, a mobile user switches on the phone, enters the number to be called on the keypad, and presses the Send button. The phone then sends the number to be called and its own identity on the access channel. If a collision occurs there, it tries again later. When the base station receives the request, it informs the MSC. If the caller is a customer of the MSC's company (or one of its partners), the MSC looks for an idle channel for the call. If one is found, the channel number is sent back on the control channel. The mobile phone then automatically switches to the selected voice channel and waits until the called party picks up the phone.

Incoming calls work differently. First, all idle phones continuously listen to the paging channel to detect messages directed at them. When a call is placed to a mobile phone (either from a fixed phone or another mobile phone), a packet is sent to the caller's home MSC to locate the intended receiver. A packet is then sent to the base station in the receiver's current cell, which then sends a broadcast on the paging channel of the form, "Unit 14, are you there?" The called phone then responds with "Yes" on the control channel. The base then sends a signal on the paging channel, for example, "Unit 14, call for you on channel 3." At this point, the called phone switches to channel 3, the access channel, and begins ringing.

Security Issues

Analog cellular phones are totally insecure. Anyone with an all-band radio receiver (scanner) can tune in and eavesdrop on a cell. Most cellular users do not realize how insecure the system is. They often give out credit card numbers and other once-confidential information this way.

Theft of air time is also a major problem. With an all-band receiver attached to a computer, a thief can monitor the control channel and record the 32-bit serial numbers and 34-bit telephone numbers of every mobile telephone he or she hears. By just driving around for a couple of hours, a thief can build a large database. Some thieves offer a low-cost telephone service by making calls for their customers using stolen numbers. Others reprogram mobile telephones with stolen numbers and sell them as phones that can make free calls. Some of these problems could be solved by encryption, but the police could not easily perform "wiretaps" on wireless criminals.

Another issue in the general area of security is vandalism and damage to antennas and base stations. These problems are quite severe and add up to hundreds of millions of dollars a year in losses for the cellular industry.

DIGITAL CELLULAR TELEPHONES

As mentioned, first generation cellular systems used analog signal access and data transmission. The second generation uses digital signals. AMPS was the first digital system in the United States. With the advent of digital telephony, more competitors have emerged, and a struggle for survival is on. Two systems appear destined for survival. The first system is backward compatible with the AMPS frequency allocation scheme and is specified in standards known as IS-54 and IS-135. The other system is based on direct sequence spread spectrum and is specified in standard IS-95. See Table 6–2 for a list of digital systems worldwide.

IS-54 is dual mode (analog and digital) and uses the same 30 kHz channels as AMPS. It packs 48.6 Kbps in each channel and shares it among three simultaneous users. Each user gets 13 Kbps; the remaining 35.6 Kbps are devoted to control and timing signals. Cells, base stations, and MSCs work the same as in AMPS. Only the digital signaling and the digital voice encoding are different.

The IS-95 system is quite novel, because in Europe the opposite happened. Five different analog systems were in use in different countries. Someone with a British phone, for example, could not use it in France, and so on. This experience led the European Public Telephony & Telecommunications (PTT) firms to agree on a common digital system, called the **Global Systems for Mobile (GSM) communications,** which was deployed before any of the competing American systems. The Japanese system is different from all of the above, as will be discussed later.

Global Systems for Mobile (GSM) communications

Because the European systems were all different, the simplest method was to make them pure digital, operating in a new frequency band (1.8 GHz) and retrofitting the 900 MHz band where possible. GSM uses both FDMA and TDMA and a unique frequency hopping process as seen in Figure 6–9. The available spectrum is broken up into fifty 200 kHz bands. Within each band TDMA is used to multiplex multiple users. Figure 6–10 provides additional information on GSM and its cellular structure.

Some GSM telephones use smart cards, that is, credit card–sized devices containing a CPU. The serial number and telephone number are contained there, not in the telephone, making for better physical security

Table 6–2
Digital Cellular
Technologies

A1-Net	Austrian name for GSM900 networks.
B-CDMA	Broadband CDMA. Now known as W-CDMA. To be used in UMTS.
CDMA	Code division multiple access. There are now a number of variations of CDMA, in addition to the original Qualcomm-invented N-CDMA (originally just CDMA, also known as IS-95; see N-CDMA). Latest variations are B-CDMA, W-CDMA, and composite CDMA/TDMA. Developed originally by Qualcomm, CDMA is characterized by high capacity and small cell radius, employing spread spectrum technology and a special coding scheme. It was adopted by the Telecommunications Industry Association (TIA) in 1993. The first CDMA-based networks are now operational. B-CDMA is the basis for 3G UMTS.
CDMAOne	First generation narrowband CDMA (IS-95). See above.
CDMA2000	The new second generation CDMA MoU spec for inclusion in IMT-2000. It consists of various iterations, including 1xEV, 1XEV-DO, and MC 3X.
CDMA2000 1xEV	1xEV (Evolution) is an enhancement of the cdma2000 standard of the TIA. The CDMA 1xEV specification was developed by the Third Generation Partnership Project 2 (3GPP2), a partnership consisting of five telecommunications standards bodies: CWTS in China, ARIB and TTC in Japan, TTA in Korea, and TIA in North America. The 1xEV specification is known as TIA/EIA/IS-856 "CDMA2000 High Rate Packet Data Air Interface Specification." It promises around 300 Kbps speeds on a 1.25 Mhz channel.
cdma2000 1xEV-DO	1xEV-DO or data-only is an enhancement of the cdma2000 1X standard. It promises around 300 Kbps speeds on a 1.25 MHz channel.
Composite CDMA/TDMA	Wireless technology that uses both CDMA and TDMA. For large-cell licensed band and small-cell unlicensed band applications. Uses CDMA between cells and TDMA within cells. Based on omnipoint technology.
CT-2	A second generation digital cordless telephone standard. CT2 has 40 carriers x 1 duplex bearer per carrier = 40 voice channels.
CT-3	A third generation digital cordless telephone, which is similar to and a precursor to DECT.
CTS	GSM *cordless telephone system*. In the home environment, GSM-CTS phones communicate with a CTS home base station (HBS), which offers perfect indoor radio coverage. The CTS-HBS hooks up to the fixed network and offers the best of the fixed and mobile worlds: low cost and high quality from the Public Switched Telephone Network (PSTN) and services and mobility from the GSM.
D-AMPS (IS-54)	Digital AMPS, a variation of AMPs. Uses 3-timeslot variation of TDMA, also known as IS-54. An upgrade to the analog AMPS. Designed to address the problem of using existing channels more efficiently, DAMPS (IS-54) employs the same 30 kHz channel spacing and frequency bands (824–849 and 869–894 MHz) as AMPS. By using TDMA instead of FDMA, IS-54 increases the number of users from one to three per channel (up to ten with enhanced TDMA). An AMPS/D-AMPS infrastructure can support use of either analog AMPS phones or digital D-AMPS phones, because the FCC mandated only that digital cellular in the United States must act in a dual-mode capacity with analog. Both operate in the 800 MHz band.
DCS 1800	Digital Cordless Standard. Now known as GSM 1800. GSM operated in the 1,800 MHz range. It is a different frequency version of GSM, and (900 MHz) GSM phones cannot be used on DCS 1800 networks unless they are dual band.

Table 6–2
Continued

DECT	Digital European Cordless Telephone. Uses 12-timeslot TDMA. This began as Ericsson's CT-3, but developed into ETSI's Digital European Cordless Standard. It is intended to be a more flexible standard than the CT2 standard, in that it has more RF channels (10 RF carriers $\times$ 12 duplex bearers per carrier = 120 duplex voice channels). It also has a better multimedia performance since 32 Kbps bearers can be concatenated. Ericsson has developed a dual GSM/DECT handset.
EDGE	UWC-136, the next generation of data heading toward third generation and personal multimedia environments builds on GPRS and is known as Enhanced Data rate for GSM Evolution (EDGE). It will allow GSM operators to use existing GSM radio bands to offer wireless multimedia IP-based services and applications at theoretical maximum speeds of 384 Kbps with a bite rate of 48 Kbps per timeslot and up to 69.2 Kbps per timeslot in good radio conditions.
E-Netz	The German name for GSM 1800 networks.
FDMA	Frequency division multiple access.
Flash-OFDM	Flash-orthogonal frequency division multiplexing is a new signal processing scheme from Lucent/Flarion that will support high data rates at very low packet and delay losses, also known as latencies, over a distributed all-IP wireless network. The low latency will enable real-time mobile interactive and multimedia applications. It promises to deliver higher quality wireless service and better cost effectiveness than current wireless data technologies.
GERAN	A term used to describe a GSM and EDGE/based 200 kHz radio access network. The GERAN is based on GSM/EDGE Release 99, and covers all new features for GSM Release 2000 and subsequent releases, with full backward compatibility to previous releases.
GMSS	Geostationary Mobile Satellite Standard, a satellite air interface standard developed from GSM and formed by Ericsson, Lockheed Martin, U.K. Matra Marconi Space, and satellite operators Asia Cellular Satellite and Euro-African Satellite Telecommunications.
GSM	Global System for Mobile Communications. The first European digital standard, developed to establish cellular compatibility throughout Europe. Its success has spread to all parts of the world and over eighty GSM networks are now operational. It operates at 900 MHz.
IDEN	Integrated Digital Enhanced Network. Launched by Motorola in 1994, this is private mobile radio system from Motorola's Land Mobile Products Sector (LMPS) iDEN technology, currently available in the 800 MHz, 900 MHz, and 1.5 GHz bands. It utilizes a variety of advanced technologies, including state-of-the-art vocoders, M16QAM modulation and TDMA. It allows commercial mobile radio service (CMRS) operators to maximize the dispatch capacity and provides the flexibility to add optional services such as full-duplex telephone interconnect, alphanumeric paging, and data/fax communications services.
iMode	Launched in February 1999, this fast-growing system from NTT DoCoMo uses compact HTML to provide WAP-like content to iMode phones.
IMT DS	Wideband CDMA, or W-CDMA.
IMT MC	Widely known as cdma2000 and consisting of the 1X and 3X components.
IMT SC	Called UWC-136 and widely known as EDGE.
IMT TC	Called UTRA TDD or TD-SCDMA.
IMTFT	Well known as DECT.
Inmarsat	International Maritime Satellite System which uses a number of GEO satellites. Available as Inmarsat A, B, C, and M.

Table 6–2
Continued

Iridium	Mobile Satellite phone/pager network launched November 1998. Uses TDMA for intersatellite links. Uses 2 GHz band.
IS-54	TDMA-based technology used by the D-AMPS system at 800 MHz.
IS-95	CDMA-based technology used at 800 MHz.
IS-136	TDMA-based technology.
JS-008	CDMA-based standard for 1,900 MHz.
N-CDMA	Narrowband code division multiple access, or the original CDMA. Also known in the United States as IS-95. Developed by Qualcomm and characterized by high capacity and small-cell radius. Has a 1.25 MHz spread spectrum air interface. It uses the same frequency bands as AMPS and supports AMPS operation, employing spread spectrum technology and a special coding scheme. It was adopted by the TIA in 1993. The first CDMA-based networks are now operational.
PACS-TDMA	An 8-timeslot TDMA-based standard, primarily for pedestrian use. Derived from Bellcore's wireless access spec for licensed band applications. Motorola supported.
PCS	Personal Communications Service. The PCS frequency band is 1,850 to 1,990 MHz, which encompasses a wide range of new digital cellular standards such as N-CDMA and GSM 1900. Single-band GSM 900 phones cannot be used on PCS networks. PCS networks operate throughout North America.
PDC	Personal Digital Cellular is a TDMA-based Japanese standard operating in the 800 and 1,500 MHz bands.
PHS	Personal Handy System. A TDD TDMA Japanese-centric system that offers high-speed data services and superb voice clarity. Really a WLL system with only 300 m to 3 km coverage.
SDMA	Space division multiple access, thought of as a component of third generation digital cellular/UMTS.
TDMA	Time division multiple access. The first U.S. digital standard to be developed. It was adopted by the TIA in 1992. The first TDMA commercial system began in 1993. A number of variations exist.
Telecentre-H	A proprietary WLL system by Krone. Range 30 km, in the 350–500 MHz and 800–1,000 MHz ranges. Uses FDD FDM/FDMA and TDM/TDMA technologies.
TETRA	TErrestrial Trunked Radio is a new open digital trunked radio standard defined by the European Telecommunications Standardization Institute (ETSI) to meet the needs of the most demanding professional mobile radio users.
TETRA-POL	Proprietary TETRA network from Matra and AEG. Does not conform to TETRA MoU specifications.
UltraPhone 110	A proprietary WLL system by IDC. Range 30 km, in the 350–500 MHz range. Uses FDD FDM/TDMA technologies. The UltraPhone system allows four conversations to operate simultaneously on every 25 kMHz-spaced channel. A typical UP twenty-four-channel WLL system can support ninety-five full-duplex voice circuits in 1.2 kHz of spectrum.
UMTS	Universal Mobile Telephone Standard, the next generation of global cellular which should be in place by 2004. Proposed data rates of < 2 Mbps, using combination TDMA and W-CDMA. Operates at around 2 GHz.
W-CDMA	One of the latest components of UMTS, along with TDMA and cdma2000. It has a 5 MHz air interface and is the basis of higher bandwidth data rates.
WLL	Wireless local loop limited-number systems are usually found in remote areas where fixed-line usage is impossible. Most modern WLL systems use CDMA technology.

Source: Perlman, Leon J. Retrieved April 14, 2002, www.cellular.co.za/celltech.htm.

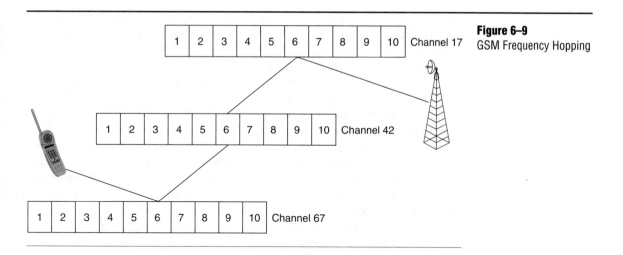

Figure 6–9
GSM Frequency Hopping

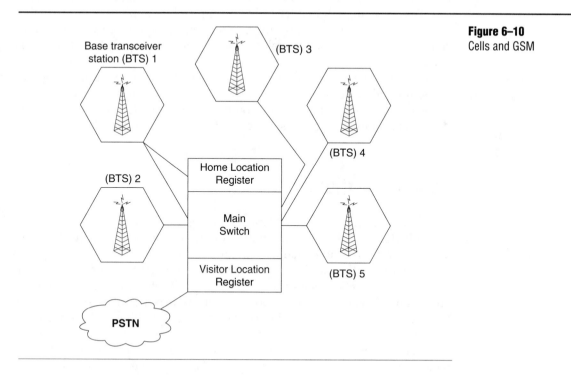

Figure 6–10
Cells and GSM

(stealing the phone without the card will not get you the number). Encryption is also used. This is a successful approach since GSM subscribers total 488.5 million.[4]

Digital transmission clearly involves great complexity. The conversion of the analog waveform into thousands of digital samples every second, the transmission and regeneration of these samples within a system that must be tightly synchronized end to end, and the faithful recreation of the voice signal from a stream of abstract 1s and 0s all involve elaborate processing at both ends and extremely challenging hardware requirements. Although the layperson thinks of the computer as the marvel of digital electronics, communication applications are much more demanding than most data processing applications, because of the complexity of communication signals and the need to handle huge amounts of data in real time. A DS-4 digital transmission processes more than 300 megabytes in 10 seconds. Historically, the availability of fast electronic components was the limiting factor in implementing digital voice systems. The first PCM systems were experimentally developed during World War II—in fact, the principles of PCM had been elaborated even before the war—but the inadequacy and high cost of digital processing circuits held back the implementation of commercial digital systems until the mid-1960s.

The cost of digital circuitry still determines which applications are suitable for digital techniques. For example, digital microwave radio was unable to compete economically with analog systems until the late 1970s in some markets. Realizing that fact, the potential of digital techniques in applications such as mobile radio will call for the development of much more sophisticated designs, more advanced materials, more complex software, faster processors, and more ingenious mathematics—by comparison with which even the T-carrier may look primitive. T-carrier services are briefly discussed in Chapter 1.

THE DIGITAL NETWORK

An analog transmitter (e.g., a telephone microphone or a radio transmitter) is basically designed to launch a signal through a transmission medium such as a copper wire or a radio channel. An analog receiver is designed to capture this signal from the medium. Particularly in radio, transmitter and receiver do not really function as a system after the initial call setup. They do not exercise control over one another. Both operate more or less independently; the transmitter blares away, and the receiver receives anything within range.

The enemy in this scenario is noise—the engineer's term for the sum of the effects of the environment upon the signal during transmission. The signal begins to deteriorate, irrevocably, as soon as it is generated. The analog system has no control over this, but it can try to ensure that the level of the

---

[4] See www.gsmworld.com.

signal relative to the noise stays above a certain threshold. As long as this signal-to-noise ratio is maintained above this threshold, the human brain can filter out the noise, and the communication will be satisfactory.

By contrast, the idea of a digital network involves the creation of an end-to-end system, which encompasses transmitter, receiver, and channel facility in a single integration. The elements of this system communicate among themselves and exercise control over one another. The system is capable of monitoring and controlling key signal characteristics during the transmission process. Transmitter and receiver, and indeed all elements of the network, are normally synchronized, or at least tightly coordinated. Both ends of the system possess sufficient intelligence to detect, diagnose, decide, and act upon various kinds of problems that may occur during transmission. In short, the system is not passive, but adaptive.

It is important to realize that the digital network is not a physical entity, but a logical entity superimposed upon a highly heterogeneous set of physical transmission facilities. In the digital network, one telephone call may encompass a copper wire local loop to the local central office, a fiber optic interoffice trunk to the toll switch, a microwave or satellite link from the toll switch to the destination toll switch, and a mobile radio-telephone link to the called party traveling in a car. Yet the digital network possesses the ability to link these disparate facilities into a single digital circuit, the performance parameters of which can be precisely monitored and controlled. It is embodied in an interrelated system of processors which originate, transmit, detect, and decode the messages at different stages in the transmission path. Such a network offers a number of broad advantages over the analog system it is gradually replacing.

Digital Networking and Quality

A digital transmission system allows for communication to be established and maintained at uniform high-quality levels in environments where analog techniques are costly and ineffective. The more difficult the communication link, the more digital communication will stand out over analog. One of the chief areas of digital application today is upgrading the quality of the long-distance segment of the network by the use of digital technology, particularly optical fiber. Long-distance circuits have always been subject to an assumption of reduced quality. With optical fiber systems, this is changing. Another particularly difficult communication environment is the field of mobile radio, where until recently digital techniques have not been applied. The robustness of digital communication is evident along several dimensions: resistance to noise, resistance to crosstalk (signals erroneously crossing over into adjoining wires), error correction, and signal regeneration.

Resistance to Noise

In part, this advantage is due to the inherent resistance of the digital signal to degradation due to channel noise, which is in turn due to the threshold effect inherent in a binary coding system. For example, a noise source that is 5 percent as strong as the desired signal may alter the amplitude of an analog signal by up to 5 percent, which is enough to create a serious distortion. The same noise will probably have no impact on a PCM-type digital transmission where each bit is detected on an all-or-nothing basis.

The gain in digital signal robustness is dramatic. In analog systems, the design standard for signal-to-noise ratio on short-haul wire telephone systems is typically 46 dB—the signal must be 40,000 times stronger than the noise. Designing an analog transmission system to this standard is often extremely costly, especially as the length of the circuit grows. A digital system will produce virtually error-free transmission, that is, with no channel noise and a signal-to-noise ratio of as little as 15 dB—a signal only 30 to 40 times stronger than the noise. In other words, a digital transmission system can withstand noise levels 1,000 times higher and still deliver a good signal.

Resistance to Crosstalk

A significant problem with analog wire systems, particularly analog carrier (frequency multiplexed) cable systems, is crosstalk. Crosstalk in cable pairs occurs when the electrical energy from one circuit is magnetically superimposed upon an adjacent circuit. This has historically been an intractable problem for wireline network designers. Crosstalk can enter the circuit at almost any place in the analog network, and even after more than a century it has not been eradicated. Many analog systems are not noise limited so much as crosstalk limited. In a digital system, crosstalk is largely eliminated by the threshold effect, and even where it does obtrude with sufficient strength to cause detection errors, it appears not as an intelligible signal, which is highly objectionable, but as random, unintelligible background noise.

Error Correction

In a sense, a digital system does not care about such things as the source of bit errors, the digital effect of noise, crosstalk, or interference. It treats all channel errors arising from whatever source in the same manner. In early digital systems, little attempt was made to deal with errors other than through the inherent robustness of the coding and signaling format. In more modern digital systems, powerful error-correction techniques have been developed and can be applied to scrub a signal and remove a large percentage of errors from any source. The application of such advanced techniques becomes a matter of weighing costs (the additional digital circuitry plus, in most cases, a bandwidth penalty) against the benefits (the ability to operate in even noisier channels or in the presence of even more severe crosstalk or interference).

Signal Regeneration

As mentioned, digital regeneration techniques allow the system designer to control the performance of the circuit to almost any desired specification; the system is no longer at the mercy of the hazards of the transmission environment in the conventional sense.

Many transmission problems involve time-variant or statistical sources of signal degradation against which an analog system is often largely helpless. Such problems can be controlled in a properly designed digital system. A digital system can be designed to monitor and adapt to changing channel conditions.

Echo Control

Another type of problem that arises because of the heterogeneity of the physical plant is the phenomenon of echo, which may be conceived of as a reflection of the transmitted signal back to the transmitter from some discontinuity in the transmission path, such as the improperly balanced interface between one segment of the physical facilities and another. Such echoes become disturbing to the talker for circuits longer than about 1,800 miles.

Channel Condition Monitoring and Decision Response Capabilities

In addition to automatic and continuous processes such as equalization and echo cancellation, a digital system is highly amenable to the implementation of decision rules, whereby countermeasures can be implemented by the system if certain thresholds are reached in the deterioration of channel conditions. A digital system can easily monitor the quality of the channel measured as the bit error rate. Other parameters such as signal strength and delay time, which in a radio system may be translated into propagation distance, can also be used as the basis for certain countermeasure decisions. In early digital systems, about the only type of decision response that was allowed for was the removal of a channel from service in the event that the bit error rate rose too high. A digital radio system offers a wide variety of opportunities for countermeasures to be implemented, and such measures can greatly improve the theoretical performance of such systems compared with their analog predecessors.

It is also possible in a digital system for the transmitter to provide a great deal of information about its own functioning and condition on a more or less continuous basis. This becomes useful when designing a large network, especially for diagnostic and maintenance purposes. Analog systems are generally less flexible in providing this type of information without disrupting the voice communication itself.

In the evolution of digital systems, the trend is strongly toward generic mainframe systems which can be upgraded over a long lifetime through installation of new software releases to maintain the system close to the leading edge of current applications. Indeed, the trend is toward a network that can be reconfigured more or less on demand, even by the users themselves.

One type of demand-configured architecture is known as packet switching. Another is based on the concept of a virtual circuit.

The digital network transmits digits. It is the end user who determines what to do with the data received. Digital voice, digital images, digital music, and digital data appear the same within the digital transmission in switching facilities. In principle, it should be possible to deliver any type of digital service over the same facility. This is precisely what the proponents of integrated services on digital networks have in mind: a single channel for digital television, telephone, and data services, delivered through a single integrated network.

The single-channel concept is perhaps the haziest area of thinking about the digital network. It is relatively easy to posit the capability; it is hard to say exactly what types of service could or should be offered in an integrated format. After all, television is available from broadcasters and coaxial cable systems which are not integrated with the telephone or data networks, and yet it is difficult to see what the advantages of integration for the end user might be in this case. However, the advantages of sharing facilities between nodes in the network are clear. The use of a single facility to transmit voice, video, and data from one collection point to another offers definite savings through the reduction in duplicative facilities.

Digital multiplexing is based on the idea of balancing the additional costs of digital circuitry against the more efficient use of the transmission channel and its associated equipment and facilities. In the case of wireline digital systems, it offers a reduction in copper costs, as well as in the logistical and overhead costs associated with the construction and maintenance of the copper plant. In the case of radio, digital multiplexing points the way to reducing the amount of transmitter-receiver radio equipment, which remains the most expensive element in any radio-based communication system. Of course, digital multiplexing also allows for more efficient use of the radio spectrum itself, which is an important goal from the standpoint of the regulatory community and the radio communications industry as a whole.

Security

The first real work on PCM systems was stimulated during World War II by the need for a truly secure radio-telephone system. Digital communication systems were utilized in military and sensitive government applications long before they became economical in commercial settings, precisely because of the benefit of communications security. Outside of the government and defense industries, privacy and security were not high concerns among telephone users, partly because we assume that the wireline telephone *is* secure. But the growth of network data services has brought security more to the forefront in facilities planning. Banks, automatic teller machines, credit agencies, and point-of-sale terminals for customer transactions may involve the transmission of sensitive information via the public switched network.

As noted, the problem of communication privacy is much more serious in mobile radio systems, where the ease of interception greatly compromises even routine cellular radio-telephone calls. Analog scrambling systems exist, but they are not accepted as having a high level of security and they typically degrade voice quality.

By contrast, a digital radio system lends itself readily to intensive encryption, which can be designed to provide almost any level of security desired and has a relatively small impact on system performance or economics. Even without encryption, the digitization process itself provides a fair degree of privacy by making inexpensive interception devices difficult for amateurs to construct.

PERSONAL COMMUNICATIONS SYSTEM

The personal communications system (PCS) uses the cellular concept, but while cellular communications utilizes frequencies in the 800 MHz range and has both analog and digital capabilities, PCS is pure digital and operates in the 1,900 MHz range. PCS only came about after it became clear that the explosive growth of cellular telephony would require the FCC to allocate another frequency range for wireless communications to avoid overloading the 800 MHz range. PCS has developed as a digital tool that not only carries voice communications but also provides paging service, fax service, and mobile connection to the vast Internet.

PCS is the collective name for the digital wireless communications technologies available in the 1,900 MHz range in North America today. Four such technologies are presently in competition for the booming digital communications market. They are GSM, IS-136 TDMA, CDMA, and personal access communications systems (PACS). In addition, the original cellular systems, which all began as analog, began converting to digital even before PCS became available. These operate in the 800 MHz range so by definition they are not PCS. However, they compete with PCS for the wireless business.

The beginnings of digital cellular telephony began in Europe, where a group formed to develop the standard for the new digital system. This group was called the Groupe Speciale Mobile and the standard they created used the same acronym, GSM. This was later changed to stand for Global System for Mobile Communications. The system that was built using this standard operated in the 900 MHz frequency band and utilized a digital technique called time division multiple access (TDMA), which tripled the capacity of the system by breaking up conversations into small bursts separated by small time intervals and then filling in those intervals with bursts from two other conversations. In this manner, three conversations could be handled simultaneously over the same frequency channel. A computer did the work of separating the bursts so quickly that the digital customers could never tell the difference. GSM also utilizes a built-in encryption and an incorporated key that eliminates the potential for piracy and fraud. The subscribers use a card (similar to a credit card) that has an identity number on it so any GSM phone can be used and charged to the subscriber's account.

In 1990, the European Telecommunication Standards Institute, which took over responsibility for GSM, granted permission for the use of the 1,800 MHz band in Great Britain. They called this the Digital Cellular System (DCS). The system remained GSM, but it used lower power transmitters than the 900 MHz system. This was also offered in the United States in 1995 as GSM-NA (GSM North America) and operated in the 1,900 MHz range set aside for PCS by the FCC.

In 1990, the FCC requested applications for digital service for U.S. aircraft-to-ground telephones, called Terrestrial Aeronautical Public Correspondence (TAPC). Licenses were granted to AT&T, GTE, and InFlight Phone for TAPC, but because the FCC did not require standardization, the three systems were incompatible with each other.

The year 1990 was also a milestone in American cellular history when IS-54B or digital AMPS was chosen to become the first standard for dual-mode cellular. It utilized TDMA technology for digital transmissions but kept the analog process on the setup. It also could use analog channels if digital channels were not available for transmission. This way it was compatible with the analog phones being sold in the booming American cellular market.

As more Americans began using cellular phones, it became apparent that the capacity of the cellular frequency channels was insufficient, even with the utilization of digital TDMA. To address the problem, in 1994, Qualcomm, Inc. introduced a system based on the digital technique of code division multiple access (CDMA) which could increase the number of conversations on a single frequency channel to ten or more. The system works by assigning a unique code to each call and then filtering out the other calls on that frequency without the particular code for that call. Although CDMA had the ability to increase the capacity of the 800 MHz cellular system, widespread utilization did not come quickly, probably because of the additional expense associated with the sophisticated equipment necessary for coding and filtering.

In the mid-1990s, densely populated areas of the United States were already beginning to reach the limits of capacity for 800 MHz cellular service. The FCC decided to open up another frequency range for wireless communications. From December 1994 to January 1997, the FCC auctioned off frequencies in the 1,900 MHz range, designated PCS, with rules that required at least two carriers in each metropolitan area. The second option became the personal communications service.

When the PCS frequencies became available, there was no single standard set, so now four technology standards are used by the carriers in the PCS market. Because the FCC's goal was to promote competition, at least two carriers were required in any market and six blocks of channels were reserved for up to six different carriers in each market. They organized the markets using the Rand McNally lists of major trading areas (MTAs) and basic trading areas (BTAs). There were 51 MTAs, each covering large portions of one or more states, and they were assigned two blocks of frequencies called the A block and the B block. There were 493 BTAs and they were centered around metropolitan areas. The BTAs were assigned frequency blocks C, D, E, and F. Because all the BTAs were contained in MTAs, six blocks were available in every market. The auction that was held by the FCC for access to the PCS frequencies generated over $7 billion just for blocks A and B. Eighteen different companies were awarded the licenses for these frequencies.

The licenses awarded granted the A block carriers the frequencies between 1,850 and 1,865 MHz. B block carriers were given 1,870 to 1,885 MHz, and C block carriers received 1,895 to 1,910 MHz. The D, E, and F blocks were only given 5 MHz of bandwidth instead of the 15 MHz awarded to blocks A, B, and C. Their frequencies were 1,865 to 1,870 MHz for D; 1,885 to 1,890 MHz for E; and 1,890 to 1,895 MHz for F. The frequencies and their locations in the overall spectrum are shown in Figure 6–11.

The carriers that won the awards for the PCS frequencies utilized all four technology standards. Of the 102 licenses of the A and B block carriers, 22 chose to use GSM-NA, 24 used TDMA, 50 used CDMA, and 1 used PAC. Five carriers had not decided which technology to use at the time of the awards.

PCS Technology Standards and Carriers

GSM, the European technology standard, had always been purely digital and had been used at 1,800 MHz in Great Britain, so it was a good fit for the new and close 1,900 MHz U.S. PCS frequency. Designated GSM-NA in North America, it allowed eight mobile phones to share one 200 kHz bandwidth channel through the use of the TDMA technique. To get full-duplex operation, the system actually operates the eight phones on two channels. GSM-NA also has several types of control channels that can carry data service information such as message paging, fax, Internet browsing, and e-mail. It is the only technology that can roam between the United States and the European countries. GSM-NA supports both digital data services and fax service. Carriers that use GSM-NA include Omnipoint, Pacific Bell, BellSouth, Powertel, Western Wireless, and Aerial. These companies or others that utilize GSM-NA can be found in nearly every market in the United States.

Another technology standard being used by the PCS carriers is time division multiple access (TDMA), designated as IS-136. TDMA had been developed for digital cellular and was originally designated IS-54. This standard was used as the basis for IS-136 in 1995. TDMA IS-136 is used in both digital cellular and PCS and has the advantage of being easily compatible with the analog phones still in use in the cellular range. Some wireless phones are built to be dual frequency so that they can be utilized in the 800 MHz range in markets where PCS is not available.

At present, TDMA technology does not support digital data services and fax. The three major carriers that use TDMA IS-136 in the PCS market are

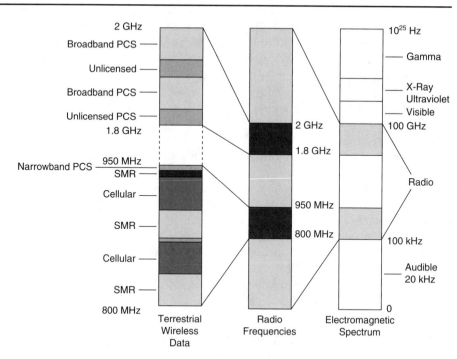

Figure 6–11
Wireless Data Frequency
Ranges in the United
States

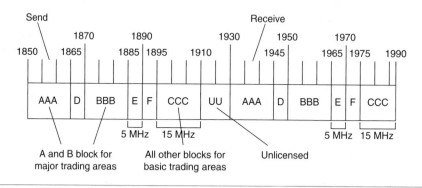

Broadband PCS Spectrum Allocation

AT&T Wireless, Southwestern Bell, and Century Tel. Although the number of companies is small, the technology is available almost everywhere because of the immense market share of AT&T Wireless.

A third major technology standard in use in PCS is code division multiple access (CDMA), designated IS-95. Like TDMA, it was originally developed for digital cellular to increase the capacity of the 800 MHz system. Although in theory CDMA can increase the capacity up to 13 times that of TDMA, it required about 4 times the investment per base station. It does, however, allow the reuse of frequencies in adjacent cells, which makes frequency planning much easier. Another unique feature of CDMA is its ability to perform a soft handoff. This means that by searching constantly for stronger signals, called raking, it can transfer the call to a different channel without breaking the call first.

Like GSM-NA, the CDMA technology supports digital data and fax and is used by a large number of carriers in the PCS market. Some of the larger carriers are Sprint PCS, PrimeCo, GTE Wireless, USWest, Nextwave, Alltel, and AirTouch.

The fourth technology standard in the PC market is personal access communications systems (PACS). This was planned as a low cost, local system using radio ports instead of base stations. It utilizes wireless local loop (WLL), which is the same technology used in cordless phones, in combination with wireless mobile service. Tall towers are not required so costs are kept low. The disadvantage is that roaming is not possible unless roaming agreements are reached between the local markets. PACS presently does not support digital data services or fax and is only used by one carrier, 21st Century Telesis. It is available in twenty-eight markets.

Consider the following practical example. In Georgia, there are fifteen BTAs, with all but one (Brunswick, which is in Jacksonville's MTA) in the Atlanta MTA. The PCS carriers who have been awarded the licenses and the technologies they use (if known) are listed here.

MTA	Block A	Block B		
Atlanta	AT&T Wireless (T)	Powertel (G)		
Jacksonville	Powertel (G)	PrimeCo (C)		

BTA	Block C	Block D	Block E	Block F
Albany	Enterprise Com.	Sprint PCS (C)	BellSouth (G)	Omnipoint (G)
Athens	General Wireless (C)	Sprint PCS (C)	Alltel (C)	Wireless Telecom (C)
Atlanta	General Wireless (C)	Sprint PCS (C)	Alltel (C)	Nextwave (C)
Augusta	Powertel (G)	Savannah Ind. PCS (C)	Sprint PCS (C)	Omnipoint (G)
Brunswick	KMTel	BellSouth (G)	Sprint PCS (C)	Mercury PCS II
Columbus	R&S PCS	BellSouth (G)	Sprint PCS (C)	Public Serv. PCS
Dalton	Southeast Wireless	Sprint PCS (C)	Alltel (C)	Troup EMC Co.
Gainesville	General Wireless (C)	Sprint PCS (C)	Alltel (C)	Wireless Telecom (C)
LaGrange	Enterprise Com.	BellSouth (G)	Sprint PCS (C)	Technicom
Macon	Georgia. Ind. PCS	Sprint PCS (C)	Alltel (C)	Omnipoint (G)
Rome	Southeast Wireless	Sprint PCS (C)	Alltel (C)	Troup EMC Com
Savannah	Southeast Wireless	Sprint PCS (C)	BellSouth (G)	Omnipoint (G)
Tifton	Enterprise Com.	Sprint PCS (C)	BellSouth (G)	Omnipoint (G)
Valdosta	Sowega Wireless	Sprint PCS (C)	BellSouth (G)	Mercury PCS II
Waycross	Savannah Ind PCS (C)	Sprint PCS (C)	BellSouth (G)	Mercury PCS II

G = GSM-NA, T = TDMA, C = CDMA

PCS and 800 MHz Cellular

PCS, unlike the 800 MHz cellular, has been 100 percent digital since its inception. There was no reason to make the system backward compatible with the huge number of analog phones, as the service providers did with digital cellular. PCS emphasizes personal service with smaller phones, greater mobility, and data service features that only digital systems could provide. The following features are not provided with analog cellular service, or at least are better provided with digital service:

- The vast resources of the Internet are now available from a laptop computer anywhere that a PCS network with data service capability can reach. Sprint PCS has announced the capability of Internet access using only the PCS phone; no type of computer is needed.
- A laptop computer and a data-enabled wireless phone are all that are needed to connect to your company LAN. Workers can access files and run computer applications on the road just as if they were sitting at their desks. No landline phone jack is required. Unlike analog cellular systems, digital PCS technology encrypts the wireless link. This provides security for transfer of confidential or proprietary data and prevents eavesdropping on voice communications.
- PCS communications (GSM-NA and CDMA technologies only at present) give the ability to fax from anywhere. Also the encryption feature of PCS makes sure that the message cannot be pirated.
- Other features of PCS phones that do not require a computer include the short messaging service (SMS). The phone's display can directly receive up to 160 characters in a message without even making a connection. The SMS feature allows your phone to work as an alphanumeric pager. The phone can also provide caller ID and voice mail and display news, traffic and weather reports, and stock market quotes.
- In addition to the special features that the PCS can provide, the quality of voice transmission is better with digital service than with analog cellular service.

Although digital cellular can provide most if not all of the features of PCS, there are some inherent advantages to the 1,900 MHz PCS digital wireless system as compared with the 800 MHz digital cellular. Because the range is shorter for the higher frequency communications, the cell sites must be closer together. The PCS system used this to its advantage, because small cells and closely proximated cell sites mean that less power is required for PCS than for cellular, and PCS provides the stronger signal. The higher frequency signal of PCS also gives better reception inside buildings, the handsets are smaller, and battery life is longer than with cellular phones.

The singular disadvantage of PCS compared with cellular is the available service area. Cellular is now in every market in the United States and

	GSM	PDC	CDMA	TDMA	GSM-1900
Sep-92					
Dec-92	201,500		1,500		
Mar-93	394,500		3,000		
Jun-93	589,720		24,000		
Sep-93	838,220		48,500	5,000	
Dec-93	1,362,990		78,150	35,000	
Mar-94	1,889,790		154,200	55,000	
Jun-94	2,592,530		276,600	122,000	
Sep-94	3,393,030		396,000	183,500	
Dec-94	4,628,790		565,600	378,000	
Mar-95	5,665,900		800,000	484,000	
Jun-95	7,951,000	1,500,000	0	1,100,000	0
Sep-95	9,884,000	2,100,000	0	1,300,000	0
Dec-95	13,034,000	3,108,000	9,000	2,055,000	40,000
Mar-96	16,343,000	5,227,000	11,300	2,750,000	60,000
Jun-96	21,148,000	7,672,000	180,000	4,500,000	105,000
Sep-96	26,150,000	10,592,500	350,000	6,000,000	150,000
Dec-96	32,878,500	13,920,000	987,000	2,700,000	301,500
Mar-97	40,200,000	16,000,000	1,100,000	3,200,000	500,000
Jun-97	48,900,000	19,000,000	2,500,000	4,000,000	700,000
Sep-97	58,145,570	23,619,000	4,300,000	5,500,000	955,570
Dec-97	71,359,000	26,772,000	5,980,000	6,900,000	1,331,000
Mar-98	83,557,000	30,074,000	10,900,000	8,000,000	1,691,000
Jun-98	98,858,230	33,006,000	12,900,000	10,928,450	2,123,000
Sep-98	114,901,660	35,728,500	16,704,640	13,976,430	2,545,000
Dec-98	138m	38m	23m	18m	2.8m
Mar-99	161m	41m	26m	21m	3.1m
Jun-99	188m	42m	29m	26m	3.3m
Mar-01	460m	65m	34m	61m	7m
Jun-01	540m	70m	40m	65m	9m
Sep-01	670m	72m	52m	80m	12m
Dec-01	720m	80m	60m	82m	16m
Mar-02	1B	82m	71m	90m	25m

Source: Perlman, Leon J. Retrieved April 14, 2002, www.cellular.co.za/celltech.htm.

Table 6–3
Market Shares of Major Cellular Systems

is still the only wireless service available in many areas. In rural areas, especially, PCS will not be available for still some time. There are, however, some dual frequency phones that will operate in the cellular 800 MHz range as well as the PCS 1,900 MHz range. See Table 6–3 for market shares worldwide.

The Future of PCS

While the 800 MHz cellular systems still have the dominant market share, PCS is gaining steadily. In 1997, there were approximately 50 million cellular subscribers and approximately 3 million PCS subscribers. PCS held a 6 percent market share. By 1999, cellular had grown to 68 million and PCS to 13 million for a 16 percent market share. In the year 2005, there will be 400 million cellular subscribers and 200 million PCS subscribers, giving PCS almost a 50 percent market share. In 1997, less than 2 percent of all calls were wireless, but in 2005 that amount will be 60 percent. This growth can be compared with other communications devices by looking at the time it took to reach 50 million customers. Whereas it took 14 years for wireless phones to reach 50 million customers, it took 75 years for landline telephones to reach the same milestone, 39 years for commercial radio, and 24 years for television. With growth projections like these it is easy to see why the FCC auction for PCS frequencies brought such high dollars.

The biggest technological advance right now in wireless telephony is packet switching. Cellular phones were developed using circuit switching, which means a dedicated radio frequency is required for data transmission at all times. With packet switching, data are sent in small bursts (or packets) and the frequency is free for other packets between these bursts. Packet switching allows the efficient data transfer required for Internet or LAN connections on wireless telephones.

With packet switching now available, the largest growth area for wireless telephony is expected to be in the mobile data market. Wireless mobile connections to the Internet allows electronic banking and e-commerce from anywhere and even en route to anywhere. This function is expected to revolutionize the way business is done. Sales can be transacted anywhere and corporate data and support functions that were only available to workers who had access to a landline connection are becoming available to mobile workers anywhere in the world. Technology will continue to improve the speed of data transfer. Rates of 1 Mbps to 2 Mbps are here now and speeds will increase to such an extent that in the future even video conferencing will be practical with wireless phones.

Teleworking is becoming increasingly a normal work process. Some states and some European countries are actually devising initiatives to promote teleworking because of the problems with automobiles of commuting workers. The new digital wireless communications will make that even easier because the teleworker will not be confined to a workstation and landline Internet connection.

Digital wireless communication such as PCS helps blend two major modern-day facets of life: the freedom of mobility and the timeliness of information. Due to modern transportation methods, we live in a highly mobile and global society. Due to the advances in communication such as television and the Internet, we are a highly and quickly informed society. If, because of the digital computer, the latter portion of the twentieth century could be labeled the information age, then the early part of the twenty-first century, because of digital wireless communications, could well be known as the mobile information age.

THIRD GENERATION

The next page in the cellular industry will be written by **third generation (3G)** technology. It is certainly true that the telecommunications world is changing as the trends of media convergence, industry consolidation, Internet and IP technologies, and mobile communications occur simultaneously ("Data on 3G—An introduction" 2000).

third generation (3G)

Third generation is a mobile phone system that became available commercially in 2002. 3G unifies the disparate standards of second generation wireless networks. Instead of different network types being adopted worldwide, the 3G plan provides a single network standard.

Significant change will occur during this rapid evolution in technology, with 3G mobile Internet technology being a radical departure from the first and even the second generations of mobile technology. Some of the changes include the following:

- People will look at their mobile phone as much as they hold it to their ear. As such, 3G will be less safe than previous generations, because television and other multimedia services tend to attract attention to themselves, so instead of hands-free kits, we will need eyes-free kits!
- Mobile communications will be similar in its capability to fixed communications. Therefore, many people will only have a mobile phone.
- The mobile phone will be used as an integral part of the majority of people's lives. It will not be an added accessory but a core part of how they conduct their daily lives. The mobile phone will become akin to a remote control or magic wand that lets people communicate what they want, when they want.

As with all new technology standards, there is uncertainty and the fear of displacement. The deployment of 3G is contentious for several reasons.

- Because the nature and form of mobile communications is so radically changed, many people do not understand how to make money in the nonvoice world, and do not understand their role in it.
- 3G licenses are being awarded around the world, necessitating existing mobile communications companies in the 2G world to think about and justify their continued existence.
- 3G is based on a different technology platform—DMA—which is unlike the TDMA technology that is widely used in the 2G world.

- The U.S., Japanese, and European mobile players have different cellular technology competencies, but now can be unified in this single standard. Separate wireless evolution paths and European wireless leadership are thereby challenged.
- Japanese network operators were the first to implement 3G networks in 2001, and Japanese terminal manufacturers, who have not had much market share outside their home market, are the first with 3G terminals.
- Many industry analysts and other pundits have questioned the return on an investment in 3G technology, questioning whether network operators will be able to earn an adequate return on the capital deployed in acquiring and rolling out a 3G network.[5]
- Many media and Internet companies have shown a strong interest in using 3G technology as a new channel to distribute their content, opening the opportunity for new entrants, partnerships, and value chains.

The following details may help to clarify 3G.

- 3G can be thought of as 2.5G services such as GPRS plus entertainment (games, video, mobile multimedia) and new terminals. 3G brings with it significantly more bandwidth. Whereas GPRS terminals will have the same range of form factors as today's 2G phones, many 3G terminals will be video centric.
- There is a clear business case for investing in 3G for existing network operators that are facing congested 2G networks. Voice traffic over 3G networks will ensure that the 3G business can pay for itself. The main positive (rather than defensive) reason for mobile network operators to secure 3G network licenses is to solve capacity issues in terms of enabling far greater call capacity than today's digital mobile networks allow.
- Nonvoice (data) traffic will also be huge, with new mobile multimedia applications such as mobile postcards, movies, and music driving new applications and services along with corporate applications. Applications and services available through the Internet, intranets, and extranets will drive the interest in and traffic on 3G networks.
- Providing that network operators adopt an open model to all Internet traffic, the business case for 3G fueled by both greater data and voice traffic is clear and so this bolsters the business case for winning and rolling out a 3G network. If the network operator insists upon a closed model in which data traffic is funneled primarily through its own in-house portal or limits access to its customers for e-commerce and other Internet services, the business case is endangered.

---

[5] The many such opinions can be found at various sites, including www.cnet.com and www.computerworld.com.

- 3G technology is essential. Think about the huge changes that will occur in the next 5 years to today's products.
- It is often assumed that early adopters will be corporate customers for 3G, but it is expected that since consumer electronics devices, as their name suggests, appeal to consumer markets they will have 3G built in.[6] Mobile multimedia games, entertainment and the like, are highly consumer oriented—so much so that mobile 3G is expected to be a consumer revolution, not a corporate one.
- Many people will not have a fixed phone at home. The preventer of this until now has been the slow speed of mobile data in 2G and even so-called 2.5G technology that has made Internet access the principal application for home phones.
- There will be many suppliers of mobile terminals as Japanese companies, mobile hand-held computer manufacturers (Palm, Microsoft), and information appliance and IT suppliers enter the global mobile terminal market. Mobile-enabled devices will proliferate as all portable consumer electronics devices have built-in short-range wireless communications technology. The successful cellular manufacturer vendors (such as Nokia or Ericcson) will be the ones to deliver new products rapidly and reliably.
- Given the fragmented market for wireless phones, alliances and mergers between Korean, Japanese, European, and American mobile phone and consumer electronics manufacturers will continue and accelerate, because few if any companies have all the enabling technologies in-house—from video to camera to mobile to interfaces. Smaller players in these sectors will continue to consolidate, as companies are acquired to gain better distribution for their technologies.
- 3G terminals, now emerging in the marketplace, are significantly more complex than today's GSM phones, because of the need to support such items as video, more storage, multiple modes, new software and interfaces, and better battery life. The biggest single inhibitor of new services such as WAP and high-speed circuit switched data (HSCSD) has proven to be a lack of handsets; however, every stage in the evolution path for GSM, from current technology to 3G, requires a new handset. Therefore, availability of 3G-enabled handsets will be a critical factor in determining exactly when 3G becomes a success.
- Partnerships will increasingly develop between (U.S. based) Internet, IT companies, traditional mobile communications vendors (from Europe and the United States), and (Japanese) consumer electronics manufacturers. Different regions have different strengths and are likely to leverage them through strategic alliances.

---

[6] See www.lifestreams.com.

- From a technical point of view, the introduction of packet data services such as GPRS to circuit switched networks is more challenging than the move from GPRS to 3G. GPRS is the first-time addition of packet capability to a circuit switched network, whereas 3G is the addition of more packet switching technology.
- From an end-user point of view, the move from GPRS to 3G is much more revolutionary than the move from 2G data services to GPRS. GPRS allows the mobile network to catch up with the data bandwidths available over fixed telecommunications networks, whereas 3G provides unprecedented bandwidth for mobile users—so much bandwidth that new applications will be developed to use it.

3G Standardization Process

In 1998, the ITU called for radio transmission technology (RTT) proposals for IMT-2000, originally called Future Public Land Mobile Telecommunications Systems (FPLMTS), the formal name for the 3G standard. Of the different proposals submitted, the DECT and TDMA/Universal Wireless Communications organizations submitted plans for the RTT to be TDMA based, while all other proposals for nonsatellite-based solutions centered on wideband CDMA. The main submissions were called wideband CDMA (W-CDMA) and cdma2000. The GSM players including infrastructure vendors Nokia and Ericsson backed W-CDMA. The North American CDMA community, led by the CDMA Development Group (CDG) including infrastructure vendors Qualcomm and Lucent Technologies, backed cdma2000.

Third Generation Partnership Project

In December 1998, the Third Generation Partnership Project (3GPP) began work, following an agreement between five standards-setting bodies around the world including ETSI, ARIB and TIC of Japan, ANSI of the United States, and TTA of Korea. This unprecedented cooperation into standards setting made 3GPP responsible for preparing, approving, and maintaining the technical specifications and reports for a 3G mobile system based on evolved GSM core networks and the frequency-division duplex (FDD) and time-division duplex (TDD) radio access technology.

3G Data Rates

The ITU established indicative minimum requirements for data speeds which the IMT-2000 standards must support. These requirements are defined according to the degree of mobility involved when the 3G call is being made. As such, the data rate available over 3G will depend on the environment of the call.

High mobility data rate standards require 144 Kbps for rural outdoor mobile use. This data rate is available for environments in which the 3G user is traveling more than 120 km per hour in outdoor environments. Let us hope that the 3G user is riding in a train, not driving at such high speeds and trying to use the 3G terminal. Full mobility data rates require 384 Kbps for pedestrian users traveling less than 120 km per hour in urban outdoor environments. Limited mobility requires at least 2 Mbps with low mobility (less than 10 km per hour) in stationary indoor and short-range outdoor environments. The maximum data rates that are often quoted when illustrating the potential for 3G technology will therefore be available only in stationary indoor environments.

Timelines for 3G

New services usually progress through several stages before becoming established. For 3G service developments these stages include standardization, infrastructure development, network trials, service and infrastructure contracts, network rollout, terminal availability, and application development. These stages for 3G are shown in Table 6–4.

3G Specific Applications

Several applications enabled by the broad bandwidth will come with 3G. These applications include audio and video, VOIP, still and moving images, virtual home environments, and UMTS.

Table 6–4
3G Timeline

Date	Milestone
1999	3G radio interface standardization and initial 3G live demonstrations of infrastructure and concept terminals
2000	• Continuing standardization with network architectures, terminal requirements, and detailed standards • Formal approval of the IMT-2000 recommendations at the ITU Radio Communication Assembly (in early May) • 3G licenses by governments around Europe and Asia
2001	• 3G trials and integration • 3G in Japan by NTT DoCoMo • First trial 3G services in Europe
2002	• Basic 3G capable terminals available in commercial quantities • Network operators launch 3G services commercially and roll out 3G. Vertical market and executive 3G early adopters use 3G regularly for nonvoice mobile communications
2002–2003	New 3G specific applications, greater network capacity solutions, more capable terminals, fueling 3G usage
2004–2005	Commercial 3G at critical mass in both corporate and consumer sectors

See www.lifestreams.com.

Audio and Video

Audio or video over the Internet is downloaded (transferred, stored, and played) or streamed (played as it is being sent but not stored). Streaming tends to be of lower quality than downloading. Content is transferred using various compression algorithms such as those from Microsoft or Real Networks or the MPEG-1 audio layer 3 (better known as MP3) protocol. In fact, MP3 is a compression/decompression (codec) algorithm. With 3G, MP3 files will be downloadable over the air directly to a phone via a dedicated server. There are numerous business models to allow both the network providers and the copyright owners of the MP3 material to benefit financially. Mobile Lifestreams, a noted leader in the cellular telephony research field, expects that the integration of mobile telephony with everyday consumer products will emerge within the next several years to the extent that users can retrieve any type of data, anytime, anyplace through the next generation of mobile devices.[7]

Mobile phones with built-in MP3 from Samsung and with add-on MP3 modules from Ericsson were demonstrated in late 1999 and are expected to be commercially available during the year 2000.

Due to current bandwidth constraints, users go online and download files to their portable device over the fixed network, which are then watched at their convenience, without real-time audio and video streaming over mobile networks. Because even short voice clips occupy large files, high-speed mobile data services are needed to enable mobile audio applications. The higher the bandwidth, the better the quality, which explains the attractiveness of 3G for mobile multimedia applications such as mobile audio and video.

Voice-Over Internet Protocol

Another audio application for 3G is Voice-over IP (VoIP), which is the ability to route telephone calls over the Internet to provide voice telephony service at local call rates to anywhere in the world. With 3G and higher rate 2.5G technologies such as EDGE, VoIP will be available for the first time on mobile phones. VoIP can be used as an alternative to cable service.

VoIP is not, however, a replacement for standard voice services, because it requires a great deal of bandwidth. In addition, it needs a high switching rate on the IP backbone to minimize the very high likelihood of delayed and lost packets. VoIP is the ability to make telephone calls and send faxes over IP-based data networks with a suitable quality of service (QoS) and superior cost benefits. Everyone is talking about VoIP and everyone wants a piece of the pie.[8]

---

[7] See www.lifestreams.com.

[8] See www.protocols.com and www.techguide.com for more information on VoIP.

- Equipment developers and manufacturers see a window of opportunity to innovate and compete. They are busy developing new VoIP-enabled equipment for their timely break into the market.
- Internet service providers see the possibility of competing with the PSTN for customers.
- Users are interested in the integration of voice and data applications in addition to the cost savings.

Although VoIP seems to be most attractive, the technology has not been developed to the point where it can replace the services and quality provided by the PSTN. First it must be clear that VoIP will indeed be cost effective. To compete with today's PSTN, there must be significantly lower total cost of operation. This savings should initially be seen in the area of long-distance calls. VoIP provides a competitive threat to providers of traditional telephone services that will clearly stimulate improvements in cost and function throughout the industry.

Another immediate application for IP telephony is real-time facsimile transmission. Fax transmission quality is typically affected by network delays, machine compatibility, and analog signal quality. To send faxes over packet networks, an interface unit must convert the data to packet form, handle the conversion of signaling and control protocols, and ensure complete delivery of the scan data in the correct order. Packet loss and end-to-end delay are even more critical here than in voice applications.

Many other applications have been identified to be implemented by VoIP. For example, voice messages can be prepared using a telephone and then delivered to an integrated voice/data mailbox using internet or intranet services. Voice annotated documents, multimedia files, and so on can easily become standard within office suites in the near future.

The main justifications for development of VoIP can be summarized as follows:

1. *Cost reduction.* As described, there can be a real savings in long-distance telephone costs, which is extremely important to most companies, particularly those with international markets.
2. *Simplification.* An integrated voice/data network allows more standardization and reduces total equipment needs.
3. *Consolidation.* The ability to eliminate points of failure, consolidate accounting systems, and combine operations is obviously more efficient.
4. *Advanced applications.* The long-run benefits of VoIP include support for multimedia and multiservice applications, an area in which today's telephone system cannot compete.

Growth in the VoIP market is expected to be considerable over the next 5 years. The annual growth rate for IP-enabled telephone equipment was 132 percent between 1997 and 2004, with a market of $3.16 billion in 2004. Annual revenues for the IP fax gateway market were over $100 million in 2000 (from less than $20 million in 1996).

Still Images

Still images such as photographs, pictures, letters, postcards, greeting cards, presentations, and static web pages can be sent and received over mobile networks just as they are across fixed telephone networks.

Two variables that affect the usability of such applications are bandwidth and time, and they are inversely related. The greater the bandwidth, the less time needed to transmit images, and vice versa. Transmission of image rather than textual information has not been a popular nonvoice mobile application until now, due to the slow data transmission speeds that were available prior to the introduction of mobile packet data. Files between 50K and 100K bits in the JPEG format can be transmitted quickly using mobile packet data.

Moving Images

Sending moving images in a mobile environment has several vertical market applications including (sensor triggered) monitoring parking lots or building sites and sending images of patients from an ambulance to a hospital. Video-conferencing applications, in which teams of distributed salespeople can have a regular sales meeting without having to go to a particular physical location, is another application for moving images. Some may argue that vertical markets do not need video, and consumers do not want it. However, with the Internet becoming a more multimedia environment, 3G devices will display those images and access web services.

The transmission of moving images is one of the applications that GPRS and 3G terminal and infrastructure vendors such as Sprint or Verizon routinely and repeatedly tout as a compelling application area that will be enabled by greater data rates. However, it must be noted that even demonstrations of 1 megabyte of data over the air using Microsoft NetMeeting to perform a video conference facility do not deliver smooth broadcast quality video images. Improving compression techniques should allow acceptable quality video images to be transmitted using 64 Kbps of bandwidth.

Although videophones have failed to gain public acceptance, this actually could be a function of the fact that a videophone is only as good as the number of other people who have one. Corporations that have several people with video-capable mobile phones could easily hold virtual remote sales meetings among regional sales representatives.

The problem is that as much as still images such as pictures and postcards will be a significant application for GPRS, moving images may not be of high enough quality initially to be of much interest. Users could spend all their time adjusting the size of the image on their screen and trying to make out what they are seeing.

The greater bandwidth of 3G telephony allows for high-quality image transmission over the mobile network. As such, we see all moving video and image transmission application migrating to the 3G bearer as soon as that technology becomes available. By 2003–2004, full-length movies will be downloadable from a plethora of Internet sites.

Virtual Home Environment and Universal Mobile Telephone Service

A universal mobile telephone service (UMTS) that is often mentioned in the vendor's brochures is the virtual home environment (VHE), a service that allows customers seamless access with a common look and feel to their telephone services from home, office, or elsewhere and in any city as if they were at home. VHE is therefore aimed at roamers (a small subset of total mobile phone users).

VHE could also allow some other more useful services by placing a universal identity module (UIM) into any terminal. Those terminals also could be something other than mobile devices if smart cards gain wider acceptability. VHE could hardly be described as a major application though, especially since e-mail and other services are increasingly available worldwide, as the Internet becomes more widespread and services migrate to the Internet and can therefore be accessed from any internet browser—with or without a smart card.

VHE is only one story in the UMTS saga; however, UMTS may well be a future technology in the cellular telephony world. UMTS can also mean universal mobile telecommunications system. UMTS, called the "third generation" mobile communications system, is intended to be the standard by which the next generation of wireless communication devices will operate. UMTS is positioned to play a key role in creating the future mass market for high-quality wireless multimedia communications. It will accomplish this standard of higher performance by employing higher frequencies and increased capacity. Simply put, UMTS is intended to be the next incarnation of the cell phone. UMTS will also play a key role in future PDA developments.

Deemed the "third generation," UMTS is advancing Europe into the emerging capability to provide wideband applications such as video telephony and video conferencing as well as interactive multimedia services at an astounding rate of up to 2 Mbps. This breakthrough will provide the ability to be virtually connected at all times. Intertwining high-speed packed and circuit switched data transmission, this design will offer data rate on demand, which will reduce the operating costs to consumers. 3G networks promise to be in demand as forecasters predict over 1 billion wireless subscribers by the year 2010. UMTS offers customers competitive prices and "customer interfacing" equipment as well as a variety of terminals to provide affordability to both the novice and the advanced user. VHE will provide full security using a mixture of radio environments or roaming to other UMTS operators regardless of satellite or terrestrial access with the help of access and core networks. Added bonuses include alternate billing tailored to the individual's unique needs whether it be per session, flat rate, pay per byte, or uplink/downlink asymmetric bandwidth. This high-quality global service promises exciting advances for the on-the-go person.

UMTS must accommodate services and features already in use by the wireless customer while allowing the addition of broadband services. The levels of these new broadband services must be comparable with fixed broadband networks. What are the services that UMTS must support? How much bandwidth is required of UMTS by consumers in cell environments? What applications are likely to be supported by UMTS? These are questions to be covered in this brief description. A considerable amount of effort has been taken to complete the migration from second generation without a paradigm shift and still make provisions for a competitive market.

The new mobile generation will be based in a mixed environment of fixed and mobile communications environments. The need to standardize services offered to mobile and fixed networks and to have mobile systems that have bandwidth on demand for applications is a must to bring this about. This should allow roaming and interworking capabilities while catering to specialized markets. Telematics is a major market for this, such as dynamic transport routing, fleet control, and tourism.

UMTS will utilize the mobile access portion of B-ISDN, to increase the standardization of services supported by both fixed and mobile networks. Mobile networks will now be able to offer services that had been previously attainable only through fixed networks. This extends to include wideband services upwards of 2 Mbps, thus allowing UMTS to be a stand-alone network option.

The implementation of UMTS on a large global scale will definitely have to rely on mobile satellite systems to provide for coverage in rural areas in developed and developing countries as well as for naval and aircraft communications. The system would allow for complete integration with its terrestrial sites, thus allowing for a seamless service that keeps the functions of the systems virtual or transparent. The interworking of terrestrial and space communications must be carefully considered as performance could deteriorate as economic and technology issues arise from the integration of equipment, networks, and systems.

ATM is slowly becoming the dominant transmission technique. UMTS must allow for the compatibility with this technique, thus allowing for variable bit rate access for different access media for mobile or fixed networks. The provision for the use of UMTS across multiple countries requires special attention to issues with regulatory implications such as spectrum, billing, security, numbering, and privacy.

UMTS is the product of the ITU. The organization set out to create a standard by which all next-generation wireless communications devices would operate. ITU's first job was to develop an acceptable framework. This framework is IMT-2000, which the ITU has touted as *the standard* for the next generation of all wireless products, cell phones, PDAs, and so forth. 3GPP was also a part of ITU's effort. 3GPP is described as ". . . a global cooperation between 5 organizational partners (ARIB, CWTS, T1, TTA, and TTC) who are recognized as being the world's major standardization bodies from Japan, China, Europe, United States, and Korea" (3GPP Market Representation 2002). These organizations bring a powerful set of knowledge and abilities that pertain to the field of wireless communications. This knowledge continues to be helpful in making UMTS a successful standard on which to build the future.

The plethora of firms involved in implementing IMT-2000 includes an organization of manufacturers called the Global Mobile Suppliers Association (GSA). According to their web page, GSA has "...a rapidly growing membership that encompasses the total supplier chain for wireless industry—over 70 leading companies are working today on GSA programs" (GSA Member Profiles 2002). This global partnership includes such well-known firms as Ericsson, Hewlett-Packard, Intel, Motorola, Nokia, and Siemens. Lesser known companies include 3G Mobilize, Cerebrum, Eplication, and Telemessage.

Why is it so important to have one universal standard? Recent history has shown that having one set standard for any device, regardless of its function, greatly facilitates and simplifies its usefulness. For example, consider the problems that have emerged from the myriad of standards imposed by the computer and wireless communications industries up to this point. UMTS and IMT-2000 are intended to be the single standard under which all companies can build and expand wireless communications worldwide. Having a single standard also will allow easier communications between individuals and companies.

The two applications for UMTS are business and individual consumer use. UMTS offers many benefits to global businesses and individuals. Cost, however, is a major deterrent to UMTS becoming a system that is commonly used by both groups. The estimated start-up cost is $2.5 billion with a like amount for operational and marketing cost. The most obvious use of UMTS is global business. UMTS offers companies global mobility and a wide range of services and applications. As mentioned, some of the services offered by UMTS are teleservices, telephony, messaging, Internet, paging, and broadband data. Allotment of these services would greatly benefit business, especially if UMTS delivers what it promises.

The services and applications have been divided into four classes by the 3GPP. The first is conversational class. This class encompasses areas important to business such as voice, telephony, and video. The second class is streaming class, and it is involved with multimedia, video on demand, and webcast. Third is a class called the interactive class, and it offers web browsing and database access. The final class is the background class, which includes e-mail, SMS, and downloading applications.

Security is an area where UMTS excels. For example, UMTS security is increased beyond 2G wireless GSM. UMTS offers better network security against attacks and eavesdropping, so sensitive business data are not at such a risk. Another advantage to UMTS over GSM is its ability to give confidentiality or network domain security for the user and the user's location. Companies can also configure the security features to fit their own needs.

The individual will need to acquire service from a UMTS service provider. Initially this service will be much more expensive compared with existing services, but as mentioned, the benefits are much greater. The services will essentially be the same as business with a couple of additional benefits. The conversational class also has a video gaming service that the individual may find beneficial. The interactive class offers network gaming for those who enjoy online gaming.

The reason for one wanting a standard such as UMTS is simple economics. There are 770 million people who own mobile radio equipment. Of those people, 500 million can be reached by the GSM standard found in Europe and 167 other countries outside of Europe. With the 1 billion mark drawing near and 85 percent of that group having Internet capability, the margin for growth is high and the untapped resource is great (Schwarz da Silva, Arroyo-Ferncindes, Barani, Pereira, and Ikonomou, 2001).

Siemens conducted an extensive market research study, during which he interviewed 11,000 mobile radio users, of which 88 percent said they would use expanded mobile services. Private and business customers were prepared to increase spending 69 percent and 42 percent, respectively, on mobile communications equipment. These numbers were slightly higher in the under 25 class, where entertainment seemed to be the main focus. Figure 6–12 shows the results of the survey.

The end of 2002 is expected to see 53 million mobile Internet users. This is roughly 14 percent of Europe's population, which would be approximately equal to 76 billion euro as long as UMTS networks and handsets are available by 2003. Mobile communications will set the global market with 68 percent of revenue being established in Europe and Asia and scheduled

Figure 6–12
Requested Services and
User Demand.
Source: Siemens End-User
Survey, 2001.

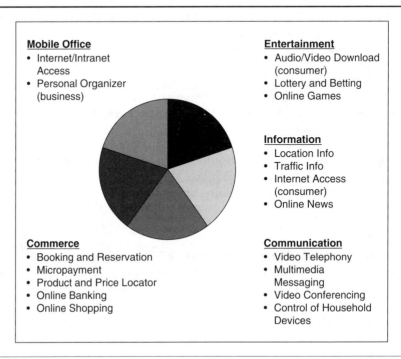

Mobile Office
• Internet/Intranet
  Access
• Personal Organizer
  (business)

Entertainment
• Audio/Video Download
  (consumer)
• Lottery and Betting
• Online Games

Information
• Location Info
• Traffic Info
• Internet Access
  (consumer)
• Online News

Commerce
• Booking and Reservation
• Micropayment
• Product and Price Locator
• Online Banking
• Online Shopping

Communication
• Video Telephony
• Multimedia
  Messaging
• Video Conferencing
• Control of Household
  Devices

to be worth $1 trillion by the year 2010.[9] However, the success of UMTS depends on it being able to reach its 2002 launch date. This launch date has now been put back to 2004–2005. Creating a regulatory framework and spectrum that is adequate also is paramount to success. The FCC is fast at work trying to get an entire consensus on the spectrum slotted for UMTS. Recently the 1,770 MHz to 1,880 MHz range was taken because of national security reasons following the September 11, 2001, attacks. The relative working frequency range extends from 1,885 MHz to 2,025 MHz and 2,110 MHz to 2,200 MHz for future use.

The availability of licenses is also a big issue when discussing the success of UMTS. Licensing will be relatively easy to obtain, so the market can become saturated with services such as application, voice, and video. Licenses have already been auctioned off to much of the eastern world including Europe and Asia. None have been issued in the United States. Canada is the only North American country to date that has won any licenses. Producing timely UMTS standards is also key in making sure UMTS is fully operational by a 2004–2005 date.

Although the global market is working to implement this 3G technology by the end of 2002, current estimates show implementation to be sometime during mid-2004 at the earliest. This is due to the massive expenditure which must be made in advanced switching and transmission equipment. The promise to consumers is high, but it will be up to the UMTS vendor community to deliver.

Electronic Agents

Electronic agents are a technology that many believe will play an important role for mobile working in the future.[10] Electronic agents are defined as mobile programs that go to places in the network to carry out their owners' instructions. They can be thought of as extensions of the people who dispatch them. Agents are self-contained programs that roam communications networks delivering and receiving messages or looking for information or services.

Certainly, 3G terminals will give their owners much more control over their lives than today's mobile phones. They will be e-assistants, e-secretaries, e-advisors, and e-administrators. This kind of control is what home automation applications anticipate. Indeed, Orange in the UK expects that within 10 years, our mobile devices will be waking us up, reading out our e-mails, ordering our groceries, telling us the best route to work, reminding us, and translating our conference calls. The key question is the extent to which these processes are human initiated or computer generated and controlled and the extent to which devices can "learn" individual preferences and act accordingly.

---

[9] For additional information, go to www.umtsworld.com.
[10] See www.cnet.com, www.lifestreams.com, and www.networkworld.com for more information.

Downloading Software

As the twenty-first century progresses, software will increasingly be downloaded electronically from the Internet rather than purchased as boxed products in stores. Downloading software has the following advantages.

- It is environmentally friendly. There is no packaging to discard or store.
- It is quick and convenient. Downloadable products are delivered direct to your computing device. It arrives in minutes, not days.
- It is value for money. You pay no delivery charges.

Download times vary depending on modem speed and application size. Typical download times vary from 10 minutes to 2 hours. Following are download times for a 5 megabyte application.

Connection Speed	Download Time
Very fast corporate type connection (e.g., T1)	30 seconds
Corporate type connection (e.g., ISBN)	12 minutes
Typical home modem (e.g., 28.8 modem)	104 minutes

Sites such as beyond.com and Mobiledatashop.com offer many software products for immediate electronic download. Additionally, the application service provision (ASP) market, in which software platforms and server software are being hosted by third parties and accessed by client software, mimics this "thin client" world in which the bandwidth is high enough for applications and files to be retrieved from the Internet on the fly whenever they are needed. Because download speed relies on the bandwidth, 3G is likely to become a key bearer for this service.

As you can see, 3G promises a great deal. If it delivers even a small portion of this, the cellular industry and people's lives will be dramatically changed. The next few years will determine to a large degree what the future holds in this regard.

RECENT EVENTS IN THE CELLULAR WORLD

This section reports recent events (within the last 18 months) in the cellular world. It is important to understand the role the cellular industry plays in our technologically oriented society.[11]

- The market for cell phones continues to expand. A total of 400 million cell phones were sold in 2001, while 1 billion are expected to be sold in 2003.

---

[11] Sources used throughout are from www.cnet.com, www.infoword.com, and www.zdnet.com (with permission).

- Kozmo.com, a home delivery service that operates in urban areas, announced a plan to offer web-enabled cellular phones free on its Internet site through a partnership with mobile phone distributor InPhonic. Kozmo delivers a variety of items, including toiletries, videos, and snacks, to customers in eleven metropolitan markets. What's in it for Kozmo? The free phones will make it easier for more customers to buy more from Kozmo, as commuters could order items and have them delivered soon after they arrive home. Why would InPhonic be involved? InPhonic customizes cell phones at no cost to the consumer, then receives a percentage of Internet sales made from those phones.
- In September 2000, Intel unveiled an architecture for wireless devices called the Personal Internet Client Architecture. By adopting the architecture, device makers would essentially receive a free blueprint for wireless hand-held devices. The blueprints, of course, depend on Intel chips.
- In another interesting development and as an example of how cell phones are expanding their role, the publishers of *Star Trek Deep Space Nine: The Fallen* announced that gamers can sign up to receive clues and cheat codes in text messages via their cell phones (Barnes, www.cnetnews.com). Though video game publishers have long posted game guides and clues on websites, and sent the information directly to people via e-mail, Simon & Schuster Interactive says it is the first game publisher to relay the information via cell phone. Simon & Schuster Interactive's vice president, Peter von Schlossberg, said he wants people to continue living the *Star Trek* game while away from their computers—by viewing their web-enabled cell phones as their own "communicators." Video game players can sign up for the service on the game's website. Simon & Schuster Interactive says it will be offering the free text messaging service through a collaborative effort with Finnish-based Add2Phone. The companies also intend to create a distinctive ring for cellular phones that will alert *Star Trek* players that they are about to receive some new information, a feature that is already available in Japan.

Increased Interactivity

A major technical improvement is that the cellular phone is experiencing a higher degree of interactivity.

- Simon & Schuster Interactive is not alone in its move toward increased interactivity. Last week, Sega announced it will develop games for Motorola's next-generation cell phones. Die-hard video game players have proven to be a receptive group to the marketing of new technology, analysts say. They tend to be more technologically savvy and eager to experience whatever is new on the market. The company said that the service will be free and that people's personal information will not be distributed.

- The top cell phone maker, Nokia, now has almost over one-third of the global handset market. In 1999, Finland's Nokia held a 27 percent share, according to Dataquest. Schaumburg, Illinois-based Motorola had 17 percent, followed by Sweden's Ericsson with 11 percent. In 2001, Nokia's share increased to 35 percent. Motorola had 25 percent with others comprising the rest.

- Consider some of the services the 15.5 million people using I-mode can tap. A passenger in a car hears a song she likes on the radio, checks into the station's website, and reads all about its singer. A theatergoer on a city-bound train picks which play has available seats and makes the reservation (source).

- AT&T Wireless said its PocketNet mobile Internet service had signed up just 300,000 customers, although it was adding 50,000 a month. A study by the research firm Yankee Group found 54 percent of Americans who own mobile phones say they do not want a mobile Internet service. Among those who now do their banking online and already have a mobile phone, only 14 percent expressed interest in accessing their account through wireless Internet links.

- DoCoMo, Hewlett-Packard, and the U.S. company's Japanese unit said they have signed an agreement to conduct joint research on "fourth-generation" cellular phone systems. So-called 4G technology is expected to support services such as video streaming to cell phone handsets, which allows users to view large image files without waiting for the entire file to download.

- Teenagers are the fastest growing market segment for cell phones due to their busy lifestyles. A study, released by the market research firm Cahner's In-Stat Group, predicts the wireless market for young people ages 10 to 24 will grow from 11 million subscribers today to 43 million in 2004. Two of every four teens will own a cell phone by 2004, and three of every four will use a cell phone, typically one borrowed regularly from a parent, the report states. The Cahner's study further found that wage-earning young people, who have opted to work rather than attend college, will make up the largest segment of the youth market. Research showed that teens mostly use phones for social purposes and want more colorful and interesting cell phone options.

- In a development sure to open up new markets for cellular, Freedom Wireless, a privately held Phoenix-based wireless technology company, was granted a U.S. patent earlier this month for its methods of completing prepaid wireless phone calls—a service that allows a customer to pay a certain amount in advance, then make wireless calls until the amount of credit is used up.

   The prepaid market, although a small percentage (10 percent) of most carriers' subscriber totals, has grown in popularity in recent years. Prepaid wireless plans are particularly popular with parents of teens, customers with poor credit histories, and other subscribers who want to ensure they are not surprised by a larger-than-expected monthly bill. About 12 million of roughly 104 million U.S. wireless customers use a prepaid service, but that is expected to increase to 27 percent by 2005, according to the Yankee Group.

Freedom Wireless developed a simple system for completing prepaid wireless calls in the mid-1990s when it operated as a cellular service reseller. The company first sought a patent in December 1994.

Previously, cellular customers had to dial toll-free 800 numbers or enter personal identification numbers (PINs) for the carriers to check their account balances. But those methods were undesirable for many customers.

In documents from the U.S. Patent and Trademark Office, Freedom Wireless outlines its patented method for checking a database of prepaid subscribers when a call is made or received to determine if there is enough credit in the account to connect the call.

Freedom Wireless also details a variety of additional features, including periodically checking the accounts of prepaid subscribers while they are on the phone to ensure customers do not overdraw their accounts.

- In December 2000, two important studies indicated that hand-held cell phones, even when used regularly for 5 years or more, do not appear to increase the risk of developing brain cancer. Some experts have suspected that radio-frequency signals, a low-power microwave radiation emitted by many of the 92 million cellular phones in the United States, may cause health problems, particularly brain cancer in parts of the head closest to the antenna. In the two studies released by medical journals, those concerns appear to be groundless. The National Cancer Institute's study, which was released in early 2001 by the *New England Journal of Medicine,* examined use of cellular phones among 782 patients with malignant tumors and 799 patients with tumors from nonmalignant conditions. The phones were not linked to malignant tumors regardless of the amount or length of phone use, and tumors were not more likely to appear on the same side of the head where the telephone was used, the researchers said.

In the second study, 469 patients with brain cancer were compared with 422 people without the disease. Patients with cancer did not use their phones substantially more than those without the disease, the study found. Moreover, people who rarely or never used cell phones were not less likely to develop cancer.

Others are concerned about this issue as well. At the behest of a major wireless industry trade group and amid heightening concerns about potential health risks, many mobile phones now carry information about radiation levels (Grice 2000).

- Long-term users of some first-generation cell phones face up to 80 percent greater risk of developing brain tumors than those who did not use the phones, a new Swedish study shows.

  The study, published in the European Journal of Cancer Prevention, looked at 1,617 Swedish patients diagnosed with brain tumors between 1997 and 2000, comparing them with a similar control group without brain tumors.

  Researchers found that those who had used Nordic Mobile Telephone handsets had a 30 percent higher risk of developing brain tumors than people who had not used that type of phone, particularly on the side of the brain used during calls. For people using the phones for more than 10 years, the risk was 80 percent greater.

  "Our present study showed an increased risk for brain tumors among users of analog cellular telephones. For digital cellular phones and cordless phones, the results showed no increased risk overall within a five-year latency period," the study said.

  Two major mobile phone manufacturers disputed the findings of an increased risk of cancer.

  The world's biggest mobile producer, Finland's Nokia, which still produces two models of phones working in the Nordic Mobile Telephone standard, said scores of other studies conducted on the health effects of cell phones showed no evidence of health hazards for users.

  "There have been close to 200 studies done on different areas of mobile phones, and in the light of those and the way the scientific evidence is, there is no health risk in using mobile phones," Marianne Holmlund, communications manager at Nokia Phones, said Thursday.

  Mikael Westmark, a spokesman for Sweden's Ericsson, which used to make Nordic Mobile Telephone handsets, said: "The study and the conclusions it reaches differs from at least three other studies in the past in several highly regarded scientific journals. None of these studies found a connection between mobile phones and cancer."

  The Nordic Mobile Telephone network was initially developed to serve the Nordic countries, starting operations in the early 1980s, but then became popular in Russia and the Baltic countries.

  It is still used in more than 40 countries, but has been overtaken in several countries by the GSM (Global System for Mobile Communications), which is due to be gradually replaced by rapid third-generation mobile networks (CNET News 2001).

- The Cellular Telecommunications Industry Association (CTIA) announced plans to require mobile phones to carry new information on the phone's package, revealing the rate of radiation emitted by certain models.

  "Voice-only devices are giving way to real heavy Internet-ready devices with a lot more guts that, at least theoretically, may not be that great to have right next to your head," said Bryan Prohm, a wireless industry analyst at the market research firm Dataquest.

The radiation information, which could be packaged with new handsets in as early as 3 to 6 months, is expected to show that cell phones are within acceptable radiation exposure ranges. All cell phones sold in the United States today have been certified to meet federal requirements for the quantity of radio-frequency energy absorbed by the body, according to industry representatives. The World Health Organization urged that more testing be conducted worldwide.

New evidence suggests the radio-frequency waves emitted by the handsets could be cause for some concern. The renewed interest in cell phone safety, including related driving risks, comes just as the industry is preparing for an explosion of growth in the next few years around new voice users and high-speed wireless Internet access.

The addition of radiation information on a cell phone box could help stave off new regulations and future lawsuits and could appease consumer groups, analysts say.

- Mobile phones are also being scrutinized on another health front. The National Highway Traffic Safety Administration today said "driver distraction" from cell phones and other devices could in part be to blame for a higher number of car crashes. Increasingly, policy makers are voicing their opposition to cell phone use while driving. Analysts estimate that about 50 percent of mobile phone use occurs in automobiles.

  After a variety of studies in recent years indicating that mobile phone users have higher accident rates than average motorists—in some cases approaching rates of those legally intoxicated—some lawmakers are now calling for curbs in the use of mobile phones while driving. The distractions of conversing while driving can lead to unnecessary accidents or unsafe driving habits, opponents say.

  Hands-free earpieces and microphones, which have gained in popularity and often come with the newest phones, not only allow motorists to drive with both hands on the steering wheel, thus improving safety, but also move the handset away from a person's head.

On Star Technology

Sun Microsystems announced a deal with the General Motors OnStar unit to make Java technology the computing standard for the automotive industry. Microsoft has its Windows CE for Automotive os, as well as a Car.Net initiative aimed at creating a common language for cell phones, hand-helds, and dashboard computers.

IBM and Intel, too, have announced plans to collaborate on a nonproprietary standard for dashboard "telematics," a term for cellular and Internet services in vehicles such as navigation systems, roadside assistance, and entertainment.

The spectrum bands, or segments of airwaves, could be used in addition to existing cellular and PCS airwaves for such services as high-speed wireless Internet access or streaming video. These 3G services are being launched aggressively in Europe and Asia, where there are fewer conflicts over spectrum. In the United States, this deployment has been slower.

Wireless carriers believe mobile Internet access will become more popular when they can offer speeds and services that compete with cable modems and digital subscriber lines (DSLs), making wireless Internet a multibillion dollar business. The technology already has been developed and hardware manufacturers are eager to sell the necessary equipment to carriers. Analysts also believe that the debut of 3G technology will allow new entrants into the wireless market, leading to more growth in an already rapidly growing market.

Much work needs to be done, however, before carriers such as Verizon Wireless and Cingular Wireless can begin offering consumers mobile broadband. The first step is for the wireless industry and current users of the identified spectrum to tell the FCC how it can best fit these new, bandwidth-heavy wireless services into a crowded spectrum. The agency must then adopt new rules for doing so, which could involve displacing incumbent spectrum users. The five bands the FCC is expected to identify for 3G services are 2,500–2,690 MHz, 1,710–1,755 MHz, 1,755–1,850 MHz, 2,110–2,150 MHz, and 2,160–2,165 MHz.

The FCC has focused on 2,500–2,690 MHz, which houses fixed wireless users such as Sprint and WorldCom that are beginning to offer broadband services to homes and offices. The 1,710–1,755 MHz previously was reserved for government use but has been opened up to commercial operation. The FCC will ask if another government band, 1,755–1,850 MHz, should also be opened up, according to agency representatives. Wireless engineers say 1,755–1,850 MHz is a prime band for 3G services, but the U.S. Defense Department operates heavily in this band and has been extremely reluctant to share or relocate, citing national security concerns. The other two bands, 2,110–2,150 MHz and its neighboring 2,160–2,165 MHz, hold various fixed service and mobile operators that had already been notified that they were subject to relocation.

The FCC in August 2002 opened the door to allow mobile telephone carriers to pull either entirely or partially out of a $15.9 billion wireless license sale, in what could be a financial boost to a sagging industry.

The agency laid out two scenarios it was willing to consider. The first was to allow carriers like Verizon Wireless and Deutsche Telekom's T-Mobile USA to walk away entirely from a controversial January 2001 auction.

The FCC tried to sell the licenses to other carriers after repossessing them for nonpayment from NextWave Telecom, which is mired in bankruptcy proceedings; however, a federal appeals court said that move violated bank-ruptcy law. The decision halted the agency's attempt to complete its auction of the licenses, but the FCC has kept carriers who participated in the auction on the hook for $15.9 billion until the legal wrangling was over.

The other scenario the FCC contemplated was allowing carriers to opt out of bidding for specific licenses they no longer wanted to pursue.

The U.S. Supreme Court is slated to hear the arguments over the licenses on Oct. 8, and a decision could take several months. However, if NextWave loses, the company has indicated that it would mount other legal challenges to hang onto the licenses.

Verizon and other carriers have gone to Congress and beseeched the FCC to allow them to withdraw from the auction without penalty because of the legal fight, arguing the financial overhang from the auction is hurting their outlook since they cannot use the licenses they supposedly bought (CNET News 2001).

OnStar

GM's OnStar division has experimented with location-based advertising beamed to vehicles through a cellular network (Konrad 2001; Ross 2001).[12] The service is part of OnStar's Virtual Adviser, an automated, cellular-based concierge service offered as an option on most GM vehicles.

In addition to cheap gas alerts, drivers who indicate they want so-called push advertising can become the target of other sales and marketing pitches. For example, a golf store might send information about a sale to golf enthu-siasts whenever they are driving within a 2-mile radius of the store. As a driver approaches the store, a computerized voice could override music on the stereo and say, "All golf bags are 50 percent off."

To receive the location-based, automated data, OnStar subscribers first complete a confidential questionnaire detailing exactly what kind of adver-tisements they want to receive. Subscribers may also enter their credit card information into an OnStar database so that they can purchase items directly from their vehicle and later stop at the store to pick up merchandise. OnStar is finalizing deals with third-party advertisers and potential advertisers—ranging from online financial service providers to sporting goods retailers—have barraged OnStar with requests to advertise.

Virtual Adviser's first advertisements will be from financial service providers, including Fidelity and Charles Schwab, that will sponsor the ser-vice's stock updates. The 4-second ads on Virtual Adviser work as follows:

---

[12] Also see New York State Internet Office, www.nydmv.state.ny.us/dmvfaqs.htm#cell for more information.

Just before the computerized voice gives the driver an update on selected stocks, the driver may hear, "This update is brought to you by Fidelity."

Experts say location-based push advertising is poised to become one of the hottest marketing trends of the next decade. Revenue from location-based wireless services in North America will increase more than 100-fold, from less than $30 million in 2000 to $3.9 billion by 2004, according to market researchers at the Strategies Group. Although hand-held computers and cellular phones are fertile ground for location-based advertisements, many say the automobile has much greater potential.

Letting advertisers subsidize the cost of onboard navigation and concierge systems is another advantage. Although navigation systems and onboard cellular services are relatively common in Japan, American customers have been far less willing to pay for the services. Advertisers may eventually subsidize a fraction of the cost of OnStar, which ranges from $199 to $399 annually.

OnStar's Virtual Adviser service is conducted through a hands-free, voice-activated phone. The driver hears a computerized voice through the vehicle's stereo speakers, while a microphone embedded in the vehicle's headliner picks up the driver's commands.

Virtual Adviser customers also are able to hear numerous types of web-based information, including updated stock prices, weather, and sports scores. Customers must first select which information they want to hear by filling out an online questionnaire. General Magic of Sunnyvale, California, provides the voice-activated user interface (VUI) for the Virtual Adviser's web content.

Roughly 800,000 people subscribe to OnStar, and the GM division is expecting dramatic growth. Toyota features optional OnStar service in its luxury Lexus vehicles, and Honda features it as an option on its luxury Acura models. OnStar equipment is currently included in roughly one of four GM vehicles, and about 5,000 new customers sign up for the service every day. It is offered as a standard or optional feature on all GM vehicles except for compact cars, such as the low-end Chevrolet Cavalier. OnStar debuted several years ago as an option on Cadillac luxury sedans.

OnStar comes in two levels of service. The safety and security package, which costs $199 per year including all per-minute cellular charges, features a single emergency button that links to OnStar customer service representatives at call centers in Troy, Michigan, and Charlotte, North Carolina. Representatives can remotely open the car doors if a person is locked out, or conduct simple diagnostic research if the car stalls. The service also sends out an emergency call if the airbags deploy, in which case a representative will either confirm that the driver is safe or dispatch an emergency vehicle to the site.

Premium service costs $399 per year including all per-minute cellular charges and is only offered on Cadillacs. It includes the safety and security features as well as routing assistance for 5 million points of interest in North America. A concierge service is also included, in which a call center representative will purchase tickets or make reservations at local entertainment events and restaurants. Location-based services are another significant feature, and OmniSky is working to add them to its mix. Such services offer information about local restaurants and entertainment, as well as maps.

On-Road Distractions

As a sign of things to come, Ford Motors has opened a $10 million driving simulator laboratory to study the dangers of dashboard computers, hand-held devices, cellular phones, and other electronic appliances that are becoming ubiquitous on U.S. roads (Konrad 2001). This will be the first automotive lab to feature a full-scale, moving-base driving simulator that tracks drivers' eye movements while using onboard gadgets and trying to maneuver curves on simulated highways.

Distractions ranged from lighting cigarettes and eating hamburgers to applying makeup and even attempting to write memos. The study also showed an alarming increase in the number of electronics-related distractions as the number of Americans with hand-held electronic devices has mushroomed.

The concern over the use of cell phones has reached a point where some counties and states have banned or are considering banning the use of mobile phones in cars. Several pieces of legislation have been proposed lately that would ban such use.

Keeping Up in the Technological Arena

The problem is that wireless data transmission rates are still too slow and that e-mail and Web surfing—the key features people want from these devices—are poor experiences. Most of the products run on CDPD networks, which are limited to 19.2 Kbps, which compares with a dial-up modem that can run at speeds up to 56 Kpbs. The current data transmission rates limit the amount of content and the level of personalization on these devices, so there is only so much that providers can do until bandwidth dramatically increases.

Why is the U.S. market so tough when other countries are flocking to the new technology in large numbers? Part of the answer lies in a combination of industry hype, consumer confusion, and cultural differences that change the way people use such technologies from country to country. The main reason, however, according to analysts is the simple fact that products on the U.S. market are more difficult to use (Borland 2001).

Interestingly, a major reason for this technological lag may be the very thing that is often touted as America's most powerful weapon in technology: competition. In effect, the increasing number of companies getting into the wireless field may have created a holdup of competing services, products, and other factors that has helped stall the overall progress of the technology. The result has created difficult choices for wireless companies facing a fragmented marketplace.

By contrast, domestically dominant companies such as Japan's NTT DoCoMo have been able to forge tighter links between hardware, software, and content companies. This has created an environment which many industry insiders say makes it easier to develop software and services, and thus makes it easier for consumers to use. This has helped the Japanese market mature more quickly, many say.

Few argue that the technology has matured quickly in other countries. Analysts from Silicon Valley to Wall Street routinely point to Japan and Scandinavia, where average consumers have welcomed even the most rudimentary access to the Internet over phones.

Worldwide, the only product to emerge as an unambiguous success is NTT DoCoMo's I-mode service. I-mode offers color screens and small graphics that make it look and act very different from U.S. versions, which largely offer only small, monochrome, text-only screens. More than 17.7 million people have signed up for I-mode service, vastly outpacing the rate of growth of any U.S. wireless Internet company.

To be fair, it is not a perfect comparison. Net connections via the computer are less common, and generally more expensive in Japan than in the United States. The I-mode technology is thus a first or only method of reaching e-mail and the Internet for many customers.

Social mores have also helped drive the use of the technology. The Japanese tend to adopt consumer electronics technologies more quickly and use them for entertainment purposes. Others have speculated that social conventions there frown more heavily on talking on the phone in public places such as buses or restaurants, driving people to use the silent Internet connection more readily.

These caveats aside, virtually everyone who has played with the I-mode phones or developed websites for the technology say it is easier to use than the American and European WAP technologies.

Americans are definitely interested in the technology. Analysts estimate that about 6 million people in the United States use some kind of wireless data access, even if only infrequently. Sprint's subscriber figures and numerous industry surveys show that consumers are open to the idea of reaching internet services—e-mail, news or stock quotes, even simple games—on the go. Sprint PCS saw 1.5 million of its subscribers try the service by the end of 2001, half of that number signing up for at least a free trial subscription.

The problem, many analysts say, has been one of unfulfilled expectations. What is actually available, for example, in the wireless Internet arena is slow, text-based access to a relatively small number of sites, which themselves vary widely in reliability. These perform adequately, but only if the consumer is expecting something comparable to the Internet circa 1994.

Defenders of the technology note that a new generation of standards is being prepared, in which the WAP text-based system and Japan's popular I-mode system will merge. Once the high-speed, next-generation networks are put in place, the speed and processing limitations that prompted the creation of three-line, text-based interfaces will disappear, experts say, and phone screens will feature graphics and multimedia.

Cell Phones, Spectrum, and Wireless Technologies

Moving along a parallel track, the distinction between hand-held computers and cellular phones is becoming clouded. Companies such as Kyosera and Samsung now offer phones with the Palm operating system that have keyboards and large screens built in. The revenue equation may be more difficult, however. The carriers themselves are still posting strong revenues from wireless voice services, but growing competition has raised the prospect of shrinking profit margins, and the carriers are hoping data services can fill in gaps.

Critics of today's U.S. mobile phones say carriers need new technology, new applications, and a better interface before consumers jump aboard enthusiastically. Although that point has yet to be reached, some look with hope to the possibility of I-mode coming to the United States.

Cell phone sales worldwide are expected to remain steady, but not increase, because consumers are patiently waiting for models with better internet capability before replacing their old phones ("Cell phone sales" 2001).

In January 2001, top wireless carriers and their allies won the bulk of the licenses in the $17 billion airwaves auction, but smaller carriers are crying foul and may tie the licenses up in court (Ross 2001b).

Big-name technology makers, including Qualcomm, Motorola, Sun Microsystems, Nokia, Microsoft, and even chip giant Intel, are looking to become to wireless devices what Windows is to the PC: a universal standard for software developers (Charney and Grice 2001).

With projections of a multibillion dollar future, many technology companies are vying to establish a de facto standard for wireless software, creating a new revenue source even if sales of cell phone handsets—the most commonly used wireless devices—slip.

In the 1980s, Microsoft and Intel established dominance in the personal computer market with the Windows os and Intel's line of processors. The winner in the wireless standards battle, analysts say, could wind up as the next "Wintel."

Indeed, worldwide sales of cellular phones have dwarfed PC sales in recent years. Many analysts and company executives, however, are now worried that handset sales are slowing, prompting research and development into entire software architectures suitable for cell phones or whatever new wireless devices come to the fore.

At least three major wireless transmission technologies—GSM, CDMA, which Qualcomm developed and licenses, and TDMA—dominate the worldwide market. At the same time, dozens of carriers and handset makers require wireless software and mobile Internet developers to tweak their applications for different phones and carriers. Analysts say this is partly to blame for the slow adoption of wireless Internet services in the United States and their higher cost here than in other areas of the globe.

In the end, the development of the wireless web market and consumers' awareness of the benefits of surfing using a phone in some regions may be the biggest impediment to the growth and success of any of the software efforts.

BellSouth, which sells local phone service in the southeastern United States, plans to get out of the payphone business by the end of 2002, partly because the boom in wireless usage has sapped its sales (BellSouth 2001). The Atlanta-based company wants to concentrate on its high-speed Internet and digital networks, its wireless data and voice business, and its operations in Latin America. Payphone usage has declined as more people carry cellular phones and wireless pagers, BellSouth spokesman David Blumenthal said. In fact, about 40 percent of people in the United States have wireless phones.

I-mode cell phones have proven that companies can make money on wireless Internet access, and with their explosive popularity, the phones could one day rival the number of PCs on the Internet (Kanellos 2001). NTT DoCoMo's I-mode cellular service—which lets consumers receive e-mail and browse the Web—is gaining 50,000 new subscribers a day, or 1.5 million a month.

The I-mode has the potential to catch up with the PC Internet market. Of course, the current number of PC users dwarfs the number of I-mode subscribers. The I-mode growth rate, however, is far more rapid. The service began in February 1999 and has since garnered 18 million subscribers, all in Japan. International expansion is underway. Mobile Internet access is expected to be a huge market, with hundreds of companies targeting the emerging sector. With the exception of I-mode in Japan, however, U.S. wireless Internet access has been something of a bust thus far.

In mid-2001, Sprint PCS and Cingular Wireless began offering advanced services, giving customers faster access to the Internet and new features such as transmitting pictures to mobile phones (Sprint 2001). Sprint PCS, a unit of Sprint Corp., said it spent less than $1 billion to start 3G service in 2001 and early 2002. Cingular, a venture of SBC Communications and BellSouth, said it offers users instant, unlimited Internet access with voice calls in California, Nevada, and Washington.

The announcements come as analysts worry the United States is trailing other nations, such as Japan, in introducing advanced services. Companies complain about insufficient airwaves and untested service software. Sprint will upgrade its network again by 2003, and by early 2004, it will be able to transmit data at speeds faster than cable modems. Cingular, the second-ranking mobile-phone company, upgraded services in former BellSouth markets where it uses the GSM standard. Features are available in the Carolinas, coastal Georgia, and eastern Tennessee.

Cost issues and concerns abound as well. The average cost of monthly cellular telephone service in twenty-five major U.S. cities dropped less than 1 percent in March 2001, according to a survey by industry watchers Econ One. Prices dropped an average of 0.3 percent compared with February (Charny 2001). Two Florida cities showed the greatest percentage increases of the twenty-five surveyed. The price of a month of cell phone service in Miami rose 2.8 percent to $38.99. Tampa increased 1.8 percent to $38.76. San Francisco had the largest decrease—2 percent to $43.34. The five most expensive cities surveyed are San Francisco, Cincinnati, Los Angeles, Boston, and New York.

In summary, when the Internet burst on the business scene several years ago, several companies were criticized for "not getting it" and not moving their businesses to the Web fast enough. Just as we are gaining a better understanding of that dislocation, a new technology has emerged asking many of the same questions: the wireless Internet.

Wireless Internet and SMS

Unfortunately, in many ways, the wireless Web is more complex and confusing than its wired brother. Let us attempt to simplify this world in two phases. First we look at an overview of the current infrastructure from a global perspective. Then we examine five important issues that confront the business executive when thinking about a wireless strategy.

Perhaps the most discussed issue with regard to the wireless Internet is the fact that the United States trails both Japan and Europe in terms of innovation and progress. Many executives in the United States, particularly those involved directly with the wireless industry, will debate this issue, but in the end the numbers speak for themselves.

Leading the world is NTT DoCoMo, with its I-mode wireless data service in Japan. The number of I-mode subscribers is quickly approaching 10 million from its start in February 1999. In addition, the features and functionality of the company's phones far outweigh anything available in the United States. Europe is probably closer to the United States than to Japan in terms of progress, but is still somewhat ahead. The primary innovation in Europe is short message service (SMS), which is a cross-carrier cellular phone version of Instant Messenger. SMS users are expected to top 50 million by year's end, and the number of SMS messages per month exceeds 1 billion. In fact, SMS may well be the future in the United States as well in this arena, so it is appropriate to spend some time discussing it here.

SMS appeared on the wireless scene in Europe in 1991. It is a globally accepted wireless service that enables the transmission of alphanumeric messages between mobile subscribers and external systems such as electronic mail, paging, and voice mail systems (Buckingham 2000). The standard in Europe for digital wireless is now referred to as GSM. This standard included short messaging services from the beginning.

North America was introduced to SMS by digital wireless networks such as BellSouth Mobility, Prime Co., and Nextel. These networks are based on GSM, CDMA, and TDMA standards. Many of these smaller networks have consolidated their systems, forming large wireless networks that are used nationwide and even internationally. Many of these providers not only carry SMS but also other forms of wireless communications. SMS, based on an intelligent network (IN) approach, provides a uniform solution, enables ease of operation and administration, and accommodates existing subscriber capacity, message throughput, future growth, and services reliably (Buckingham 2000).

SMS messages must be no longer than 160 alphanumeric characters and contain no images or graphics (Webopedia 2002). Some phones allow 224 characters if using a 5-bit mode (Smith 2001). Once the message has been sent, a short message service center (SMSC) receives it and launches it to the correct wireless device. For an SMSC to send a message, it must send a message to the home location register (HLR) to find the roaming customer. Once the HLR has acknowledged the request, it responds to the SMSC with the subscribers' status (active or inactive) and where the subscriber is roaming.

If the HLR sends a response that says the subscriber is inactive, the SMSC holds the message for a period of time. Once the subscriber becomes active again, the HLR sends an SMS message notifying the SMSC to attempt the delivery of the message again (Webopedia 2002). The SMSC transfers the message in a short message delivery point-to-point format to the serving system. The system pages the device, and if the device responds, the message is sent and delivered to the intended recipient. The SMSC then receives a verification message that the end user received the short message. The SMSC then classifies the message as sent and will not try to send the message again (Ragaza 2000).

The number of cellular phone users has increased rapidly over the past 10 years. By 2003, the expected number of mobile phone users is 500 million worldwide. With the help of SMS, 75 percent of all cellular phones will be Internet allowable (Webopedia 2002). Figure 6–13 shows the rapid growth of SMS technology worldwide.

SMS has become predominantly popular with teenagers in Europe, but also is starting to catch on with teens in the United States. Many teenagers have developed their own distinctive shorthand to keep others, such as teachers and parents, from reading their messages. Some companies are offering this service for free. However, in areas where SMS is more popular and heavily used, companies are charging a small fee per message after a certain amount of messages have been used (Ragaza 2000). This fee is similar to most minute plans for cellular-voice use. This service also is similar to paging, but is more reliable, because SMS messages do not require the mobile phone to be active or within range, and it will be held for many days until the phone becomes active or within range. Most pagers do not offer this service.

The many uses of SMS, especially in the business sector, include the following (Smith 2001):

- Informing a mobile phone owner of a voice-mail message
- Notifying a salesperson of an inquiry and contact to call
- Alerting a doctor of a patient with an emergency problem

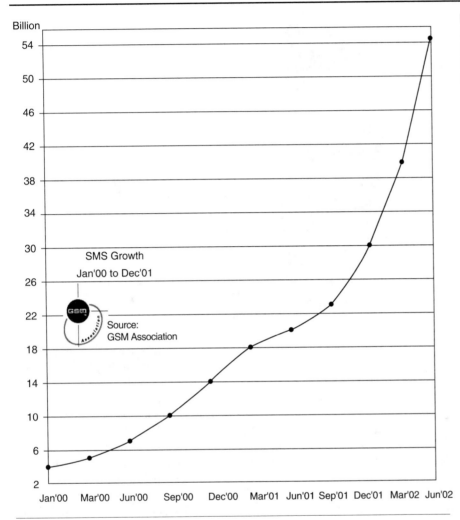

Billion

SMS Growth
Jan'00 to Dec'01

Source:
GSM Association

Jan'00 Mar'00 Jun'00 Sep'00 Dec'00 Mar'01 Jun'01 Sep'01 Dec'01 Mar'02 Jun'02

Figure 6–13
SMS Growth
(Source: GSM
Association,
www.gsmworld.com)

- Reporting to a service person the time and place of his or her next call
- Reporting needed during times of emergency
- Notifying a driver of the address of the next pickup

Cellular service providers that offer SMS also offer public SMS gateways, which allow one to compose and send messages from the service provider's website. Users need only a computer and an Internet connection to send a short message. A number of independently operated message gateways also exist on the Internet. Some gateways have more features than others. A few let you compose messages to more than one recipient, create group lists, manage messages, and send preset or customized replies (India Mobile 2002).

Figure 6–14
Nokia's SMS (A link to this screen can be found at Nokia's main website: www.nokia.zimismobile. com/nokiasms/ nokiasms.html)

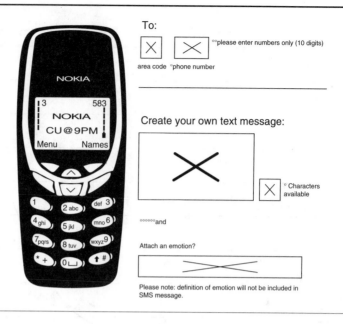

Users can send a message to a device on another network using a public gateway, and friends or business associates without SMS-equipped phones or pagers can send messages back. If the user does not have access to an e-mail account or an SMS-equipped phone, a public gateway (available from any computer with Internet access) is a convenient way to keep in contact. Figure 6–14 shows an example of Nokia's Mobile Messaging Service.

As you can see from the figure, the service is very easy and user friendly. Regrettably, one drawback to the use of public gateways is file security. Dissimilar to web transmissions that occur as secure transactions, any third party could monitor the cell phone number and message contents when a message is sent through a public gateway. Messages sent from phone to phone through SMSCs are more secure; however, the possibility does exist that others could also observe these transmissions (India Mobile 2002).

The most common devices used with SMS are digital cell phones equipped with one-way receive SMS capability. A number of digital pagers also have this feature. Analog cell phones, or cell phones that are several years old, are unlikely to be compatible with SMS.

Currently wireless service providers catering to the hand-held market do not offer SMS. These providers depend on networks that use a different system than SMS. Messages to and from hand-helds work in conjunction with standard e-mail, but images are often stripped out.

The greater part of SMS usage is accounted for by consumer applications. It is not uncommon to find 90 percent of a network operator's total SMS traffic being accounted for by the following consumer applications.

1. *Simple person-to-person messaging.* Cellular phone users use SMS to send messages via their cell phones, typically to arrange meeting times or just say hello. These messages are typed with the phone's number keypad, which is not appropriate for long conversations. This use of SMS is very popular with younger generations who are more willing to learn new technologies.

2. *Voice and fax mail notifications.* This is the most common use, mainly used for alerting phone users that they have new voice or fax messages waiting for them. Because SMS is routinely used for this particular function, this application is and will remain one of the largest generators of short messages.

3. *Unified messaging.* This service combines the different types of messages from various services into one message box. Whether it is fax, e-mail, or text messages, they all come into the same box. Once accessed, the SMS tells the user what type of message it is and then the user can process it however he or she wishes.

4. *Internet e-mail alerts.* This service is used by linking the SMS to the user's Internet e-mail box. Each time a new e-mail is received, a short message is sent to the user. This is a time-saving tool, because logging on to the Internet can be time consuming especially in an area with slow connection speeds.

5. *Ring tones.* Many cellular phones have approximately fifteen tones from which to choose. With SMS, different rings can be created and downloaded to the user's phone, so that the tone is unique from other users. Ring tones that are commercial tunes must be licensed before use.

6. *Chat.* This service is used the same way as chat services are used via the Internet; however, a drawback is that SMS chatting can become rather expensive. Some providers charge by how many individual times a message is sent, which can get pricey when having a conversation with another person.

7. *Information services.* This service allows users to view a wide range of information, such as stock prices, sports scores, weather readings, light information, news headlines, lottery results, jokes, and horoscopes. Any information that can be written in 160 characters or less can be considered for information services.

Currently, the corporate applications that use SMS are scarce, mainly due to the age of the corporate mobile phone users. Corporate users tend to prefer using voice as their primary communication method.

The various uses of SMS are operated on the same system. The basic network architecture is displayed in Figure 6–15. Communications with the HLR and the MSC occurs through the signal transfer point (STP).

Many different elements belong to the SMS network. First is an external short messaging entity (ESME). This may be part of a fixed network, a mobile device, or another service center. The voice message center (VMC) receives,

Figure 6–15
Network Architecture

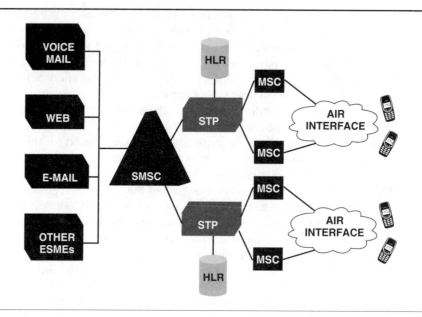

stores, and plays voice messages (Webopedia 2002). It also sends voice mail notifications to SMSC users. The SMSC is a combination of hardware and software that is responsible for the relaying, storing, and forwarding of a short message between an SME and mobile device (India Mobile 2002). The STP is a network element normally available on IN deployments that allows IS-41 interconnections over signaling system 7 (SS7) links with multiple network elements (Smith 2001).

The HLR is a database used for permanent storage and management of subscriptions and service profiles. The visitor location register (VLR) contains temporary information about users who are homed in one HLR but roaming in another HLR. The MSC performs the switching of calls and messages. The air interface is defined in each one of the different wireless technologies (GSM, TDMA, and CDMA). See Table 6–5. The base station system (BSS) houses all of the transmissions of electromagnetic radio signals between the MSC and the different mobile devices. The mobile device (MD) is the user's mobile device, or receiver (Webopedia 2002). A pictorial overview of the network infrastructure is shown in Figure 6–16.

As usual, whether SMS will remain as it is, grow, or die is up to the users of the system. The more SMS is used, the more widespread and popular it will become. The next few years should see this outcome, at least in the United States.

Japan is even more ahead in the cellular telephony arena than Europe and certainly the United States, because Japan was first to implement a

Interface	Modulation Type	Channel Bandwidth	Widely Adopted?	Comment
Omnipoint Composite TDMA/CDMA	CDMA between cells, TDMA within cells	Various	No	May be used for applications using unlicensed PCS frequencies
Upbanded TDMA (IS-136)	TDMA	30 kHz	Yes	Interoperability with existing cellular
Upbanded CDMA (IS-95)	CDMA	1.25 MHz	Yes	Interoperability with existing cellular
PCS-1800 (Upbanded GSM)	TDMA	200 kHz	Yes	Potential international roaming
PACS	TDMA/TDD	30 kHz	No	Not adopted by major carriers
Oki W-CDMA	CDMA	5 MHz	No	Not adopted by any carrier
DCT-1800	TDMA	Variable	No	Only for use in unlicensed PCS band

Table 6–5
Summary of BPCS Air Interface Standards

packet switched infrastructure (as opposed to circuit switched). In the United States and Europe, data transmission is accomplished by connecting via a normal phone call to a bank of modems that handle data transmission. The differential is almost identical to the difference between using a dial-up modem and connecting directly to the Internet via a corporate LAN. Packet-based networks are ideally suited for bursty data, or burstable packets of data, and result in a better "always-on" user experience and a lower cost.

Although Europe remains circuit switched, it still has advantages over the United States. First, the carriers in Europe have been far more accepting of common industry standards. The carriers use the same underlying GSM technology, which results in much better interoperability and allows for equipment innovations to be shared. This standardization philosophy also enables intercarrier functionality such as SMS. The United States still is without interoperable instant messaging on the wired Internet. The other primary European advantage is cell phone penetration, which is well above 50 percent in some countries.

Another interesting difference pertains to e-mail. In Europe and Japan, many people obtain their first personal e-mail addresses through their cellular phones. We all know that e-mail was one of the original "killer apps" of the Internet. (A killer app is an application that drives customer adoption.) Well, that same phenomenon is driving the acceptance of the wireless Internet overseas. The cell phone has become the primary e-mail device for most consumers.

If you read about wireless, you are likely to encounter the letter *G* which is used to describe different evolutions in wireless Internet technologies.

Figure 6–16
SMS Network
Infrastructure

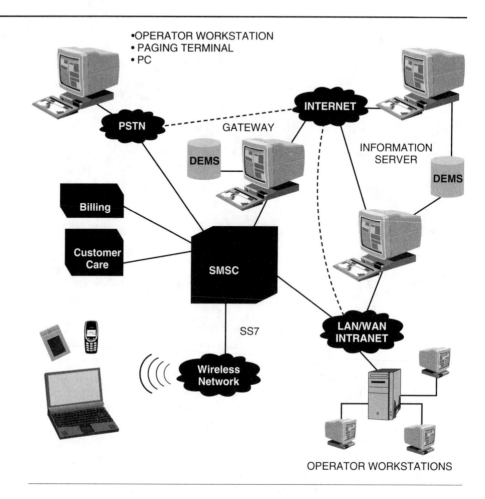

This can be much simpler than it sounds. 2G, or second generation, refers to today's technology. 2.5 G is an interim step to advanced cellular data networks. These advanced cellular data networks are called 3G.

When thinking about the wireless data infrastructure, keep these two things in mind. First, the move to packet switched is more important than raw speed; therefore, 2.5G is more important than 3G. (Watch out also for the phrase "up-to," because the first 2.5G implementations are well below 100 K.) Second, for all practical purposes, it looks as if Japan will extend its lead, and the United States will continue to follow. NTT DoCoMo is clearly the first to implement 3G and is well in front in terms of understanding business models and customer net present values (NPVs). The United States must overcome interoperability issues and gain help from the government, which has been slow to clean up spectrum issues that could hinder 3G rollout locally.

Most wireless Internet users in Europe and the United States access their wireless data through WAP, which is a set of standards that describes how a cell phone accesses data over the Internet. WAP supporters argue that a cell phone's latency, interface challenges, and low processing power require a separate set of technologies than those used on the Internet (HTTP and HTML, for example). The anti-WAP forces argue that these same three things will go away, particularly once we move to packet switched networks (2.5G+). Supporting this argument is the fact that I-mode, the only packet-based network operating today, relies on HTML, not WAP.

Another interesting question is which gadget will be the primary Internet access device for most people. The PC seems like the natural choice, primarily because of the large installed base, the richness of the experience, and the relatively high connectivity speeds. However, the installed base of cell phones is about twice that of PCs, and some argue that being "always with you" is a critical advantage. The PDA has a richer user interface than the phone, but the installed base numbers are much lower. One interesting question is, will the PDA become a phone, or will the phone become a PDA?

Everyone likes to talk about the killer app. You will see a lot of articles and press releases talking about corporate applications, contact information or "m-commerce," but the applications driving adoption in Japan are fundamentally consumer focused. Games, daily screensavers, daily rings, and instant messaging are the most popular apps. We should ignore the market's current distaste for business-to-consumer as we plan the future of the wireless Internet. This is a social medium more than anything else.

The best way to answer the question about PDA is to look at I-mode. NTT has been successful in charging people per-bit fees as well as subscriptions for extra services. One key to this has been using the billing relationship with the customer and using this billing infrastructure as a one-click payment alternative for content providers. In other words, they piggyback on DoCoMo's bill. Of course, on the broader Internet, competition has driven many services to free or near free, and the same could happen here. Many start-ups will point to advertising and e-commerce, but those streams will likely be limited to those with "portal power."

Portal power can be briefly defined as any company that has a large number of users and can use those relationships to extract rent from others who want to reach those customers. Everyone from cellular carriers and hardware manufacturers to portals such as America Online and Yahoo and new mobile portals are counting on having this position. If the carriers can overcome the mistakes of the internet service providers, they should be successful here, particularly if they use the billing relationship. Otherwise, the current portals are likely winners. Just remember this, in the words of the sci-fi thriller *Highlander,* "There can be only one."

It is surprising that even with the rise of the Internet as a proxy, many unanswered questions remain with regard to the wireless Web. The good news is that we get to watch NTT DoCoMo in action and learn from its model—a unique position for American companies.

The wireless Internet has obvious advantages. Mobile employees from doctors to salespeople are already tapping into data from the field via Internet-ready cell phones and PDAs. But wireless devices currently suffer from slow data transfer speeds, dropped connections, and limited interactivity. Before wireless computing becomes a business staple, these issues must be addressed, which leads us to 3G technology (Graven 2001).

As discussed, today's wireless service is delivered over 2G wireless systems. The three network standards—CDMA, GSM, and TDMA—are intended for voice services. Even under ideal conditions, these networks transfer data at only 19.2 Kbps, tops.

However, 3G will offer data transfer rates from 144 Kbps to 2 Mbps and an always-on connection. With increased bandwidth, 3G will support large multimedia files and applications, not just the basic apps available on today's wireless devices. You will be able to access multimedia-rich sites or even videoconference. Because 3G will move to two standards—WCDMA and cdma2000—roaming capabilities will improve. With fewer competing standards, service will be accessible nearly anywhere in the country.

One serious hurdle is the federal government. To achieve more bandwidth and speed, carriers need more spectrum (the radio frequencies that carry data). On October 13, 2000, President Clinton signed a memorandum mandating that the FCC allocate more spectrum for the future needs of wireless technology. Now the FCC and National Telecommunications and Information Administration (NTIA) are struggling to figure out how to distribute the limited amount of new spectrum that will be allocated for G3. The FCC currently plans to announce the 3G bands and auction off spectrum licenses by the end of 2002.

Cost is also a hurdle. Telecommunication carriers will have to spend billions of dollars. In the United Kingdom, they spent $32 billion recently on just five spectrum licenses, which was only a portion of the cost. Operators also must install cellular transmitters and radio towers. Most carriers do not yet know the cost of implementing 3G systems, says Forrester Research (Graven 2001).

Still, carriers are likely to embrace 3G, because their former cash cow—basic cellular—is becoming a low-cost commodity. Wireless services could be an increasingly important revenue stream for these companies. "The motivation for building 3G networks is that operators want a way to differentiate themselves and to have another revenue source," says Michael McMahan, Texas Instruments fellow, director of research and development, and wireless communications.

Wireless usage in the United States will nearly double by 2006 from 2001 levels as more and more consumers use their cell phones to make calls that they previously made from traditional telephones, a study said.

According to a study by the Yankee Group, wireless subscribers are expected to increase their monthly minutes of use to 641 by 2006 from 356 in 2001 and 109 in 1994.

"Although only 3 percent of U.S. consumers use their mobiles as their only phone. . . 26 percent of mobile users' minutes are already being displaced from wireline to wireless and 45 percent of mobile users indicated at least some substitution," the research firm said in a study, citing a survey it conducted.

Wireless analyst Keith Mallinson said in a report that mobile phones were perceived by many to be cheaper, more indispensable, more permanent, and more accessible as wireless service prices fall to levels similar to traditional telephone service.

Users also see wireless phones as a more reliable communication tool to reach someone regardless of the time of day or his or her location.

These findings support evidence already seen by traditional telephone companies, who have reported a decline in access lines as consumers choose cell phone service over second or third lines in the home.

Mallinson said he sees strong market demand for wireless services, with subscribers increasing to nearly 200 million by 2006 as 70 percent of Americans come to own cell phones. Nearly 50 percent of Americans own cell phones currently (Reuters 2002).

■ SUMMARY

Chapter 6 explains how cellular telephony evolved. It provides the important dates and events that helped shape cellular telephony, as well as the basic concepts of both analog and digital networking. You learned how a typical cell works, and how the immediate future of cellular is shaping up with the introduction of 3G technology. The text also presents the recent events surrounding cellular telephony.

In the next chapter, we explore the uses of wireless in the public services arena. This arena has been in the forefront of wireless for many years, from deploying dispatch systems for emergency personnel to the use of GPS technology for locating missing persons.

REVIEW QUESTIONS

1. The function of a repeater is to _____.
   a. extend a network
   b. increase the transmitted power of a network
   c. merge cells in a cellular network

2. An MSC will _____.
   a. control a cell's activities
   b. extend a network
   c. cover temporary cellular service for a special event

3. AMPS is a _____ technology.
   a. digital
   b. analog
   c. satellite-based

4. The mobile telephone is technically a _____ transceiver.
   a. half-duplex
   b. unidirectional
   c. full-duplex

5. The number of channels required for a given traffic load is measured in _____.
   a. decibels
   b. erlangs
   c. furlongs

6. Cellular telephony's beginnings started before _____.
   a. World War I
   b. World War II
   c. Vietnam War

7. Cell splitting is necessary due to _____.
   a. number of subscribers
   b. topographical features
   c. COLTS

8. The first proposal for a large-capacity mobile telephone system was put forward by Bell engineers in _____.

   a. 1955
   b. 1909
   c. 1947

9. The capture effect is _____.
   a. uniquely an FM feature
   b. only available on analog systems
   c. not present on cellular systems, only microwave systems

10. By reducing the coverage areas and creating a large number of small cells, it became possible (in theory) to _____ the same frequencies in different cells.
    a. reduce
    b. reuse
    c. eliminate

11. Microcells are deployed in a network for a number of reasons such as _____.
    a. increasing system capacity, improving signal reception, and providing inconspicuous coverage
    b. reducing handoffs and relieving traffic congestion at certain points or time of day
    c. economics
    d. all of the above

12. It is possible in a _____ system for the transmitter to provide a great deal of information about its own functioning and condition on a more or less continuous basis.
    a. digital
    b. analog
    c. AMPS

13. GSM is the most agreed-upon _____ cellular telephony standard.
    a. Japanese
    b. European
    c. U.S.

14. Analog cellular systems are _____ than digital ones.
a. less secure
b. more secure

15. The MTSO is also known as a _____.
a. PSTN
b. COW
c. MSC

16. The base station is at the center of a _____.
a. cell
b. cell split
c. antenna

17. By _____ cell sizes, more and more circuits could be created.
a. increasing
b. reducing
c. doubling

18. COLTS are _____.
a. temporary cell sites used during natural disasters

b. temporary cell sites used for special events
c. housed in Indianapolis

19. IMTS, an early cellular system, was created in the _____.
a. 1940s
b. 1950s
c. 1960s

20. The cost of _____ still determines which applications are suitable for digital techniques.
a. electrical power
b. digital circuitry
c. precise antennas

21. 3G is _____.
a. the other name for a base station in a macrocell
b. the next technology in the cellular world
c. propagation effects

HANDS-ON EXERCISES

1. Using your browser and a search engine such as Google, conduct a quick research activity and find out information about PCS. Write a two-page paper describing PCS.

2. AMPS is the older analog cellular standard. Conduct research to determine who still uses this standard.

3. Using your browser, go to www.whatis.com and enter "amplitude modulation" and "frequency modulation". Follow the links found there. Can you analyze the differences between the two?

4. Using your browser, conduct some research on the Web. Can you find articles/references that describe the PSTN → MTSO linkage?

5. Using drawing software such as Photopaint, draw a typical cellular telephony network. Include a linkage to the PSTN. Show all MSCs. Also show a person driving in a car and how the handoff process would work.

6. Conduct research to find out what cellular providers are available in your area. Create a chart listing these with their respective service plans and coverage areas, along with any extras such as free nighttime minutes, and so forth.

7 Public Services

In the previous chapter we discussed cellular telephony. In this chapter, we address a cousin, namely wireless technology in the public service arena such as fire and emergency services.

OBJECTIVES

After reading this chapter and completing the exercises, you will be able to:

- Learn the importance of public wireless-based services
- Understand the technologies used in this arena
- Learn what types of wireless devices are used in this arena
- Become aware of how public wireless-based services have proved beneficial

WIRELESS EMERGENCY SERVICES—PANICPENDANT.COM

PanicPendant.com is a company owned by Boss Security Systems, which is based in Leonia, New Jersey. PanicPendant provides wireless emergency notification services for individuals who may be homebound or in some other way incapacitated. The cornerstone of the company's services, the one button panic pendant is used to wirelessly notify someone in case of an emergency. These devices transmit a signal to the in-home receiver, which in turn sends the message on to the central monitoring station. When the signal is received there, a representative assesses the situation and notifies the proper authorities. This service is just on the edge of explosive growth with a great amount of our population aging and feeling safer with a service like PanicPendant. The key to this service though is the wireless capability it has. If an elderly person falls and breaks a hip and cannot get to the phone, he or she may be stuck there for days, but with a wireless service like PanicPendant that same person could activate the wrist device and obtain help. Another neat device that PanicPendant has integrated into this wireless system is a wireless-connected smoke detector. The smoke detector is installed and used like any other detector, except that when it sounds because of a fire, it also sends a wireless signal to the in-home receiver. The in-home receiver forwards the alarm to the nearest fire station and help is on the way. These devices seem very simple, but they are a great use of wireless technology and could definitely save many lives.

Source: www.panicpendant.com

MOBITEX—ERICSSON WIRELESS EMERGENCY SERVICES

Gothenburg, Sweden, is considered one of the best places to have a heart attack—if that's a good thing—because their ambulances are equipped with special mobile data communications. These devices allow doctors at the hospital to begin the process of treating and diagnosing the patient while the patient is still being transported by the ambulance.

Two ambulance or emergency vehicles in Gothenburg carry this special equipment intended for accidents and heart attacks. EMTs place the patient in the ambulance and attach electrodes to the patient's chest. They can then press a SEND button on the PC, which starts the transmission of a complete twelve-channel EKG to the hospital's coronary intensive care unit. The mobile data communication (Mobimed) system also alerts personnel at the hospital via their minicall pagers. This technology allows the doctor to examine the patient's EKG on a monitor and receive a print out. The doctor can the determine the next step in treatment.

Thanks to Ericcson, Gothenburg Eastern Hospital is the first, but hopefully not the last, to introduce mobile medical lifesaving technology for cardiac patients and accident victims.

Source: www.mobitex.com

INTEROPERABILITY IN COMMUNICATIONS

When a family is trapped in the fiery wreckage of an automobile accident, the seconds it takes to respond are measured in lives. Local, county, and state police officers rush to the scene. Nearby fire and rescue personnel are quickly dispatched to aid in the rescue efforts. Emergency medical technicians care for the injured en route to local hospitals.

None of these public safety agencies works in isolation. Their joint response is key to a successful rescue. In fact, the ability of the public safety community to provide a coordinated response to criminal activities, fires, medical emergencies, or natural disasters can mean the difference between life and death.

To provide immediate and coordinated assistance, the nation's public safety workers must communicate with each other effectively, swiftly, and securely. In the mobile environment where public safety personnel do their work, radio communication is the lifeline. Without it, both life and property are put at significant risk.

interoperability

To be successful, public services must have interoperable communications. What is **interoperability**? It is the ability of public safety personnel to communicate by radio with staff from other agencies, on demand and in real time. Public safety agencies require three distinct types of interoperability—day to day, mutual aid, and task force.

Day-to-day interoperability involves coordination during routine public safety operations. Interoperability is required, for example, for county firefighters to join forces to battle a structural fire or for neighboring law enforcement agencies to work together during a vehicular chase.

Mutual aid interoperability involves a joint and immediate response to catastrophic accidents or natural disasters and requires tactical communications among numerous groups of public safety personnel. Airplane crashes, bombings, forest fires, earthquakes, and hurricanes are all examples of mutual aid events.

Task force interoperability involves local, state, and federal agencies coming together for an extended period of time to address a public safety problem. Task forces lead the extended recovery operations for major disasters, provide security for major events, and conduct operations in response to prolonged criminal activity.

Two 1998 surveys of more than 2,000 public safety agencies document major interoperability problems (Public Safety Wireless Network 2001). The law enforcement, fire, and emergency medical service agencies that were surveyed rated radio wave spectrum and funding limitations as their biggest obstacles to interoperability. They identified incompatible technologies and the lack of adequate systems planning as additional obstacles.

Public safety radio wave spectrum refers to the array of channels, like those on a television, available for communications transmissions. These

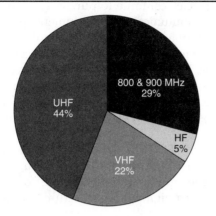

Figure 7–1
Division of Frequency
Use in the Public
Services Arena

channels are a finite natural resource, meaning they cannot be created or discovered. In many communities, not enough spectrum is available for public safety use in general; even less is available for interoperability purposes. Scarce spectrum results in congested radio channels and increased interference, limiting the ability of public safety personnel to communicate. Additional spectrum is needed to meet current communication needs and to support the deployment of new technologies (FCC 2001).

The current public safety channels are located in several portions of the radio spectrum, as Figure 7–1 demonstrates, resulting in separate spectrum islands that isolate public safety operations and jurisdictions. This fragmentation of spectrum impedes interoperability and joint public safety operations. Because no single radio can span all of the public safety channels, agencies using different portions of spectrum cannot talk with each other. Agencies often resort to using multiple radios or other ad hoc means of linking communications.

Many existing public safety communications systems cannot support modern technologies that are needed for interoperability (Public Safety Wireless Network 2001). Replacement of outdated systems or system expansions is expensive. Inadequate funding for upgrades often prevents a public safety agency from purchasing the technology and equipment that can enhance interoperability and improve organizational effectiveness. To obtain the necessary funding, public safety agencies must convince public officials and concerned citizens of the critical need for modern radio communications.

A variety of new radio technologies are becoming increasingly popular as agencies plan to replace or upgrade their existing systems. Despite these new technologies, competing equipment vendors continue to manufacture, and public safety agencies continue to purchase, equipment that is not interoperable. Radio communications equipment produced by multiple vendors uses proprietary and incompatible technology schemes. These incompatibilities prevent interoperability even when the radios operate in the same spectrum bands. Without technical standards, vendors are producing "closed systems" that create significant barriers to interoperability for the public safety community. Industry and the public safety community must work together to foster the development of standards and compatible equipment.

In January 1998, the FCC reallocated an additional 24 MHz of spectrum for public safety use. A portion of that spectrum has been designated to support nationwide interoperability among local, state, and federal agencies. The process of assigning this spectrum to public safety is moving forward.

On February 14, 2002, public safety communications took a major step forward. During a full meeting, the FCC unanimously voted to allocate 50 megahertz (MHz) of spectrum to public safety in the 4.9 gigahertz (GHz) band. The Commission also adopted conservative rules to protect public safety communications from interference caused by emerging ultra wideband (UWB) services. These two decisions are expected to usher in a new era of growth and improvement for communications by improving both the quality and range of services available to the public safety community.

Recognizing the critical need for additional spectrum to meet established domestic security priorities, the Commission ordered the 50 MHz spectrum allocation for public safety use on WT Docket 00-32. It is the largest single public safety spectrum allocation in history. The allocation is specifically designed to support broadband data and video services, and will require a significant rulemaking period to develop rules and standards for the new band.

Also during 1998, the Attorney General convened an interagency working group to develop a program that would provide federal seed money to the states to plan statewide public safety wireless communications and provide money for demonstration projects. President Clinton's fiscal year 2000 budget contained $80 million for this program and the Bush Administration has doubled that amount. The standardization efforts are making incremental progress on addressing difficult issues such as sharing proprietary information and developing standards that improve interoperability and ensure competition among multiple vendors. The **Public Safety Wireless Network (PSWN)** program, a joint initiative of the Department of Justice and the Treasury Department, fosters communications interoperability among local, state, and federal public safety agencies. The PSWN also receives guidance from the Federal Law Enforcement Wireless Users Group and an executive committee of prominent local and state public safety officials.

Public Safety Wireless Network (PSWN)

Since its inception, these groups have promoted partnerships among public safety agencies, conducted case studies in several regions of the nation, initiated pilot projects to test and refine interoperability solutions, addressed spectrum policy and funding issues important to public safety, and investigated issues associated with system security as well as standards and technology development.

Their plan provides the vision, mission, and strategic goals and objectives for the program. It describes their strategies for meeting these goals, and it depicts how these strategies interact to determine annual activities.

Shortly after the September 11, 2001, terrorist attacks in New York, a man trapped where a courtyard previously existed between the twin World Trade Center towers was rescued after he called for help on his mobile phone.

To locate more victims in the WTC, Lucent Technologies and Verizon Wireless used directional antennas to pinpoint the location of cell phones in the crumbled buildings (Batista 2001).

In the early hours of the tragedy, there were reports from concerned citizens that cellular telephony calls were being heard from within the rubble. It is immediately determined that an intensive effort was needed to locate any survivors. But, how to locate them? Well, cutting-edge cellular telephony locator equipment from Lucent Technologies was immediately shipped in and used to locate any turned-on wireless devices in the area.

Employees from several wireless firms in the area would then use this equipment to isolate cell phone signals from still-turned-on cell phones (Werner 2001).

VECC primarily is funded by a 9-1-1 surcharge on telephone bills that pays for calltaking by a user's charge allocated on dispatch usage. The center began operating at the same budget levels as existing communications allocations. It has not required a budget increase since 1989, except for radio equipment purchases. The actual price of a call has declined while call volume has significantly increased and has been handled without requiring additional staff.

Planning for the center took 9 months from start to initial request for proposal (RFP). The planning was conducted by an evaluation team composed of technical employees from each agency. Initially, the center operated using telephone lines and existing radio transmission sites. Shortly after the first 18 months of operation, VECC pooled existing radio frequencies and was able to justify an upgrade from simplex to repeater operations. A second RFP then was released for radio and microwave equipment purchase. This expansion permitted VECC to provide greatly enhanced service to the community while eliminating the possibility of landline problems that could result from a natural disaster or other disruption.

To accommodate microwave transmitters, receivers, and VHF and UHF repeaters, a 160-foot communications tower was erected about 12 miles west of the center at a leased site in the Oquirrh Mountain range. Another site already in service a few blocks southwest from VECC headquarters also was utilized. This site is on an unused industrial smokestack that rises 500 feet near the center of Salt Lake County. It offers a line-of-sight view of the entire valley, which is an important and required component for line-of-sight microwave service.

Additional microwave relays and radio repeaters are at Sandy City, a municipality about 6 miles southeast of the center; a Utah Highway Patrol station adjacent to the Utah State Prison on the border of Salt Lake and Utah Counties (the southern boundary of the center's service area); and a western valley site on the American Stores headquarters building near the Salt Lake International Airport.

VECC employs sixty-two people, ten of whom are supervisors and clerical workers and fifty-two of whom are dispatchers. They rotate among various shifts scheduled to serve peak traffic times. The key to maximum efficiency is the computer aided dispatch (CAD) system that provides front-end central processing unit (CPU) control of the communications consoles. This CAD system incorporates Motorola CentraCom II console electronics and IBM AS400 computers running Chiefs software. US West, a regional Bell operating company, provides enhanced 9-1-1 (E9-1-1) service that uses AT&T controllers. With E9-1-1 service, video monitors display callers' names and addresses based on telephone company records. Calls are routed via a system 75 private branch exchange (PBX) connected to V-band operator sets. Dictaphone recorders handle logging functions; call checks are handled by Veri-Trac recorder.

Control room

Equipment components are housed in custom-built consoles that accommodate nine operator positions, six calltakers, and a supervisor station. The communications control room incorporates raised computer flooring to hold miles of interconnecting cable. The ability to coordinate effectively the communications traffic for six municipalities and the county has improved emergency response time greatly.

For example, the first major event was in 1991, when the center actively participated in the nationally covered Alta View Hospital incident in Sandy, Utah, during which a gunman held several medical workers and patients hostage for most of a day. Because the center had sophisticated resources available, operators were able to keep the original caller on the line for more than 6 hours and maintain open communications directly with the hospital. Backup special weapons and attack team (SWAT) and other law enforcement team members in Sandy were completely occupied with the incident, as were teams from several other agencies. VECC dispatchers coordinated all call traffic with the various agencies participating on and off the site. The center has assisted in thousands of rescue and police operations since.

Paramedic Assistance Telemetry System

VECC carries the Paramedic Assistance Telemetry System (PATS), which provides switching for twelve medical receivers throughout the Salt Lake Valley to fourteen hospitals. PATS links paramedics in the field with a particular hospital and allows them to transmit specific hospital employees.

Another important medical communications service routed through the center is the statewide hospital emergency alert radio (HEAR). HEAR is carried via the Utah state microwave network. It is accessed by VECC through an interface at the Utah Highway Patrol relay site. HEAR carries voice and medical telemetry data to destination hospitals from ambulances and helicopters that transport patients anywhere in the state. The center maintains close ties with the county government and its paramedic services and municipalities, and their fire and emergency medical services.

Automatic aid

The center's radio system provides immediate communications among communities. An automatic aid agreement allows medical units to respond to destinations outside their jurisdictions when logistics indicate they can answer calls faster than other units. This arrangement has shortened the time it takes to respond to medical and fire-related emergencies.

Since the Loma Prieta earthquake in San Francisco in 1989, communities everywhere are more disaster conscious. The Salt Lake County government is no exception. A geographical fault threatens the entire valley. Although minor tremors are common, the area has not been subjected to a large earthquake in centuries. Thus, Salt Lake County is considered at high risk for such a phenomenon. In the face of the earthquake risk, VECC has focused much attention on disaster preparedness. The facility uses earthquake protection features and fire safeguards such as double-wall constructions and reinforced steel beams. When the communications tower at Oquirrh Mountain was erected, the tower specification was upgraded to replace the original design with a tower built to meet the earthquake code.

Three-site capabilities

Transmitters for law enforcement, fire, and medical emergency agencies are at three sites to ensure the communication system's continued operation should any of the transmitter sites be disabled. VECC facilities incorporate provisions for disaster communications networking between the Salt Lake County and Salt Lake City government agencies. The two jurisdictions include a population of more than 762,000 people. The center's long-term plans focus on further streamlining disaster communications handling and mass evacuation procedures.

New projects at VECC include establishing a records computer networked to allow participating agencies to pool information and to access the system for comparative data. The existing computer network allowed data to be shared with outside sources, such as federal agencies and law enforcement agencies in adjacent states.

Base stations are authorized to use the mobile transmit frequencies for access to fixed stations (e.g., repeaters). Mobile stations are permitted to use fixed station transmit frequencies for talk-around communications. The following restrictions apply.

- The minimum ERP necessary to support the intended use shall be employed.
- The maximum base or mobile transmitter output power shall not exceed 125 Watts.
- The gain of the base station antenna shall not exceed 6 dBi.

ASPECTS FOR THE LAW ENFORCEMENT COMMUNITY

Frequencies 167.0875 MHz and 414.0375 MHz are designated as national calling channels for initial contact and will be identified in the radio. Initial contact communications will be established using analog FM emission. The agency in control of the incident will assign specific operational channels as required for incident support operations.

The interoperability frequencies will be identified in mobile and portable radios with continuous tone-controlled squelch systems (CTCSS) frequency 167.9 Hz and/or network access code (NAC). CTCSS was invented by Motorola and involves developing a way to get more than one land mobile customer on the same frequency at almost the same time.[1] The theory is that different customers can coexist on the same frequency if they do not have to listen to each other routinely. CTCSS is simply a plug-in circuit board or chip. CTCSS can be added to virtually any transceiver and is smaller than a postage stamp. The system is based on a subaudible tone injected or encoded into the transmitter, and the tone is detected in the receiver. The decoder switch is then used to perform some function, usually to unmute the receiver when the tone is decoded.

In the commercial equipment, the audio bandwidth tends to be narrower than amateur equipment, and there are circuits installed to filter out the tones so they are truly subaudible. Most amateur equipment transmits and receives a much broader audio bandwidth and has no special tone filters, so most hear the tones. The lower the tone frequency, the less audible it tends to be.

---

[1] For additional information, go to www.eecis.udel.edu/~dra/pl.html.

The system is designed around a set of relatively low-frequency tones ranging from 67.0 Hz to 250.3 Hz. In the receiver, the tone is detected immediately before any audio processing takes place and then decoded allowing the receiver to unmute. Commercial radios filter out the tone, but amateur radios do not, so the tone is usually noticeable. It is sometimes mistaken for a power supply hum.

It should be noted that CTCSS does not alleviate RF interference. If two FM signals are on the same frequency at the same time, there will still be a **heterodyne** or beat note (unless one is 6 dB stronger than the other). A heterodyne is a note of two different tones, that is, dum dee. . . dum. . . dee. If CTCSS is being utilized and both systems use different CTCSS tones, however, they will not have to listen to the other system's traffic. A heterodyne is produced via the process, in a radio receiver, of combining a received signal with a locally generated signal of slightly different carrier frequency. The two combining frequencies produce a heterodyne.

heterodyne

With the advent of commercial repeater stations, several customers can use the same repeater without listening to each other's transmissions. In a commercial installation, the microphone hanger is grounded and when the microphone is hung up, the decoder is turned on, thus muting the receiver. When the operator picks up the microphone, the decoder is disabled and the receiver becomes "carrier squelch," hearing everything within range. If nothing is heard, the call is made. If another user is heard, the caller is supposed to monitor until the traffic clears and then make the call. Base station microphones have a monitor button next to the PTT button to disable the decoder, allowing the operator to check for traffic. With carrier squelch, the receiver repeater responds to anything on the listening frequency, including intended signals, unintended signals, and strong static. This can make the repeater noisy and annoying to monitor during idle periods. It is possible that a user, even with emergency or priority traffic, will be unable to get anybody to respond, particularly during noisy periods, because the noise interferes with other activities while monitoring.

Amateur radios do not have this automatic feature because the CTCSS system allows users to restrict what they want to listen to, not to allow several fleets of radios to operate on the same frequency. Most hand-held radios that can be factory equipped for full CTCSS encode and decode have a monitor button, usually around the PTT bar. Unfortunately, amateur mobiles have to manually turn off the tone to monitor the channel in the carrier squelch mode.

INCIDENT RESPONSE PLANS

The frequencies are reserved for public service offices to use in cases of incidents. Frequencies 169.5375 MHz, paired with 164.7125 MHz, and 410.2375 MHz, paired with 419.2375 MHz, are designated as the calling channels for initial contact. Initial contact will be established using analog FM emission (11KF3E). CTCSS will not be used on the calling channels to ensure access by stations from outside the normal area of operation. The agency in control of the incident will assign specific operational channels as required for incident support operations.

DESIGN AND PRINCIPLES OF OPERATION

There are several types of systems used in the public services arena. Many are discussed in this section.

Conventional Land Mobile Radio (Two-Way)

two-way radio systems

Two-way radio systems dedicate a single radio channel to a particular group of users who then share the channel; for example, a trucking firm dispatches all its trucks using this system—all truckers hear all the transmissions. It is the typical broadcasting scenario experienced in the Ethernet environment. Of course, anyone else can hear all the conversation once they know the channel being used. The problem with this system is that users must wait to see if anyone else is speaking before they can use the channel. Typically, conventional systems service fifty users per channel. As an example, if you have a ten-channel system, usually around 500 users can be adequately serviced. Figure 7–2 shows this system.

Trunked Radio

trunked radio

Trunked radio allows for the automatic sharing of multiple radio channels. What this means is that a group of channels is essentially assigned to a group of users, then they share the channels. When a user attempts to make a call, the trunked system actively searches for an available channel and then assigns it to the call. The user, then, has complete use of the channel with no one else listening. Obviously, this enhances security. As the day progresses, a different radio channel may be assigned each time a user makes a call. Users are not even aware of this occurring, and why should they be? As long as the call gets made, then the user is content. If the user attempts to make a call and no channels are available, then the system produces a busy signal. Trunked radio systems typically serve a user base of 100 times the number of channels. Figure 7–3 shows this system.

Commercial Trunked Radio Systems

Commercial trunked radio systems consist of wireless radio communications to achieve two-way mobile radio communications. They provide one-to-many

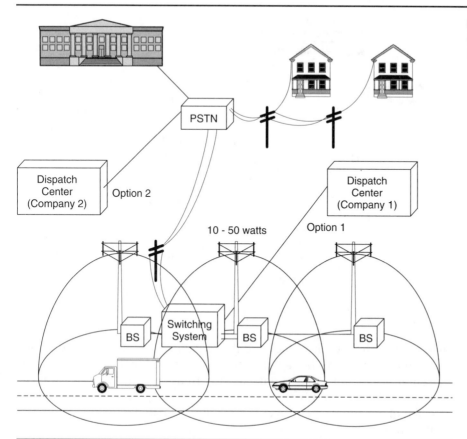

Figure 7–2
Conventional Land
Mobile Radio System

and many-to-one mobile wireless voice communications services. These are often called mobile dispatch services. For example, a commercial trunked radio system can communicate with all units in a mobile fleet or direct transmissions to a single radio or subgroup of radios. Many of these systems also offer integrated services such as voice mail and data and fax capability. Commercial trunked radio systems are typically operated by service providers who supply and then resell their services to other organizations. This differs from trunked radio systems, which are usually owned and operated by government agencies or private companies who use it for their own internal use.

Cellular Telephony and Public Service

In most areas of North America, citizens have basic or enhanced 9-1-1 service from their landline, or wireline, phones in their homes or workplaces. Basic 9-1-1 means that when the three-digit number is dialed, a calltaker/dispatcher in the local public safety answering point (PSAP), or 9-1-1 center,

Figure 7-3
Trunked radio system

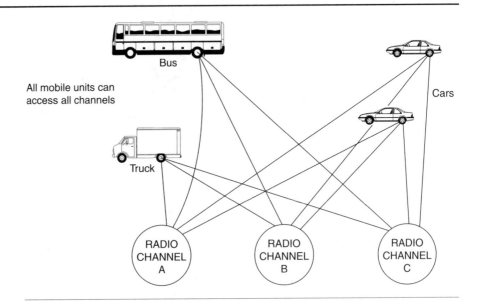

Bus

All mobile units can
access all channels

Cars

Truck

RADIO
CHANNEL
A

RADIO
CHANNEL
B

RADIO
CHANNEL
C

answers the call. The emergency and its location are communicated by voice between the caller and the calltaker. In areas serviced by E9-1-1, the local 9-1-1 center has equipment and database information that allow the calltaker to see the caller's phone number and address on a display. This enables rapid dispatch of emergency help, even if the caller is unable to communicate the location or nature of the emergency.

When 9-1-1 calls are made from wireless phones, however, the call may not be routed to the PSAP, and the calltaker may not automatically receive the callback phone number or the location of the caller. This presents life-threatening problems due to lost response time, if callers are unable to speak, do not know their location, do not know their wireless phone callback number, or the call is dropped.

There are three phases involved in implementing wireless 9-1-1. In wireless phase 0, the calltaker may be at a state highway patrol PSAP, a city or county PSAP hundreds of miles away, or a local PSAP, depending on how the wireless 9-1-1 call is routed. Wireless phase I is the first step in providing better emergency response service to wireless 9-1-1 callers. When phase I has been implemented, a wireless 9-1-1 call will come into the PSAP with the wireless phone callback number automatically displayed. This is important in the event the cell phone call is dropped, and may even allow PSAP employees to work with the wireless company to identify the wireless subscriber. However, phase I still does not help calltakers locate emergency victims or callers. Phase II must have been implemented in the area by local 9-1-1 systems and wireless carriers to locate wireless 9-1-1 callers. Phase II allows calltakers to receive both the caller's wireless phone number and location information.

It is estimated that of the 150 million calls that were made to 9-1-1 in 2000, 30 percent of them were made by wireless telephone users. This is a tenfold increase from nearly 4.3 million wireless 9-1-1 calls just 10 years ago, and the number will more than double to 100 million calls in the next few years. It is anticipated that by 2005, the majority of 9-1-1 calls will be from wireless callers (FCC, 2001).

Traditional Land Mobile Radio Systems

Paging

A paging system with dial-in capability consists of a terminal and a radio distribution network. The terminal, which contains computer logic, answers the incoming call and, based on the telephone number dialed, matches that number to the corresponding pager address stored in its own memory. If the service is other than tone only, the terminal prompts the caller for additional information (e.g., a voice message if it is a tone-voice service, a telephone number if it is a numeric display service, or free-form text for an alphanumeric display service).

In its simplest form, the radio distribution network may consist of a single transmitter and associated antenna mounted on a tower, building, or mountain (or a combination of these). In this simple configuration, the terminal merely puts the message into the correct format or protocol for transmission and conveys it to the transmitter where it is broadcast over the coverage area. The pager address may be broadcast as a sequence of bits (1s and 0s) using digital modulation. The address is followed by the message itself if it is other than the tone-only service. The paging receivers (pagers) in the coverage area listen for their address and, if they "hear" their unique address, they are activated and the trailing message is delivered. As explained in more detail below, a modern paging system may employ a network of multiple transmitter locations allowing an extension of the geographic service area.

Single Frequency Dispatch

The simplest type of two-way dispatch system uses a single frequency slot (i.e., an unpaired channel) for both transmitting and receiving. The parties at each end of the conversation (e.g., a dispatcher and a mobile unit in the field) take turns talking. This is sometimes referred to as the push-to-talk/release-to-listen (PTT/RTL) mode. In this mode each unit must push a button to talk and release the button to listen. The base station consists of a simple transmitter/receiver combination (a transceiver), an antenna mounted on the building or tower, a transmission or feedline (normally coaxial cable) for connecting the transceiver to the antenna, and perhaps some accessories such as a dispatch console. These simple dispatch systems typically operate in the analog mode and employ FM.

These conventional, single frequency dispatch systems are still used extensively at both low band and high band. Three fundamental disadvantages of such systems are (1) they mix higher power base stations and lower power mobile units in the same frequency slot, which exacerbates interference problems when the channel is shared among multiple systems, as is often the case; (2) they do not permit the use of repeaters (as described below); and (3) they do not allow for full-duplex operation.

Two-Frequency Dispatch/Community Repeater

As explained earlier, over the years, as low band and high band became more congested, the FCC regularly allocated (or reallocated) additional spectrum at higher frequencies for use by the land mobile radio services. This meant that a simple, single frequency dispatch system with the antenna on the user's premises often did not provide adequate range, especially for communicating with much lower power portable radios that were gaining in popularity (higher frequencies mean shorter range unless antenna dimensions are changed). This led the FCC to organize the bands to permit the assignment of paired frequency slots for each channel. Thus, other than for one-way paging, paired frequency assignments are available for dispatch systems in the UHF and 800/900 MHz radio bands.

This arrangement allows the higher power stations to transmit on one frequency of the pair, and the lower power mobile units on the other, thus eliminating the power disparity and reducing interference. It also allows the deployment of mobile relay or community repeater systems. In community repeater systems, a relatively high power transceiver and associated antenna are placed on a tall tower, building, or mountain top that offers good coverage over a wide area. The mobile units transmit on one frequency, which is picked up by the receiver in the unit on the mountain/ building top and then simply retransmitted or repeated on the other frequency of the pair. This special type of transceiver with the receiver and transmitter hooked back-to-back is called a repeater. With such a system, the dispatcher can communicate with mobile units via the repeater using a low-power transceiver located on the premises. The latter is known as a control station. Thus, the repeater relays all of the low-power signals, enabling both control stations and mobile units to communicate with one another over a wide area.

These systems typically operate in the PTT/RTL mode, although full-duplex operation is possible when communicating to and from the repeater (e.g., when telephone interconnection is provided at the repeater site). In addition to wide coverage, repeater systems offer other advantages. For example, because a repeater simply retransmits what it "hears," it is very easy for several customers with different fleets and their own control stations to share the use of a single repeater. This means that individual users with a small number of mobile units do not have to construct their own dedicated, high-power/high-antenna site facilities. The systems also allow the use of low-power mobile and portable units with corresponding reductions in cost. Like the single frequency dispatch systems, these two-frequency repeater systems operate in the analog mode and employ FM.

Disadvantages of Single Frequency and Paired Frequency Repeater Systems

The two types of two-way dispatch systems discussed thus far exhibit the following limitations.

Undisciplined Access to a Shared Channel.   As explained earlier, there are not enough channels available to allow each user a dedicated channel. Thus, except in the case of the very largest users, the channels must be shared among a number of such users. For example, a particular UHF channel might be shared among five companies, each with a control station and five, ten, or even more mobile/portable units each. Because each individual user may not know when another user is going to transmit (i.e., push the PTT button), users will occasionally interfere with one another. As congestion grows, the situation gets worse. There may be several users, each waiting to transmit and each unaware of the others. Moreover, some less polite users may get impatient, and transmit even if someone else is on the channel. Thus, the inevitable consequence is delays, interference, and lost messages, especially when there are unaffiliated users on the channel. In short, access to the channel is not disciplined.

Limited Addressing Capabilities and Lack of Privacy.   Both of the described systems operate as giant party lines in which each user is generally able to hear the conversations of other users. Some systems are equipped with relatively simple devices that allow messages to be directed to a particular mobile or group of mobile units. However, these addressing systems are rather rudimentary and, although they help reduce the amount of chatter to which a particular user has to listen, they do nothing to protect the user against casual eavesdropping by other people sharing the channel or by individuals (including competitors) employing simple scanning receivers. Moreover, they typically lack enough unique addresses to allow the creation of large, networked systems.

Severe Channel Congestion in Some Areas/Services.   Even with the creation of new land mobile radio bands, channels are often congested during peak periods, especially in major urban areas such as New York and Los Angeles. This often produces excessive delays in accessing the channel.

Inefficient Use of the Spectrum Resource.   Despite the heavy loading, at any instant of time, there may be some channels in the area that are unused or lightly used because, for example, the peak usage of the different channels may not coincide. However, with the simple systems just described, access is limited to only one channel. Clearly, it would be more efficient for many channels to be put into a common pool and then drawn from the pool to carry conversations on an as-needed basis. In more technical terms, the channels are not efficiently used because they are not "trunked."

Moreover, because they use high-power and high-antenna sites, a single conversation—say between a control station and a mobile unit a few miles apart—precludes the use of a channel over a very wide geographic area. For example, a system providing coverage over a 20-mile radius may preclude the assignment of the same channel for some 70 miles. In short, the systems are not spectrally efficient because the channels are not reused intensively in a given area. Finally, the channels are not efficiently used because a single voice conversation with a nominal bandwidth of 3 kHz occupies a 25 kHz frequency slot in the radio spectrum.

Multichannel Trunked Radio Systems

Multichannel trunked systems, such as those operated by very large private organizations on a private basis and by third-party providers on a commercial basis, are designed to provide more disciplined access to the channels and to allow for the message-by-message sharing of a pool of channels. They operate on an FDMA basis (see Chapter 3).

In a modern, multichannel trunked mobile radio system, this is accomplished through the use of computer logic, which assigns channels from a pool and recovers them at the end of a transmission or message. Thus, a modern trunked mobile radio system consists of a collection of repeaters, each operating on one of a pool of multiple channels (typically from five to twenty), and under computer logic control.

Because the FCC did not require the standardization of trunked radio systems, a number of proprietary systems have emerged. In one popular system, one of the available channels in the pool is set aside as a digital signaling or control channel. All of the end-user mobile units and control stations monitor the control channel and make requests and receive instructions on it. When a user sends a message by pushing the PTT button, the control station sends out a burst of digital information on the control channel, which identifies the individual mobile unit or fleet of mobile units with which the user wants to communicate. The computer logic at the repeater site finds an idle channel in the pool and sends back a burst of digital information telling the individual mobile unit or fleet of mobile units to all move to the selected idle channel. Once the control station/mobile units have arrived on the idle channel, the user who originated the transmission can begin to talk. In a modern system, this whole process takes less than 0.5 second.

If all channels are busy, the call requests are placed in a queue and handled on a "first-in, first-out" basis. At the end of the conversation, the channel is returned to the pool and mobile units and control stations in the fleet go back to monitoring the control channel. All of this is done automatically and all channels are available to all users. Note that the conversation channels are only used for the duration of the call.

Trunked mobile radio systems can be used for placing and receiving ordinary telephone calls by interconnecting the transceivers at the repeater site with the PSTN. Thus, if interconnection is provided on the system, when a user wants to engage in a telephone call rather than a dispatch call, the conversation is appropriately routed through the PSTN.

Trunked mobile radio systems overcome many of the disadvantages associated with the operation of individual repeaters such as community repeaters. First, they provide disciplined access to the channel which prevents users of the system from intentionally or unintentionally causing interference to other users. Second, because the audio output of the receiver portion of the mobile/portable radio is only activated when the unit's individual or group address is received, casual eavesdropping is eliminated. Moreover, because the assigned channel jumps from conversation to conversation (or even from transmission to transmission), it is also more difficult for others to monitor the conversations of a particular user. Third, the waiting time to access a channel is greatly reduced because, unlike in a single channel system, if any channel in the pool is idle, it can be immediately assigned for use in a conversation. Fourth, the radio spectrum is used much more efficiently because more mobile units can be accommodated per channel. For example, a modern trunked system providing dispatch service on a pool of, say, twenty channels can provide excellent service (i.e., short waiting times) with an average of well over 100 users per channel. As a rough estimate, a trunked system with a reasonable number of channels can provide about three times the capacity of untrunked channels for the same grade of service (i.e., average delay to access a channel).

The principal disadvantage of a trunked radio system of the type just described is that a single, point-to-point conversation between a control station and a mobile unit or between a mobile unit and the PSTN via the repeater site occupies a valuable radio channel over a very wide geographic area. In other words, little frequency reuse is employed, which means that the systems are spectrum inefficient for one-to-one calling. Another disadvantage in congested areas is that each end of the voice conversation occupies typically a 25 kHz channel, because the systems use ordinary FM. In addition to these spectrum efficiency concerns, the wide coverage provided by the high-power repeaters can cause difficulties in serving small, very-low-power portable units operating from within buildings and other difficult-to-serve locations. In other words, the imbalance in transmitter power may make it difficult for the portable unit to talk back to the receiver at the repeater site. These disadvantages are being addressed in third generation trunked mobile radio systems.

SPECIALIZED MOBILE RADIO

Commercial service providers have developed a number of services to meet the needs of mobile customers. One of these services is specialized mobile radio (SMR). This commercial wireless service provides mobile dispatch and data communications services. Dispatch service allows users to communicate with a single radio or simultaneously with all radios, or with a subgroup of radios in a group.

Traditional SMR service is similar to private land mobile radio (LMR), because it offers primarily voice dispatch service within a local area. Recently, new SMR technologies have led to the development of advanced SMR systems that offer cellular, dispatch, and paging services in a single handset. These networks provide features comparable to those currently offered to cellular subscribers, such as messaging, caller identification, and voice mail. The advanced networks use digital technologies and cover large geographic areas.

Important SMR considerations are summarized as follows:

- Identifies whether SMR users can access and use services during congestion or network disruption
- Identifies whether SMR services can be acquired from a carrier in a given region
- Describes the level of inherent security of the service and the capability to add security measures
- Describes the end-to-end data speed
- Identifies whether SMR calls can reach users in a given service area
- Characterizes the typical costs of SMR services

Key SMR Characteristics

Availability

There are approximately 8,000 SMR systems in the United States. Local and regional carriers operate most of these systems and cater to specific markets. Because SMR does not use a common standard or protocol, the systems operate using different technologies. SMR is also offered in three different frequency bands: 220, 800, and 900 MHz. The various standards and frequency bands make it difficult for users to operate on networks other than their own. Therefore, users are limited in the range they can travel outside their home service area.[2]

However, some service providers have joined to form regional or nationwide networks, either through acquisition or by entering into agreements with other carriers. In addition, the SMR industry is slowly expanding its coverage. Before obtaining SMR, users who require service should be sure providers offer service in those areas.

---

[2] See www.fcc.gov/wth/publicsafety/plans.html for more information.

Coverage

SMR coverage is limited compared with other commercial services because only about half of all SMR subscribers are interconnected to the PSTN. Often, communications are used strictly for the internal needs of an organization and access to the PSTN is not required. As a result, many small to midsize regional operators offer dispatch services only. Users on these networks cannot place or receive phone calls to other users on the PSTN.

Large operators offer an integrated service that combines dispatch and PSTN interconnection. If a network connects to the PSTN, users may require integrated radios, which combine dispatch and cellular services into a single handset.

SMR users may also experience coverage gaps similar to those in cellular telephony. This problem may occur in remote areas or away from major roads where carriers have not yet built service areas. Coverage gaps may also be caused by terrain or buildings that interfere with the signal. These gaps form dead spots within the region where the carrier's signal is nonexistent or too weak to communicate. In these areas, SMR users cannot send or receive phone calls.

Reliability

SMR customers compete with other users for access to the network. If the system is fully loaded and all channels are in use, users either receive a busy signal or calls are queued until a channel is free. Some SMR systems offer a priority access capability. Users with responsibility for mission-critical tasks may choose to consider a provider that offers priority access.

Unlike LMR, SMR services do not allow radios to communicate directly with each other, which is a type of communication called talk-around service. SMR radios must first communicate through the base station. Therefore, if the SMR network is damaged, users will not be able to directly communicate.

Transmission Speed

Data transmission using SMR is available only on a limited basis. Modems can be attached to a mobile unit, and then data transmitted over the network. Data services include two-way messaging, paging, and facsimile. Although SMR is capable of transmitting data at up to 9.6 Kbps, the actual data speed, or throughput, is limited in most cases to 4.8 Kbps, because data require more transmission capability than voice service.

Security

Different security risks and vulnerabilities are associated with all commercial wireless services. Analog SMR does not offer secure communications. Although digital technologies offer more privacy features than analog, they are

Figure 7–4
A Typical SMR Network
(From www.pswn.gov)

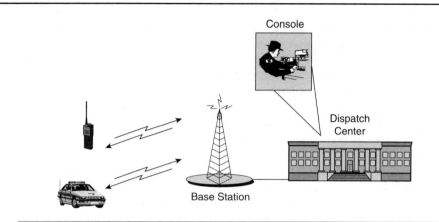

not considered secure. SMR currently does not have an encryption scheme imbedded in its network. However, some systems support encryption. Users with strict security requirements should determine, when requesting service, if encryption can be supported.

SMR from a Network Level

A traditional SMR system consists of one or more base stations, one or more antennas, and end-user radio equipment. Either a mobile user or a dispatcher can originate a call. The call is routed from the radio or dispatcher to the base station. The base station transmits and receives signals to mobile units within its coverage area.

Mobile users can be connected to the PSTN through the SMR operator (the network bearer) or through the dispatch center. Figure 7–4 illustrates a typical SMR network. An example would be CommNet Communications. Based in Dallas, Texas, CommNet Communications is the largest independent SMR service provider offering wide area, flat rate dispatch service in the Dallas/Fort Worth area.[3]

Cost

SMR customers pay equipment and service costs. Pricing structures vary by carrier and type of service. A typical SMR monthly rate plan for a large digital service carrier with PSTN connection is shown in Table 7–1.

---

[3] CommNet provides more information at www.commnetcomm.com/commnet.htm.

Cellular Service Only	Plan 1	Plan 2	Plan 3
Monthly Access	$39	$69	$99
Digital Cellular Minutes Included	60	150	300
Additional Minutes	$0.30/min	$0.25/min	$0.23/min
Numeric Pages	Unlimited	Unlimited	Unlimited
Text Messages	25	25	25
Long Distance	$0.15/min	$0.15/min	$0.15/min
Enhanced Voice Mail	$5.00	$5.00	$5.00
Enhanced Service Plan	**Plan 1**	**Plan 2**	**Plan 3**
Monthly Access	$69	$89	$109
Digital Cellular Minutes Included	60	150	300
Additional Minutes	$0.30/min	$0.25/min	$0.23/min
Talk Group Minutes Included	150	150	150
Additional Talk Group Minutes	$0.10/min	$0.10/min	$0.10/min
Numeric Pages	Unlimited	Unlimited	Unlimited
Text Messages	25	25	25
Long Distance	$0.15/min	$0.15/min	$0.15/min
Enhanced Voice Mail	$5.00	$5.00	$5.00

Table 7–1
Representative Basic and Enhanced SMR Service Fees

The average cost for dispatch-only service ranges from $14.60 to $15.50 (Carlson 1997).

In addition to service costs, subscribers usually have to buy end-user radio equipment. SMR radios vary in appearance and often resemble a cellular phone or sophisticated citizen-band radio. The radios typically operate in a push-to-talk fashion, although dial-up is becoming more common. Prices can range from less than $200 to $1,000, depending on what features the equipment includes and whether it operates using analog or digital technology (FCC 2000).

SMR Considerations

Users must think carefully about which commercial services may meet their operational requirements. Table 7–2 lists some considerations in selecting SMR services. SMR service packages and billing structures are likely to vary among carriers. Before acquiring SMR service, potential users may choose to employ the checklist in Table 7–3 to assist in determining if CDPD meets their needs.

- **Direct talk group capability**—SMR enables users to talk on an individual, talk group, or fleet basis.

- **Bundled wireless services**—Some SMR systems provide numeric paging, text messaging, PSN interconnection, mobile dispatch, and talk group functions.
- **Call setup**—SMR dispatch service takes about 1/10 of the time that cellular requires to set up a call.

- **High radio equipment costs**—The cost of SMR end-user equipment is generally higher than most commercial wireless services because SMR providers have targeted primarily mobile business users.
- **Lack of PSN interconnection**—Many SMR networks do not offer access to the PSN.

- **Coverage**—Because SMR lacks a standard protocol, users are restricted where they can roam outside of their home service areas. The majority of SMR service providers offer service in a limited area.

Table 7–2
SMR Considerations

Table 7–3
SMR Checklist

✓ Do I need dispatch, voice, and data services?
✓ Where do I need these services? Locally? Regionally? Nationally?
✓ Will SMR service work in my operational environment?
✓ Will it support mission-critical requirements?
✓ What is the coverage area of the SMR service, including extended calling areas provided through roaming agreements?
✓ Are there known dead spots?
✓ What is the full cost of the service package, including dispatch communications, PSN interconnection charges, long-distance charges, and roaming fees?
✓ Are there government rates that may reduce overall charges?
✓ Do I need encryption? If so, does the system support encryption?
✓ Is there a priority access capability?
✓ Does the carrier offer data services? If so, what type and speeds?
✓ Does the carrier offer access to the PSN?

■ SUMMARY

In this chapter we learned the importance of public wireless-based services. Topics included the technologies and types of wireless devices used in this arena. Examples of how public wireless-based services have proved beneficial provide a more complete understanding of public service communications.

In the next chapter, we discuss wireless LANs, a technology rapidly advancing in both use and performance.

1. Interoperability refers to _____.
   a. public safety personnel and PCs being interchangeable
   b. public safety personnel being able to communicate with other agencies
   c. mainframes being able to exchange information with PCs

2. A spectrum "island" is an example of _____.
   a. fragmentation
   b. scalability
   c. interoperability

3. A PSAP is another name for a _____.
   a. 9-1-1 center
   b. national geographic survey center
   c. Red Cross office

4. The PSWN stands for _____.
   a. Public Safety Awareness Network
   b. Public Service Wireless Network
   c. Public Safety Wireless Network program

5. The PSWN was formed to _____.
   a. promote effective safety communications
   b. coordinate disaster calling centers
   c. create SMR technology

6. True/False: Paging systems are used in public wireless-based service systems.

7. The simplest type of two-way dispatch system is _____.
   a. single frequency dispatch
   b. two-frequency dispatch
   c. bi-directional paging

8. Enhanced 9-1-1 service refers to the capability of _____.
   a. a joint 9-1-1 center and a Red Cross office
   b. seeing the caller's phone number and resulting address
   c. seeing all 9-1-1 calls go directly to an ambulance

9. Channels are most congested during _____.
   a. evening hours
   b. peak periods
   c. daytime hours

10. VECC stands for _____.
    a. Valley Emergency Communications Center
    b. Very Equalized Control Center
    c. Virtual Emergency Communications Center

11. Mutual aid interoperability refers to _____.
    a. exchanging medical supplies
    b. school systems exchanging public safety personnel
    c. joint and immediate responses to disasters

12. True/False: There are not enough channels available to allow each user to have a dedicated channel of his or her own.

13. Simple dispatch systems typically operate in _____ mode.
    a. digital
    b. two-way
    c. analog

14. VECC is funded by _____.
    a. local bonds
    b. federal excise taxes
    c. 9-1-1 surcharges

15. Multichannel trunked radio systems use _____.
    a. TDMA
    b. FDMA
    c. CDMA

16. SMR stands for _____.
    a. Standardized Medical Response
    b. Specialized Mobile Radio
    c. Special Modulated Radio

17. The transmission speed for SMR is only _____ Kbps.
    a. 144
    b. 4.8
    c. 56

18. The principal disadvantage of a trunked radio system is that a single, point-to-point conversation between a control station and a mobile unit or between a mobile unit and the PSTN via the repeater site occupies a valuable radio channel over a very wide geographic area. In other words, _____ frequency reuse is employed.
    a. little
    b. a significant amount of
    c. no

HANDS-ON EXERCISES

1. Using your browser and a search engine, locate at least two emergency service organizations that extensively employ wireless systems.

2. Using your browser and a search engine, plot the recent legislative history of wireless-based public services.

3. Using a graphics software package, draw an illustration of how a typical wireless-based public service system might look.

4. Using a product such as Shockwave and/or Flash, develop a short expose on a wireless-based public service system assisting in the saving of lives and/or property.

5. Go to your local emergency organization such as the local Red Cross office, a police station, or a fire station and find out how they use wireless in their day-to-day operations.

6. Conduct research and prepare a report on trunked versus untrunked systems to include characteristics and the advantages and disadvantages of each.

8 Wireless LANs

In the last chapter we discussed public services and how they utilize wireless technologies to enhance many lives. Now we address a rapidly advancing technology. Until recently, wireless LANs were a little used wireless technology due to poor performance; however, recent advances in this arena have stimulated the interest and deployment of wireless LANs worldwide.

OBJECTIVES

After reading this chapter and completing the exercises, you will be able to:

- Discuss the benefits of wireless LANs
- Describe the various uses for wireless LANs
- Describe the design and principles of operation of wireless LANs
- Discuss the current wireless LAN standard

SMOOTH SAILING FOR P&O PRINCESS CRUISES WITH PROXIM STRATUM

P&O Princess Cruises

Cruising is one of the fastest growing travel segments, with increasing numbers of people travelling to diverse locations—from Alaska to the Caribbean—enjoying the pleasure of relaxing on luxury liners. P&O Princess Cruises PLC, headquartered in London, is the third largest cruise company in the world. It includes Princess Cruises, P&O Cruises, Aida Cruises, and Swan Hellenic. The current fleet of 18 ships offers 27,360 berths and in the year 2000, P&O Princess Cruises carried over 900,000 passengers, generating revenue of more than $2 billion.

Southampton is the primary port for P&O Cruises in the UK, in addition to accommodating some of its main offices and A&P Appledore, a shipyard that the company sometimes uses. When passengers depart for their cruise on the Aurora, Oriana, Oceana, Arcadia or Victoria cruise ships, they do so from a departure lounge at the Southampton docks. Employees who work at the check-in desks and baggage reclaim need to be able to swiftly and securely transfer passenger information back to the main office by accessing the corporate local area network. This, however, proved difficult because cable communications lines down to the dock were expensive to lease. P&O were also aware that they would soon be re-fitting the passenger terminal and needed a solution that would be flexible and could adapt to this temporary situation.

Implementing a Building-to-Building Wireless LAN

There were several options open to P&O Cruises to solve their communication problems. Satellite was ruled out because it was too expensive and did not offer the high speeds that the company required. ISDN lines were considered too cumbersome, inflexible and unsuitable for a shipping environment. It was decided that the best solution would be to implement a building-to-building wireless LAN environment and, having sampled various vendors, the company realized that Proxim would best suit their needs. P&O Cruises worked with Wave Access to implement Proxim Stratum MP.

Stratum MP is a multipoint bridge product within Proxim's Stratum family of wireless Ethernet bridges for outdoor applications. Stratum MP uses the license-free 2.4 GHz ISM band to deliver 10 Mbps half duplex wireless connectivity and can be configured for point-to-point and point-to-multipoint topologies. Wave Access placed one multidirectional aerial on the roof of Duke's Keep, P&O Cruises' largest office in Southampton, and omnidirectional aerials at the passenger terminal and the shipyard, A&P Appledore, to create a complete building-to-building communications infrastructure. P&O now get 10 Mbps data throughput in the Southampton area without having to worry about security, since the wireless data is not only encrypted, but is also travelling on a certain frequency to a specific address.

A Buoyant Infrastructure to Bring a New Office on Board

P&O have been able to quickly and successfully set up a cost effective communications infrastructure with no real problems. "Many major banks and enterprises use microwave technology to securely transfer data, but this is incredibly expensive. Proxim has made long-distance line-of-sight communication affordable to other businesses," said Ken Miller, Technical Support Manager Ships, P&O Cruises. "The sites we currently have connected are up to 3 miles apart, but we have not had any issues with distance and our Proxim implementation just keeps growing. We haven't had any downtime with the network either; there have been no breaks in transmission—even in the severest weather conditions."

"Proxim Stratum gives us a permanent connection rather than having to lease lines every time that we need some kind of communication in the Southampton area, which can become expensive," explained Miller. "A wireless LAN implementation can bring return on investment in one move. We are refitting our passenger arrivals and baggage reclaim areas and I simply have to move the aerials to our temporary location and we have our communications up and running. With Proxim it is all simple: connecting the wireless building-to-building solution with our Ethernet was as easy as plugging it in to a data socket."

The network is about to be extended to Town Quays, which can be done within a matter of hours. "Soon we will be moving some of our employees from the London office to Southampton," said Miller. "Opening a new office is simplified using Proxim Stratum because we need not worry about the communications infrastructure. All we need to do is place an aerial in the window at that location and all our employees in Town Quays can instantly access the local area network. Proxim has provided the flexibility we need to grow our business."

Launching The Golden Princess

The final test for the new building-to-building wireless LAN came in the form of the launch of a new ship. When Princess Cruises was launching its luxury 15 deck,

£330 million vessel, the Golden Princess, the company decided to hold the inaugural event at the Southampton dock. Princess Cruises was anticipating 25,000 visitors to attend the event, including representatives from international media. It was imperative that the Quayside reservations office had a secure and rapid communications link with the main office of Princess Cruises in Los Angeles, so that tickets could be purchased on that day. Considering customers were spending up to £ 10,000 on a cruise ticket, it was important that they were not made to wait unnecessarily for data to be transferred.

Therefore, P&O Cruises simply added another aerial to allow the reservations office access to the network. The WAN connection was then routed from Duke's Keep. "We had a highly successful launch day that ran very smoothly, without any hitches," commented Miller. "I firmly believe that our stable communications infrastructure played an important part in this achievement."

As the company continues to expand it will be extending the wireless implementation to other buildings and locations. Proxim Stratum MP has made long distance wireless networking plain sailing for P&O Princess Cruises and has ensured that the company will be able to meet all its future communications needs—whatever the weather.

WLANs at Pine Crest School

This case study took place at Pine Crest School in Ft. Lauderdale, Florida. This school was founded in 1934 and helps children of winter tourists keep up with their studies. It is now a boarding school. The school needed anytime, anywhere network connectivity, but it had to be simple and affordable. This is where 3Com comes into the picture. They wanted to offer support for the students in the areas of completion of assignments, access to learning resources, and e-mail. They decided to use 3Com AirConnectA 11 Mbps wireless LAN, switch 4007a, SuperStackA II, switch 3300. The school launched a $10 million campaign toward this academic excellence. Thanks to 3Com, the AirConnect wireless LAN allows 400 students and teachers to transmit assignments and work from seven classrooms. Eventually, they hope to include 1,000 users on the campus. The cost savings that 3Com offered the school enabled it to fulfill other programs it had only hoped to complete. This has allowed increased student productivity and provided otherwise impossible opportunities.

Source: www.proxim.com/learn/library/casestudies/pdf/princesscruise.pdf.

INTRODUCTION

A wireless LAN (WLAN) is a flexible data communications system implemented as an extension to, or as an alternative for, a wired LAN within a building or campus. Using electromagnetic waves, WLANs transmit and receive data over the air, minimizing the need for wired connections. Thus,

WLANs combine data connectivity with user mobility, and, through simplified configuration, enable movable LANs.

Over the last seven years, WLANs have gained strong popularity in a number of vertical markets, including health care, retail, and manufacturing. Also, training sites at corporations and students at universities use wireless. These industries have profited from the productivity gains of using handheld terminals and notebook computers to transmit real-time information to centralized hosts for processing. Today, WLANs are becoming more widely recognized as a general-purpose connectivity alternative for a broad range of business customers. The U.S. wireless LAN market is rapidly approaching $1 billion in revenues.[1]

Wireless LANs frequently augment rather than replace wired LAN networks, providing the final few meters of connectivity between a backbone network and the in-building or on-campus mobile user. The following list describes some of the many applications made possible through the power and flexibility of wireless LANs.

- Doctors and nurses in hospitals are more productive because handheld or notebook computers with wireless LAN capability deliver patient information instantly.
- Consulting or accounting audit teams or small work groups increase productivity with quick network setup.
- Network managers in dynamic environments minimize the overhead of moves, adds, and changes with wireless LANs, thereby reducing the cost of LAN ownership.
- Training sites at corporations and students at universities use wireless connectivity to facilitate access to information, information exchanges, and learning.
- Network managers installing networked computers in older buildings find that wireless LANs are a cost-effective network infrastructure solution.
- Retail store IS managers use wireless networks to simplify frequent network reconfiguration.
- Trade show and branch office workers minimize setup requirements by installing preconfigured wireless LANs needing no local MIS support.
- Warehouse workers use wireless LANs to exchange information with central databases and increase their productivity.
- Network managers implement wireless LANs to provide backup for mission-critical applications running on wired networks.
- Restaurant waitresses and car rental service representatives provide faster service with real-time customer information input and retrieval.
- Senior executives in conference rooms make quicker decisions because they have real-time information at their fingertips.

---

[1] Courtesy of www.proxim.com.

BENEFITS OF WLANS

The widespread strategic reliance on networking among competitive businesses and the meteoric growth of the Internet and online services are strong testimonies to the benefits of shared data and shared resources. With wireless LANs, users can access shared information without looking for a place to plug in, and network managers can set up or augment networks without installing or moving wires. Wireless LANs offer the following productivity, service, convenience, and cost advantages over traditional wired networks.

- Improved productivity and service. Wireless LAN systems can provide LAN users with access to real-time information anywhere in their organization. This mobility supports productivity and service opportunities not possible with wired networks.
- Installation speed and simplicity. Installing a wireless LAN system can be fast and easy and can eliminate the need to pull cable through walls and ceilings.
- Installation flexibility. Wireless technology allows the network to go where wire cannot go.
- Reduced cost of ownership. Although the initial investment required for wireless LAN hardware can be higher than the cost of wired LAN hardware, overall installation expenses and life cycle costs can be significantly lower. Long-term cost benefits are greatest in dynamic environments requiring frequent moves, adds, and changes.
- Scalability. Wireless LAN systems can be configured in a variety of topologies to meet the needs of specific applications and installations. Configurations are easily changed and range from independent networks suitable for a small number of users to full infrastructure networks of thousands of users that allow roaming over a broad area. Wireless LANs provide all the functionality of wired LANs, but without the physical constraints of the wire itself. Wireless LAN configurations include independent networks, offering peer-to-peer connectivity, and infrastructure networks, supporting fully distributed data communications.

Point-to-point local area wireless solutions, such as LAN for LAN bridging and personal area networks (PANs), may overlap with some WLAN applications, but fundamentally address different user needs. A wireless LAN to LAN bridge is an alternative to cable that connects LANs in two separate buildings. A wireless PAN typically covers the few feet surrounding a user's workspace and provides the ability to synchronize computers, transfer files, and gain access to local peripherals.

Wireless LANs also should not be confused with wireless metropolitan area networks (WMANs), packet radio often used for law enforcement or utility applications, or with wireless wide area networks (WWANs), wide area data transmission over cellular or packet radio. These systems involve costly infrastructures, provide much lower data rates, and require users to pay for bandwidth on a time or usage basis.

DESIGN AND PRINCIPLES OF OPERATION

Wireless LANs use electromagnetic airwaves (radio and infrared) to communicate information from one point to another without relying on any physical connection. Radio waves are often referred to as radio carriers, because they simply perform the function of delivering energy to a remote receiver. The data being transmitted are superimposed on the radio carrier so they can be accurately extracted at the receiving end. This is generally referred to as modulation of the carrier by the information being transmitted. Once data are superimposed (modulated) onto the radio carrier, the radio signal occupies more than a single frequency, since the frequency or bit rate of the modulating information adds to the carrier.

Multiple radio carriers can exist in the same space at the same time, without interfering with each other, if the radio waves are transmitted on different radio frequencies. To extract data, a radio receiver tunes in (or selects) one radio frequency while rejecting all other radio signals on different frequencies.

access point (AP)

In a typical WLAN configuration, a transmitter/receiver (transceiver) device, called an **access point (AP),** connects to the wired network from a fixed location using standard Ethernet cable. At a minimum, the access point receives, buffers, and transmits data between the WLAN and the wired network infrastructure. A single access point can support a small group of users and can function within a range of less than a hundred feet to several hundred feet. The access point (or the antenna attached to the access point) is usually mounted high, but may be mounted essentially anywhere that is practical so long as the desired radio coverage is obtained.

End users access the WLAN through (1) wireless LAN adapters, which are implemented as PC cards in notebook computers, (2) ISA or PCI adapters in desktop computers, or (3) fully integrated devices within hand-held computers. WLAN adapters provide an interface between the client network operating system (NOS) and the airwaves (via an antenna). The nature of the wireless connection is transparent to the NOS.

WLAN CONFIGURATIONS

Independent WLANs

independent WLAN

The simplest WLAN configuration is an **independent** (or peer-to-peer) **WLAN** that connects a set of PCs with wireless adapters. Any time two or more wireless adapters are within range of each other, they can set up an in-

Figure 8–1
Independent WLAN

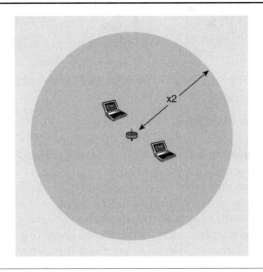

Figure 8–2
Extended-Range
Independent WLAN
Using Access Point as
Repeater

dependent network (Figure 8–1). These on-demand networks typically require no administration or preconfiguration.

Access points can extend the range of independent WLANs by acting as a repeater (see Figure 8–2), effectively doubling the distance between wireless PCs.

Infrastructure WLANs

In **infrastructure WLANs,** multiple access points link the WLAN to the wired network and allow users to efficiently share network resources. The access points not only provide communication with the wired network but also mediate wireless network traffic in the immediate neighborhood. Multiple access points can provide wireless coverage for an entire building or campus (see Figure 8–3).

infrastructure WLANs

Figure 8–3
Infrastructure WLAN

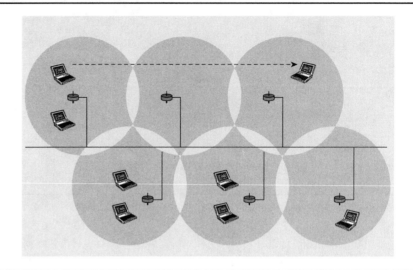

MICROCELLS AND ROAMING

Wireless communication is limited by how far signals carry for given power output. WLANs use cells, called microcells, similar to the cellular telephone system to extend the range of wireless connectivity. At any point in time, a mobile PC equipped with a WLAN adapter is associated with a single access point and its microcell, or area of coverage. Individual microcells overlap to allow continuous communication within a wired network. They handle low-power signals and hand off users as they roam through a given geographic area, as seen in Figure 8–4.

TYPES OF WLANS

The two technologies implemented in the wireless LAN arena are spread spectrum and infrared.

Spread Spectrum WLAN

Most wireless LAN systems use spread spectrum technology, a wideband radio frequency technique originally developed by the military for use in reliable, secure, mission-critical communications systems. Spread spectrum is designed to trade off bandwidth efficiency for reliability, integrity, and security.

In other words, more bandwidth is consumed than in the case of narrowband transmission, but the trade-off produces a signal that is, in effect, louder and thus easier to detect, provided that the receiver knows the pa-

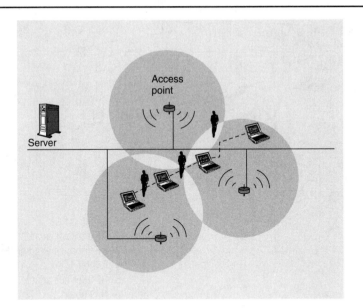

Figure 8–4
Handing off the WLAN
connection between
access points

rameters of the spread spectrum signal being broadcast. If a receiver is not tuned to the right frequency, a spread spectrum signal looks like background noise. There are two types of spread spectrum radio: frequency hopping and direct sequence.

Infrared WLAN

The second type of WLAN is infrared. Infrared LANs are rapidly being phased out, because they only operate at speeds less than 10 Mbps due to the limitation of infrared. Infrared requires line-of-sight or it will travel only short distances. Have you ever pointed your TV remote away from but close to your set? It still activated your television; however, the distance limitation of infrared is severe. Let us look at a few practical implementations of these two types of wireless LANs.

Implementation of Wireless Networking

Wireless networks can be implemented in a variety of situations. One can use wireless networks in a home office, small business, corporate buildings, and even in bridging building to building/LAN to LAN. The following will offer a brief description of each and also specific plans on the implementation of each scenario.

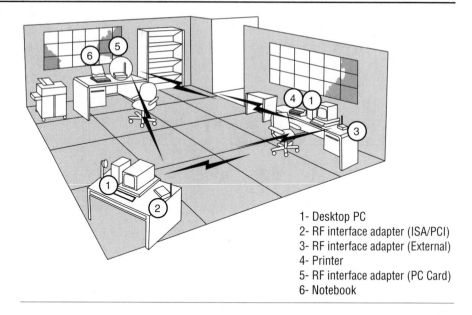

Figure 8–5
Nonaccess Point–
Based RF LAN
(Courtesy of www.wlana.org,
www.solveit.com)

1- Desktop PC
2- RF interface adapter (ISA/PCI)
3- RF interface adapter (External)
4- Printer
5- RF interface adapter (PC Card)
6- Notebook

Installing a wireless network within one's home is an efficient and convenient solution to networking needs. It offers a viable alternative to traditional wired networks within the home and provides the same benefits without the hassle or cost of setting up a wired network. Small businesses located in older buildings find that wireless LANs are efficient and they can be freestanding or connect to existing LANs to extend the network area. There are a few different methods of implementation when dealing with a home or small network. One example is a nonaccess point–based radio frequency LAN (see Figure 8–5). This provides a LAN environment for all the devices within range of each other, and allows for sharing of resources within the LAN without wires. Each device connects to the LAN in a peer-to-peer arrangement with either an internal or external interface adapter.

The RF interface adapter (ISA/PCI) is a device that fits into a PC or desktop-sized ISA/PCI slot and acts as the interface between that device and an access point or other interface adapter. The interface occurs through radio frequency transmissions and the speed and operating frequencies vary. The RF interface adapter (external) is an apparatus that connects to one device and acts as the interface between that device and an access point or other interface adapter. Like the ISA/PCI type adapter, the operating frequencies vary. The RF interface adapter (PC card) is the same as the other two, but it plugs into the PC card slot. The benefits of this setup are that it allows coverage through walls, it is easy to add workstations to the network, and it increases mobility within the office. However, this network is not without flaw. This LAN is subject to RF interference and unwanted reception, and the bandwidth can be affected by the number of users and amount of traffic on the network.

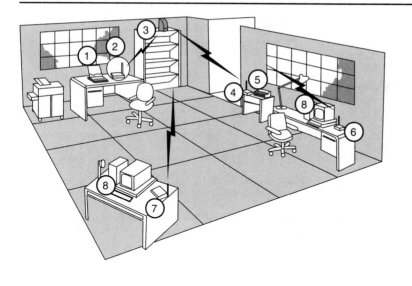

1- Notebook
2- RF interface adapter
3- RF access point
4- Specialized bridge
5- Printer
6- RF interface adapter (external)
7- RF interface adapter (ISA/PCI)
8- Desktop PC

Figure 8–6
Access Point–Based RF LAN
(Courtesy of www.wlana.org, www.solveit.com)

Another option is an access point–based radio frequency LAN. As shown in Figure 8–6, this arrangement utilizes an RF access point, which is a wireless radio frequency transceiver that connects a WLAN segment with a WLAN segment. Advantages include easy additions of workstations and increased mobility within the office. Disadvantages are it is subject to RF interference and the bandwidth can be affected by heavy network traffic.

To keep workers connected to the network and give them real-time access to information whether they are at or away from their desks, large corporations implement wireless LAN technology. Whether they are extending a preexisting LAN by adding on a wireless network or just setting up a freestanding wireless network, large businesses are finding the mobility and instant access a convenient and productive way of allowing their employees to do business.

By setting up a wireless private branch exchange (WPBX), one can extend the current private branch exchange (PBX), which is a telephone switching system (some are wireless) that connects extensions to each other as well as to the public telephone network, and extend its coverage to wireless PBX handsets. See Figure 8–7. Mobile handsets are able to initiate and receive calls on the regular voice network from anywhere within range of the base station.

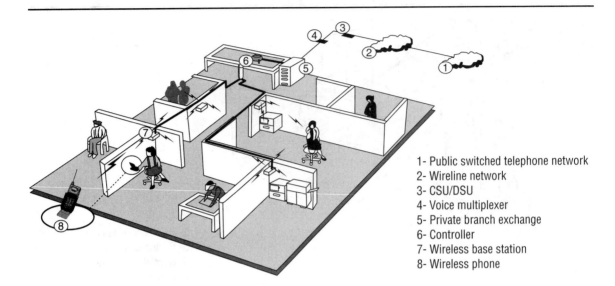

1- Public switched telephone network
2- Wireline network
3- CSU/DSU
4- Voice multiplexer
5- Private branch exchange
6- Controller
7- Wireless base station
8- Wireless phone

Figure 8–7
WPBX
(Courtesy of www.wlana.org, www.solveit.com)

This method uses a hardware controller and wireless base stations. Base stations are fixed-location devices (some are wireless) used to transmit and/or receive radio waves. The hardware controller is the system or circuitry that controls functions required for communication between a host computer or system and external devices or networks.

If the business is a retail environment, wireless technology can be utilized to provide an increased level of customer service. This implementation uses RF technology to complete customer transactions by using wireless point-of-sale (POS) devices. See Figure 8–8. These have interface adapters to access information on the company LAN, and these adapters communicate with access points wired to the LAN. The portable and mobile wireless POS devices are able to roam throughout the sales floor. In addition to providing increased customer service, they also lower theft, because the salesperson roams the floor as opposed to being stuck behind a counter.

Like the smaller offices, larger business offices can utilize both the nonaccess point–based RF LAN and the access point–based RF LAN. These are simply implemented on a larger scale and carry with them the same advantages and disadvantages as with the smaller versions.

Along with the uses previously mentioned, wireless technology can be implemented to bridge LAN to LAN or building to building. Point-to-point and point-to-multipoint connections allow users in different locations to

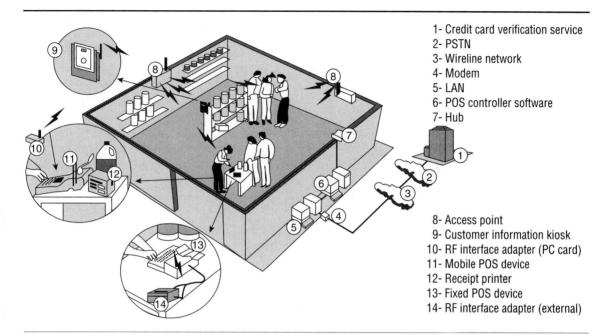

1- Credit card verification service
2- PSTN
3- Wireline network
4- Modem
5- LAN
6- POS controller software
7- Hub

8- Access point
9- Customer information kiosk
10- RF interface adapter (PC card)
11- Mobile POS device
12- Receipt printer
13- Fixed POS device
14- RF interface adapter (external)

Figure 8–8
Point-of-sale using RF
(Courtesy of www.wlana.org, www.solveit.com)

share files, access network resources, and access the Internet without wires. One type of setup for such an application uses a radio frequency wireless multipoint bridge. See Figure 8–9. This connects multiple LANs in different buildings. Modems, bridges, or routers transfer information from the LAN through an amplifier to an antenna that is used to communicate with other locations. The antennas are mounted outdoors and lightning arrestors are used to protect the wired LAN.

An advantage of this type of implementation is that network connectivity is maintained when transmission is obstructed. A disadvantage is that one has limited bandwidth without a license.

The other type of bridge is the radio frequency wireless point-to-point bridge. See Figure 8–10. This allows LANs in separate buildings to be connected over a distance ranging from several hundred yards to about 50 miles. As with the multipoint bridge, modems, bridges, or access points transfer information from the LAN through an amplifier to an antenna that is used to communicate with other locations. Lightning arrestors are used to protect the wired LAN here as well and the advantages and the disadvantages are generally the same as with the multipoint bridge.

Figure 8–9
RF Wireless Multipoint
Bridge
(Courtesy of www.wlana.org,
www.solveit.com)

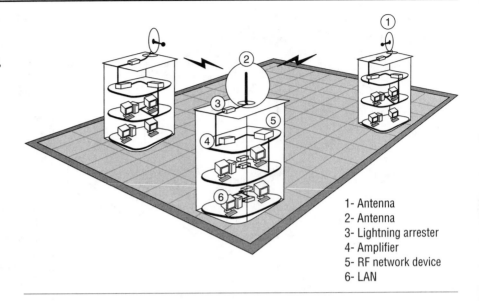

1- Antenna
2- Antenna
3- Lightning arrester
4- Amplifier
5- RF network device
6- LAN

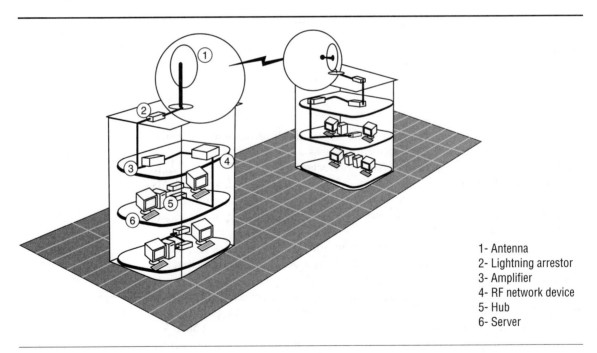

1- Antenna
2- Lightning arrestor
3- Amplifier
4- RF network device
5- Hub
6- Server

Figure 8–10
RF Wireless Point-to-Point Bridge
(Courtesy of www.wlana.org, www.solveit.com)

Wireless networks offer many solutions to networking problems and are providing the much desired convenience and efficiency to many different types of situations. These are just a few of the ways one can implement a wireless network in a variety of settings.

WLAN CUSTOMER CONSIDERATIONS

Compared with wired LANs, wireless LANs provide installation and configuration flexibility and the freedom inherent in the network mobility. Potential wireless LAN customers should also consider some or all of the following issues.

Range/Coverage

The distance over which RF waves can communicate is a function of product design (including transmitted power and receiver design) and the propagation path, especially in indoor environments. Interactions with typical building objects, including walls, metal, and even people, can affect how energy propagates, and thus what range and coverage a particular system achieves. Most wireless LAN systems use RF, because radio waves can penetrate many indoor walls and surfaces. The range (or radius of coverage) for typical WLAN systems varies from under 100 feet to more than 500 feet.

Throughput

As with wired LAN systems, actual throughput in wireless LANs is product and setup dependent. Factors that affect throughput include airwave congestion (number of users), propagation factors such as range and multipath, the type of WLAN system used, as well as the latency and bottlenecks on the wired portions of the WLAN. Typical data rates range from 1 Mbps to 11 Mbps.

Users of traditional Ethernet LANs generally experience little difference in performance when using a wireless LAN and can expect similar latency behavior. Wireless LANs provide throughput sufficient for the most common LAN-based office applications, including e-mail exchange, access to shared peripherals, and access to multiuser databases and applications.

Integrity and Reliability

Wireless data technologies have been proven through more than 50 years of wireless application in both commercial and military systems. While radio interference can cause degradation in throughput, such interference is rare in the workplace. Robust designs of proven WLAN technology and the limited distance over which signals travel result in connections that are far more robust than cellular phone connections and provide data integrity performance equal to or better than wired networking.

Simplicity/Ease of Use

Users need very little new information to take advantage of wireless LANs. Because the wireless nature of a WLAN is transparent to a user's NOS, applications work the same as they do on tethered LANs. WLAN products incorporate a variety of diagnostic tools to address issues associated with the wireless elements of the system; however, products are designed so that most users rarely need these tools.

WLANs simplify many of the installation and configuration issues that plague network managers. Since only the access points of WLANs require cabling, network managers are freed from pulling cables for WLAN end users. Lack of cabling also makes moves, adds, and changes trivial operations on WLANs. Finally, the portable nature of WLANs lets network managers preconfigure and troubleshoot entire networks before installing them at remote locations. Once configured, WLANs can be moved from place to place with little or no modification.

Security

Because wireless technology has roots in military applications, security has long been a design criterion for wireless devices. Security provisions are typically built into wireless LANs, making them more secure than most wired LANs. It is extremely difficult for unintended receivers (eavesdroppers) to listen in on wireless LAN traffic. Complex encryption techniques make it impossible for all but the most sophisticated to gain unauthorized access to network traffic. In general, individual nodes must be security enabled before they are allowed to participate in network traffic. Security issues are covered extensively via WLANA at their site (www.wlana.org).

Interoperability with Wired Infrastructure

Most wireless LAN systems provide industry-standard interconnection with wired systems, including Ethernet (802.3) and Token Ring (802.5). Standards-based interoperability makes the wireless portions of a network completely transparent to the rest of the network. Wireless LAN nodes are supported by network operating systems in the same way as any other LAN node—via drivers. Once installed, the NOS treats wireless nodes like any other component of the network.

Interoperability with Wireless Infrastructure

Several types of interoperability are possible between wireless LANs. This will depend both on technology choice and on the specific vendor's implementation. Products from different vendors employing the same technology and the same implementation typically allow for the interchange of adapters and access points. The goal of industry standards, such as the IEEE 802.11 specifications, is to allow compliant products to interoperate without explicit collaboration between vendors.

Interference and Coexistence

The unlicensed nature of radio-based wireless LANs means that other products that transmit energy in the same frequency spectrum can potentially provide some measure of interference to a WLAN system. Microwave ovens are a potential concern, but most WLAN manufacturers design their products to account for microwave interference. Another concern is the colocation of multiple WLAN systems. While colocated WLANs from different vendors may interfere with each other, others coexist without interference. This issue is best addressed directly with the appropriate vendors.

Cost

A wireless LAN implementation includes both infrastructure costs, for the wireless access points, and user costs, for the wireless LAN adapters. Infrastructure costs depend primarily on the number of access points deployed; access points range in price from $800 to $2,000. The number of access points typically depends on the required coverage region and/or the number and type of users to be serviced. The coverage area is proportional to the square of the product range. Wireless LAN adapters (see Figure 8–11) are required

Figure 8–11
Linksys Wireless LAN
Adapter
(Courtesy of M. Lipstein,
Linksys)

for standard computer platforms, and range in price from $200 to $700. The cost of installing and maintaining a wireless LAN generally is lower than the cost of installing and maintaining a traditional wired LAN, for two reasons. First, a WLAN eliminates the direct costs of cabling and the labor associated with installing and repairing it. Second, because WLANs simplify moves, adds, and changes, they reduce the indirect costs of user downtime and administrative overhead.

For example, the following table shows the costs involved in a typical home wireless network compared with a nonwireless setup.

Four-node wireless home network

3Com home wireless router	$389.00
3Com AirConnect network adapter PC card 11 Mbps	$189.00
	× 4
	$756.00
Total	$1,145.00

Four-node wired home network

Linksys EtherFast four-port cable/DSL router (Figure 8–12)	$97.88
3Com OfficeConnect Fast Ethernet NIC network	$24.45
Adapter PCI 100 Mbps	× 4
	$97.80
Total	$195.68

A diagram of a typical home-based wireless LAN is shown in Figure 8–13.

Figure 8–12
Linksys Wireless Router
(Courtesy of M. Lipstein,
Linksys)

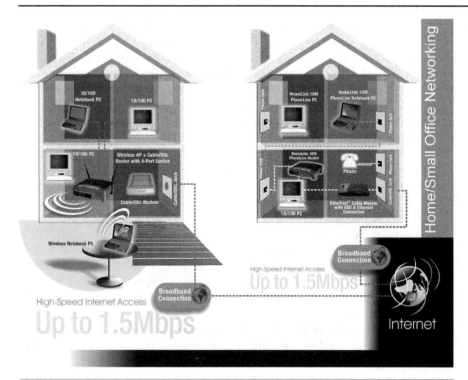

Figure 8–13
Linksys Diagram
(Courtesy of M. Lipstein,
Linksys)

Scalability

Wireless networks can be designed to be extremely simple or quite complex. Wireless networks can support large numbers of nodes and/or large physical areas by adding access points to boost or extend coverage.

Ad Hoc WLAN

In the ad hoc mode, several wireless stations (e.g., laptops) in a local area (e.g., conference room) can form a single wireless network. Through an ad hoc mode, the users can share documents and devices, creating a basic peer-to-peer network.

WIRELESS LAN STANDARD IEEE 802.11, 802.11B, AND 802.11A

In 1990, the Institute of Electrical and Electronics Engineers (IEEE) formed a committee to develop a standard for wireless LANs (WLANs). Of course, at that time, wireless LANs were only operating at 1 Mbps to 2 Mbps. Due to the consensus-driven approach of the standards process, it took them 7 years to finish and produce the official wireless LAN standard titled IEEE 802.11. This standard defines a wireless LAN that can have either fixed or mobile workstations operating at up to 2 Mbps. Another important aspect is it specifies that the features of this wireless LAN be totally transparent to the upper layers, which means that any network operating system such as Windows NT or Novell will successfully run on a wireless LAN with no special modifications.

This was all very good, except that the nonwireless LAN standard Ethernet was readily available and Ethernet operates at 10 Mbps. It was, therefore, apparent that immediate work was needed on a faster wireless LAN. This began and, in an expedited process, a higher rate was added to the original 802.11 standard, called 802.11b. This added two higher speeds of 5.5 Mbps and 11 Mbps. This new addition affected only the physical layer.

WLAN is a standard offering a limited coverage for LAN users. Cell radius is usually between 20 to 30 meters and several hundred meters. IEEE 802.11 wireless LAN is a locally situated network—a local area network. The coverage area consists of small islands and the purpose is certainly not to offer a large coverage network like the typical cellular telephony scenario. The coverage area is often tailored according to the user's own need and can also be temporary.

The specifications of wireless LAN outline two possible modes of operations: client-server and ad hoc mode WLAN. In the client-server WLAN, terminals communicate with base stations or access points, which form the coverage area. The access points are further connected to the wired network.

The coverage area of the client-server type WLAN network is usually bordered by a building or a campus and can therefore be comparable to a single cellular telephony cell. The main difference between the typical cellular telephony network and WLAN technologies are that, to cover the whole building with the WLAN technology, there has to be several access points depending on the building architecture, wall materials, and so forth. In the cellular telephony solution, the coverage area can be built with a single base station by distributing the signal into antennas located in different rooms by using power splitters.

The specifications of WLAN also define an ad hoc mode. In this mode mobile terminals by themselves build the network. The coverage area is built by the help of wireless adapters and is limited. In the ad hoc mode, the whole network is seen as movable, and it is independent of any infrastructure, unlike a typical cellular telephony or client-server type WLAN. It is also isolated, because it has no interface to the wired network.

Actually, the standardization process of IEEE 802.11 began from the need to connect wirelessly to the wired Ethernet-based data communications network and thus to offer mobility for the LAN users within a rather small area. One of the main advantages of WLAN is that it provides LAN users access to real-time information anywhere in the organization. Further, one aim was to reduce installation costs and thus achieve short- and long-term cost savings.

Installing a wireless system can be fast and the need to pull cable through walls and ceilings can be eliminated. Wireless technology also allows the network to go where the wire cannot go.

As mentioned, the specification work of the standard IEEE 802.11 started in 1990 and ended in 1997 by IEEE. The first phase of the standard IEEE 802.11 supports only 1 Mbps and 2 Mbps data rates. The first phase standard was followed by an extension IEEE 802.11b, which supports data rates up to 11 Mbps with the radio frequency technology direct sequence spread spectrum. However, the user does not have to purchase the radio operator's license to use the frequency band 2,4-2,483 GHz, which is dedicated to WLAN use, and is also known as the industrial, scientific, and medical (ISM) band. This means that anyone who has access to a WLAN can buy and install an access point, which can be problematic from the network planning point of view.

IEEE 802.11 Protocol Layers

The specifications of IEEE 802.11 define two layers. Layer 1 is the physical layer (PHY); layer 2 is the media access control (MAC) layer. Layer 1 specifies the modulation scheme used and signaling characteristics for the transmission through the radio frequencies, whereas layer 2 defines the means for accessing the physical layer. The specifications of the IEEE 802.11 layer 2 also define services related to the radio resource and the mobility management.

The specifications of IEEE 802.11 define three different physical layer characteristics for WLAN: one infrared and two RF transmission methods. We concentrate only on the two RF methods: direct sequence spread spectrum (DSSS) and frequency hopping spread spectrum (FHSS). Due to the operation in an unlicensed RF band, however, the spread spectrum modulation must fulfill the requirements set by each country. For example, not all countries use the same frequency as their neighbors.

For wireless LAN devices to be interoperable, they must have the same physical layer standard. Therefore, DSSS equipment is not capable of communicating with FHSS-based equipment.

As stated in Chapter 3, the use of the ISM band requires no operator license. Therefore, the actual capacity limit is not defined according to the available bandwidth, but is influenced by the interference level. Further packet mode traffic, which is used in WLAN, is not affected by the specification of congestion. Therefore, the available transmission capacity depends strongly on the interference level. The problem of a WLAN is that there could be interference not only caused by the other access points located in their own network, but also caused by neighboring WLAN or other equipment using the same frequency band. There is no means for dealing with the interference level in the latter case.

Association is a basic service that enables the connection between the station (STA) and the access point (AP) in an infrastructure WLAN. An access point is basically a radio base station that covers an area of about 30 meters

roaming

to 300 meters depending on the environment. An access point and its associated clients form a basic service set (BSS). Though no handover mechanism is specified in the standard, the standard introduces a service called reassociation, which is related to roaming from one BSS to another. Two adjoining BSSs form an extended service set (ESS) if they share the same ESS identity (ESSID). This is the case when **roaming** is possible. Thus, the parameter ESSID is analogous to the concept of the neighboring cell in a GSM network.

An independent basic service set (ad hoc mode, IBSS) is the most basic type of IEEE 802.11 WLAN. At minimum, it consists of two stations. The network is often formed without preplanning, and remains until either station is moved away from the other's coverage area.

WLAN is a symmetrical system, which means that stations act as access points in the ad hoc mode. Thus, power control functions similarly in the stations and in the access points. In the DSSS WLAN, the maximum transmission power is specified to be 100 milliwatts (mW) (EIRP), whereas the minimum power level shall be no less than 1 mW. Four different power levels are defined. For example, in an infrastructure network, the station selects its power management mode by a parameter of which the station informs the access point through a successful frame exchange. In a case when there is nothing to be received or transmitted, the station is able to fall into a sleep mode and then consumes very low power. In this mode, the station is unable to transmit or receive any data; however, in a low-power mode, the station listens periodically for beacons sent by the access point. In this way the access point is announcing to the stations its existence.

In the FHSS WLAN, the power level is specified as follows. The EIRP of the maximum transmission power is 100 mW, and the smallest transmission level is 10 mW. The specified power levels are regulated by local regulators and are different from continent to continent. The values presented here are the values used in Europe.

It could be assumed that the speed of the mobile also affects the user's ability to recover from the fading point. Thus, the mobile station moving at high speed recovers from the fading caused by the multipath propagation more quickly than the slow-moving mobile. Thus, the power control can be regarded as more critical to the slow-moving user.

Now in case of infrastructure WLANs, the situation is that the user who has access to the LAN and wants to make that wireless buys an access point and installs it. The user makes a coverage plan by simply installing the equipment (e.g., to the wall) and, if the place is not optimized, the user probably tries another place. The user also makes a frequency plan by randomly selecting the channel; if there is a great amount of interference, that will probably change as well. Therefore, the coverage and frequency planning steps can be found in the WLAN planning process. The planning step of capacity, however, can be omitted in the WLAN environment, because there is always room for another WLAN packet user. The information is transmitted in packet mode, so the user experiences no congestion, only a variable transmission rate. The user also may consider using several access points to guarantee a certain transmission rate, especially with hundreds of concurrent users, even though one access point provides sufficient coverage. Overall, a distinctive capacity design principle can be found within the GSM world, not the WLAN environment.

Another problem with WLAN planning is that the user may not be familiar with technical equipment. The user may install an access point and try to make a wireless connection to the LAN, but is unsuccessful. Not knowing what the problem is, the user searches for answers: Is the used frequency too interfered or is the installation place for the access point not good enough? Further, what is making the interference? The biggest problem the user has is probably the interference caused by other equipment (not WLAN) using the same frequency band. Microwave ovens, Bluetooth equipment, or different electrical equipment using 2.4 GHz (e.g., door-opening equipment) could cause interference.

A relevant requirement for the IEEE 802.11 infrastructure network planning would be that the intelligence of the frequency planning be in the equipment itself, to minimize the level of interference. The other demand would be a larger frequency band dedicated to the WLAN use than is available today, so the interference would be spread into a larger area and thus lower the average interference level. Both demands are taken into account in the next generation WLAN called HiperLAN2, which uses higher frequency (5 GHz) and supports interfunction with third generation cellular standard UMTS. In HiperLAN2, the access point automatically selects an appropriate channel and dynamically changes it according to the level of interference.

PLANNING FOR A NEW NETWORK

The steps for planning a WLAN are similar to those for planning a wired network. Aspects of assessment include the purpose of the organization, any expected organizational changes and/or expansions or contractions, current network uses, existing networks such as existing Ethernet LANs, and any changes in the capabilities of the organization. Costs can vary widely depending on the projected size of the WLANs, the number of users, and whether you select standards based (i.e., 802.11-based products) or proprietary (i.e., HomeRF). The planning must also document the current network in exhausting detail. This brings a positive result, because it can point out deficiencies in the current network and provide a paper trail for the network as a whole.

SELECTING A WLAN

When selecting a WLAN, several crucial issues must be addressed. First, consider the joint issues of compatibility and interoperability. If an existing wired LAN is in place, then the issue of compatibility between this LAN and the new WLAN must be taken into account. This is particularly critical in the wireless arena because of the standards-based (802.11) and nonstandards-based (HomeRF) approaches.

Overall, however, compatibility generally should not be a problem, because almost all access points can connect to an Ethernet network and some to a Token Ring network. This connection will allow the WLAN to seamlessly mesh into the existing network, as shown in Figure 8–3.

Compatibility is also not an issue as most network operating systems recognize WLAN NICs just as they would the typical Ethernet NICs attached to the workstations.

A real problem can arise, however, when adding a WLAN to an existing WLAN. For example, different WLAN technologies cannot interoperate. An FHSS WLAN cannot communicate with an infrared WLAN or a DSSS WLAN. Also, WLAN configurations that use different radio frequency bands cannot communicate, even if they are both DSSS systems or both FHSS systems.

Proprietary versus Standard

Determining proprietary or standard is an issue of critical importance. It may be assumed that the standards-based approach would be the only viable alternative; however, as mentioned, HomeRF, a proprietary WLAN standard, has a large audience. This is due to its lower cost and because it can be tailored for home applications. The proprietary approach also may be more suitable if you are installing a wireless LAN that will communicate with another proprietary wireless LAN.

Radio-Based versus Infrared

The 802.11b standard approved by IEEE is radio based; therefore, selecting a WLAN based on that standard means choosing a radio-based network. There are several excellent advantages to radio-based WLANs. Radio waves do not have the limitations of light and heat waves, and thus make an excellent medium for data transmission. Radio waves can travel great distances and can also penetrate nonmetallic objects. Because infrared light does not interfere with other communications signals and is itself not affected by other signals, an infrared WLAN may be the network of choice if the stations are near sensitive scientific or medical equipment or in the case of factory locations where there may be large EMI-emitting machinery. Because infrared signals do not penetrate walls and the signals are kept inside the room, an infrared WLAN also may be suited for a network that handles sensitive data, such as in government or military applications. For most WLAN applications, however, a radio-based system is the best solution.

After deciding which type of WLAN to install, the organization must consider several issues about the layout of the stations, including the topology of the WLAN, the coverage area, the number of users, network utilization, the number and type of physical obstructions, and noise (relating to wiring) levels.

Coverage Area

Area coverage size is a crucial factor as well. This aspect can affect how you should design the floor layouts and even which floors of a building can be utilized for WLANs. The following table provides guidance in this arena for DSSS WLANs.

Physical Layout	Comments	Maximum Distance
Open	Outdoor setting	1,000 feet
Partially open	Modular office layout	375 feet
Closed indoor	Typical office layout	300 feet

Flexibility and mobility make WLANs both effective extensions and attractive alternatives to wired networks. Wireless LANs provide all the functionality of wired LANs, but without the physical constraints of the wire itself. WLAN configurations include independent networks, suitable for small or temporary peer-to-peer configurations, and infrastructure networks, offering fully distributed data connectivity via microcells and roaming. In addition to offering end-user mobility within a networked environment, WLANs enable portable networks, allowing LANs to move with the knowledge that workers need them.

A wide range of WLAN products are now available. By evaluating the strengths and differences of each of these offerings, savvy network managers and users can choose a WLAN solution that best meets their business and application objectives. Customers who want to learn more about wireless LANs and WLAN technology can look to the Wireless LAN Alliance for assistance (www.wlana.com or www.solveit.com). Figure 8–14 provides a list of wireless LAN vendors.

Appendixes A though D provide WLAN solutions.

Battery Life for Mobile Platforms

End-user wireless products are capable of being completely untethered, and operate using the battery power from their host notebook or hand-held computer. WLAN vendors typically employ special design techniques to maximize the host computer's energy usage and battery life.

Safety

The output power of wireless LAN systems is very low, much less than that of a hand-held cellular phone. Because radio waves fade rapidly over distance, very little exposure to RF energy is provided to those in the area of a wireless LAN system. Wireless LANs must meet stringent government and industry regulations for safety. No adverse health effects have been attributed to wireless LANs.

1stWAVE Wireless International
2Wire Inc.
3Com
Accton Technology Corporation
Achako Corporation
ACMA Computers Inc.
Acrowave Systems
Acrowave Systems Co., Ltd.
Actiontec Electronics Inc.
ACTiSYS Corporation
Addtron Technology Co., Ltd.
ADTRAN Inc.
AeroComm Inc.
Aerotron-Repco Systems Inc.
Agere Systems Inc.
Alcatel
Allied Telesyn International Inc.
Alvarion Inc.
Ambicom Inc.
Ambit Microsystems Corporation
AnyOne Wireless
AOpen Inc.
Apple Computer Inc.
ARC Wireless Solutions Inc.
ARESCOM Inc.
ARtem GmbH.
Asante Technologies Inc.
Askey Computer Corporation
Avaya Inc.
Axis Communications Inc.
Bluesocket Inc.
Broadband Gateways
BroadTel Communications
Buffalo Technology (USA) Inc.
California Amplifier Inc.
Canadian Digital Systems
    Technologies Inc.
Cayman Systems Inc.
CC&C Technologies Inc.
Cirronet™ Inc.
Cisco Systems Inc.
Clarion Corporation of America
Colubris Networks Inc.
Communication mbH

Compaq Computer Corporation
Compex, Inc.
C-SPEC Corporation
Data-Linc Group
Dataplex Pty Ltd.
Dell Computer Corporation
Digital Networks
Direct Network Services
D-Link Systems Inc.
ELSA Inc.
EMTAC Technology Corporation
Enterasys Networks
Ericsson Inc.
Extended Systems
fSONA Communications Corporation
Furtera Inc.
Gabriel Electronics Inc.
Gateway Inc.
Gemplex Technology Co., Ltd.
GemTek Technology Co., Inc.
GigaFast Ethernet
GoC Gesellschaft fuer optische
GRE America
GVC Corporation
Harris Corporation
Hewlett-Packard Company
HighSpeed Surfing Inc.
IBM Corporation
Inficom Inc.
Infrared Communication Systems
    Inc.
Intel Corporation
Intermec Technologies Corporation
Intersil Corporation
ioWave Inc.
Karlnet Inc.
Katharsis Ltd.
Lantronix Inc.
LaserWireless Inc.
LeArtery Solution Inc.
LightPointe
Linksys Group Inc.
LSA Inc.
Lucent Technologies

LumenLink Corporation
LXE Inc.
Matsushita Electric Industrial Co.,
    Ltd.
MaxGate Inc.
Maxrad Inc.
MaxTech Corporation
Microhard Systems Inc.
MiLAN Technology
MiTAC International Corporation
Mobile Mark Inc.
Multi-Tech Systems Inc.
National Datacomm Corporation
NDC Communications Inc.
NETGEAR Inc.
Netopia Inc.
Nokia
Nomadix
Nortel Networks
Novatel Wireless
Omnispread Communications, Inc.
Optel
OTC Wireless Inc.
Padcom Inc.
PAV Data Systems Ltd.
Pinnacle Communications Inc.
Plaintree Systems Inc.
Possio
Proxim Inc.
Psion Teklogix Ltd.
RadioLAN Marketing Group Inc.
Raylink Inc.
Raytheon
RFTNC Co., Ltd.
Samsung Electro-Mechanics Co. Ltd.
Seavey Engineering Associates Inc.
SENAO International Co., Ltd.
Siemens Corporation
Silcom Technology
SMC Networks
Socket Communications Inc.
SoftHill Technologies Ltd.
SOHO Wireless
Solectek Corporation

Figure 8–14
Wireless LAN Vendors
(Courtesy of www.wlana.org, www.solveit.com)

Sonik Technologies Corporation
Sony Electronics Inc.
Southwest Microwave Inc.
SpectraLink Corporation
Spectrix Corporation
SuperPass Company Inc.
Symbol Technologies
TECOM Industries Inc.
Tekram Technology Europe GmbH
Teletronics International Inc.
Telex Communications Inc.

TIL-TEK Antennas
TRENDware International Inc.
TROY Wireless
U.S. Robotics
UC Wireless
V3 World Ltd.
Wave Wireless Networking
WaveRider Communications (USA)
   Inc.
Western Multiplex Corporation
Wi-LAN

WIMAN Systems Inc.
WinMate Communication, Inc.
Wireless Scientific Inc.
Wireless Solutions Sweden AB
Wireless Inc.
Wistron NeWeb Corp.
Young Design, Inc.
Z-Com Inc.
Zcomax Technologies Inc.
Zoom Telephonics
ZyXEL Communications Inc.

Figure 8–14
Continued

MICROWAVE LANS

Today's remote site interconnections can be considered as an extended LAN and as such require higher bandwidth solutions to handle the immense traffic transported between separate locations (as much as 10 Mbps, the native Ethernet LAN speed). There are a growing number of wireless solutions to interconnect remote sites such as spread spectrum radios, laser, and microwave. All have their respective advantages and disadvantages.

Of the many solutions available to connect remote sites, microwave can be considered for most applications. Microwave systems can connect remote distances up to 20 miles or more, but most applications fall under 4 miles. Microwave provides a solution that offers full bandwidth Ethernet LAN connectivity, high reliability, and reasonable payback periods from 6 months to 2 years when compared with traditional wired solutions. Microwave links are absolutely transparent, acting as an extension of the Ethernet backbone or segment. Because it is fully compatible with the IEEE 802.3 Ethernet standard, microwave supports all Ethernet functionality and applications.

A typical microwave system can consist of three major components: an indoor data interface unit to connect to the network, a radio unit, and a parabolic antenna. The indoor data interface unit will usually offer an IEEE 802.3 interface to the local area network. The indoor data interface unit generally operates as an Ethernet LAN transceiver interface, but may sometimes offer bridging or routing functions. The radio unit is most often placed outside near the antenna. The radio unit modulates the information from the indoor data interface unit to a higher microwave frequency. The parabolic antenna functions primarily as a component that focuses the microwave signal into a narrow beam that radiates toward the opposite receiving antenna.

Microwave radios have traditionally had a perception of being reliable and fast, but also complicated to install and maintain. Some of today's newer microwave systems have made significant improvements. They are lightweight, compact, and easy to install and maintain. In fact, some of the smaller microwave systems require only a 2.5 inch post mounted on the roof or inside an office window to operate effectively. These factors reduce the time it takes to install, and they cut costs. In most cases, all that is required to physically install a microwave system are details such as placement of the antennas, type of mount to be used, and placement of the cable runs from the outdoor radio unit to the indoor data interface unit.

Microwave systems require a clear line of sight between the two locations. This implies that there can be no obstacles or obstructions in the path of the microwave signal. To accomplish this, microwave antennas are mounted as high as possible, generally on roofs of buildings, radio towers, or other tall structures. In the event that a clear line of sight cannot be established, systems are sometimes repeated over objects (i.e., other buildings) or reflected around an object using specific microwave passive repeaters. In many cases it is necessary to perform a site survey to ensure a clear line of sight.

Microwave system performance can be affected by environmental and atmospheric factors. The most prevalent factor that affects microwave signals is rain, especially in the higher frequency bands (above 18 GHz). Rain has the adverse property of absorbing microwave energy and attenuating the signal path. Other factors such as snow, fog, smog, and temperature inversions have minimal effects on microwave performance.

Even with these environmental factors, microwave communications can offer greater than 99.995 percent availability. To ensure reliable communications, microwave systems are designed to accommodate the rain factor in several ways. First, the worst-case rain rate is identified for a specific region. Most often, microwave path designers plan for rainfalls that exceed the 0.01 percent rate. This amount can range anywhere from 1 inch per hour in Arizona to 4 inches per hour in Florida. Second, the maximum path distance is calculated with the worst-case rainfall. This means that systems that can operate up to 3 miles in Arizona may only be able to operate up to 1 mile in Florida. Third, antenna size is determined to achieve the desired reliability factor. The larger the antenna size, the higher the focusing factor, or gain. A 99.995 percent availability translates into about 26 total minutes of statistical outage in a year.

Most microwave LAN systems operate in the 23 GHz band and require licensing by the FCC. This licensing procedure is a simple process and assures the user of interference-free operation. The licensing process consists of two steps. First, a frequency coordination is performed to determine an available frequency. Many search firms are available in the United States that can perform this task. They basically have access to a regularly updated database that identifies other existing microwave users in various geographic areas. The geographic coordinates of your intended site are submitted to one of these search firms to allow them to accurately recommend a unique frequency that has minimal interference potential. Site coordinates are generally obtained from a site survey. Second, a License Request Form 402 is filed with the FCC along with the frequency coordination report. After reviewing your submittal, the FCC grants your station the license to operate that specific microwave system at that specific site using the requested frequencies. The process takes about 2 weeks for frequency coordination and 60 days for the FCC license approval. This entire process can be handled by most search firms.

RECENT EVENTS

Recent events in the wireless LAN arena include a new substandard of the initial IEEE 802.121 standard and a proprietary wireless LAN standard specifically developed for the home networking arena. The IEEE has developed a new specification called 802.11a that represents the next generation of enterprise-class wireless LANs.[2] Among the advantages it has over current technologies are greater scalability, better interference immunity, and significantly higher speed, up to 54 Mbps and beyond, which simultaneously allows for higher bandwidth applications and more users. Following is an overview in basic terms of how the 802.11a specification works, and its corresponding benefits.

Physical Layer

5 GHz Frequency Band

802.11a utilizes 300 MHz of bandwidth in the 5 GHz Unlicensed National Information Infrastructure (U-NII) band. Though the lower 200 MHz is physically contiguous, the FCC has divided the total 300 MHz into three distinct 100 MHz domains, each with a different legal maximum power output. The low band operates from 5.15 to 5.25 GHz and has a maximum of 50 mW. The middle band is located from 5.25 to 5.35 GHz with a maximum of 250 mW. The high band utilizes 5.725 to 5.825 GHz with a maximum of 1 W. Because of the high-power output, devices transmitting in the high band will tend to be building-to-building products. The low and medium bands are more suited to in-building wireless products. One requirement specific to the low band is that all devices use integrated antennas.

Different regions of the world have allocated different amounts of spectrum, so geographic location will determine how much of the 5 GHz band

---

[2] Information courtesy of Proxim Corporation, a leader in wireless LAN technology at www.proxim.com.

is available. In the United States, the FCC has allocated all three bands for unlicensed transmissions. In Europe, however, only the low and middle bands are free. Though 802.11a is not yet certifiable in Europe, efforts are currently underway between IEEE and the European Telecommunications Standards Institute (ETSI) to rectify this. In Japan, only the low band may be used. This will result in more contention for signal, but will still allow for very high performance.

The frequency range used currently for most enterprise-class unlicensed transmissions, including 802.11b, is the 2.4 GHz ISM band. This highly populated band offers only 83 MHz of spectrum for all wireless traffic, including cordless phones, building-to-building transmissions, and microwave ovens. In comparison, the 300 MHz offered in the U-NII band represents a nearly fourfold increase in spectrum—all the more impressive when considering the limited wireless traffic in the band today.

OFDM Modulation Scheme

802.11a uses orthogonal frequency division multiplexing (OFDM), a new encoding scheme that offers benefits over spread spectrum in channel availability and data rate, as seen in Figure 8–15. Channel availability is significant because the more independent channels that are available, the more scalable the wireless network becomes. The high data rate is accomplished by combining many lower speed subcarriers to create one high-speed channel. 802.11a uses OFDM to define a total of eight nonoverlapping 20 MHz channels across the two lower bands; each of these channels is divided into fifty-two subcarriers, each approximately 300 KHz wide. By comparison, 802.11b uses three nonoverlapping channels.

A large (wide) channel can transport more information per transmission than a small (narrow) one. As described, 802.11a utilizes channels that are 20 MHz wide, with 52 subcarriers contained within. The subcarriers are transmitted in parallel, meaning they are sent and received simultaneously. The receiving device processes these individual signals, each one representing a fraction of the total data that, together, make up the actual signal. With this many subcarriers comprising each channel, a tremendous amount of information can be sent at once.

Figure 8–15
OFDM
(Courtesy of
www.proxim.com)

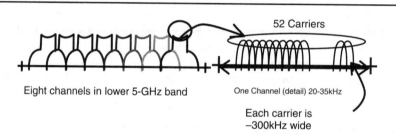

Eight channels in lower 5-GHz band

52 Carriers

One Channel (detail) 20-35kHz

Each carrier is
−300kHz wide

With so much information per transmission, it obviously becomes important to guard against data loss. Forward error correction (FEC) was added to the 802.11a specification for this purpose (FEC does not exist in 802.11b). At its simplest, FEC consists of sending a secondary copy with the primary information. If part of the primary information is lost, insurance then exists to help the receiving device recover (through sophisticated algorithms) the lost data. This way, even if part of the signal is lost, the information can be recovered so the data are received as intended, eliminating the need to retransmit. Because of its high speed, 802.11a can accommodate this overhead with negligible impact on performance.

Another threat to the integrity of the transmission is multipath reflection, also called delay spread. When a radio signal leaves the "sending" antenna, it radiates outward, spreading as it travels. If the signal reflects off a flat surface, the original signal and the reflected signal may reach the "receiving" antenna simultaneously. Depending on how the signals overlap, they can either augment or cancel each other out. A baseband processor, or equalizer, unravels the divergent signals; however, if the delay is long enough, the delayed signal spreads into the next transmission. OFDM specifies a slower symbol rate to reduce the chance that a signal will encroach on the following signal, minimizing multipath interference.

Multipath interference occurs when reflected signals cancel each other out. 802.11a uses a slower symbol rate to minimize multipath interference, as seen in Figure 8–16.

Data Rates and Range

Devices utilizing 802.11a are required to support speeds of 6, 12, and 24 Mbps. Optional speeds go up to 54 Mbps, but will also typically include 48, 36, 18, and 9 Mbps. These differences are the result of implementing different modulation techniques and FEC levels. To achieve 54 Mbps, a mechanism called 64-level quadrature amplitude modulation (64QAM) is used to pack the maximum amount of information possible (allowable by the standard) on each subcarrier.

As an 802.11a client device travels farther from its access point, the connection will remain intact but speed decreases (falls back) just like with 802.11b. However as Figure 8–17 illustrates, at any range, 802.11a has a significantly higher signaling rate than 802.11b, and is always faster than 802.11b, at any range.

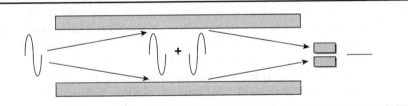

Figure 8–16
Minimized multipath interference of 802.11a

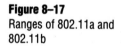

Figure 8–17
Ranges of 802.11a and
802.11b

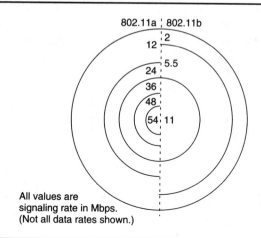

MAC Layer

802.11a uses the same media access control (MAC) layer technology as 802.11b, carrier sense multiple access with collision avoidance (CSMA-CA). CSMA-CA is a basic protocol used to avoid signals colliding and canceling each other out. It works by requesting authorization to transmit for a specific amount of time prior to sending information. The sending device broadcasts a request to send (RTS) frame with information on the length of its signal. If the receiving device permits it at that moment, it broadcasts a clear to send (CTS) frame. Once the CTS goes out, the sending machine transmits its information. Any other sending devices in the area that "hear" the CTS realize another device will be transmitting and allow that signal to go out uncontested.

Relation to HiperLAN2

HiperLAN2 is a wireless specification developed by ETSI, and has some similarities to 802.11a at the physical layer. It also uses OFDM technology, and operates in the 5 GHz frequency band. The MAC layers are different, however. While 802.11a uses CSMA-CA, HiperLAN2 utilizes time division multiple access (TDMA).

Because the 5 GHz U-NII equivalent bands have been reserved for HiperLAN2 systems in Europe, 802.11a is not yet certifiable in Europe by ETSI. In an effort to rectify this, two additions to the IEEE 802.11a specification have been proposed to allow both 802.11a and HiperLAN2 to coexist. Dynamic channel selection (DCS) and transmit power control (TPC) allow clients to detect the most available channels and use only the minimum output power necessary if interference is evident. The implementation of these additions will significantly increase the likelihood of European 802.11a certification.

Compatibility with 802.11b

While 802.11a and 802.11b share the same MAC layer technology, there are significant differences at the physical layer. 802.11b, using the ISM band, transmits in the 2.4 GHz range, while 802.11a, using the U-NII band, transmits in the 5 GHz range. Because their signals travel in different frequency bands, one significant benefit is that they will not interfere with each other. A related consequence, therefore, is that the two technologies are not compatible. There are various strategies for migrating from 802.11b to 802.11a, or even using both on the same network concurrently. These strategies are beyond the realm of this chapter.

802.11a represents the next generation of enterprise-class wireless LAN technology, with many advantages over current options. At speeds of 54 Mbps and greater, it is faster than any other unlicensed solution. 802.11a and 802.11b both have a similar range, but 802.11a provides higher speed throughout the entire coverage area. The 5 GHz band in which it operates is not highly populated, so there is less congestion to cause interference or signal contention. The eight nonoverlapping channels also allow for a highly scalable and flexible installation. 802.11a is the most reliable and efficient medium by which to accommodate high-bandwidth applications for numerous users.

Another recent development, HomeRF, was introduced in mid-2001.[3] Although 802.11 WLANs can be used for home networking, HomeRF is a proprietary standard that is specifically designed for home networking purposes. See the later section on HomeRF.

The home marketplace for networking is expanding rapidly. Internet usage, now at more than 80 million users worldwide, has quadrupled since 1996. Nearly 40 percent of all U.S. households are online, and the use of PCs for computing, communicating, and entertaining is now so pervasive that more than 20 million U.S. households have more than one PC (Parks Associates). In addition, the mobility of laptop computers gives users an easy way to take work between home and office, and as many as 60 percent of all laptops routinely travel between these locations (Intel 2001). What is more, the role of the traditional office is changing as evidenced by the large and growing number of home-based businesses and home offices. These trends are driving the need to share Internet access, PCs, phone lines, peripherals, and information in the home. It is this need that is fostering the tremendous growth in the home networking market, and broadband deployments with bundled services should fuel continued growth.

Home networks offer users a way to share information and resources by interconnecting PCs and other devices within the home. With easy and affordable solutions now available, the home networking market is predicted to reach $1.4 billion by the end of 2003 (Intel 2001). See Figure 8–18.

---

[3] Courtesy HomeRF Working Group at www.homeRF.org., 2002.

Figure 8–18
Projected Growth of
Home Networking in the
United States
(Courtesy Dataquest
March 2000)

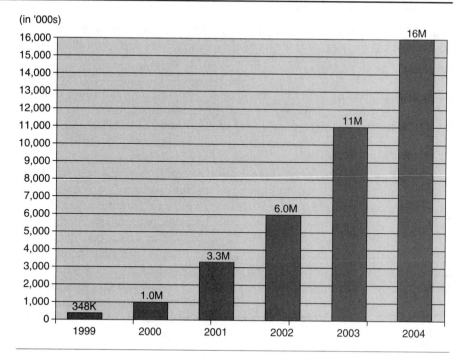

(in '000s)

IEEE 802.11g

In December 2001, IEEE determined key specifications that would essentially double the current data transmission rates over the much-ballyhooed WLAN environment, enabling true wire-free multimedia content streaming (Liu 2001). "Data rates are extremely critical," said Kurt Scherf, vice president of research at Parks Associates, a Dallas-based market research and consulting firm. "As 802 applications move from the office to the home, it means multimedia applications become more realistic for the home." Parks Associates estimated that, while 5 percent of U.S. households currently have a PC network in place, as many as 15 percent will have one in five years. Of that, wireless networking will account for 40 percent of all those home networks.

"I think [the IEEE decision is] very important. It's surprising to me how little attention it's gotten," said Navin Sabharwal, vice president of residential and networking technologies at Allied Business Intelligence (ABI), an Oyster Bay, New York, think tank.

Much of the reason for the heightened interest in the IEEE vote is related to the limitations of current 802.11b technology. Since the 802.11b specification was finalized by IEEE in 1999, many networking companies from Intel to Compaq have quickly adopted it—only to find its theoretical transmission speeds of 11 Mbps a bit of a misnomer. Due to legacy synchronization issues, 802.11b only nets about 7 Mbps of throughput.

The 802.11 working group is also creating other specifications such as .11e (for voice transmission and security) and .11a (which promises data rates of 54 Mbps). But as a result of the nonprofit organization's glacial pace of standard adoption only via consensus, no other technologies will make their way to chipsets sooner than 802.11g.

Every specification must comply with current FCC rules for operating in the unlicensed 2.4 GHz spectrum space. As a general condition, devices operating in the unlicensed spectrum may not cause harmful interference with authorized services and must work around any interference that may be received from phones, microwaves, or other RF devices.

To operate within the 2.4 GHz spectrum, the FCC mandates that a device must operate in one of two ways.

- Frequency hopping spread spectrum (FHSS) transmits bits of data by hopping along various frequencies. These data signals are transmitted and received using the same algorithm, which allows the sender and recipient to follow the signal along the various frequencies.
- Direct sequence spread spectrum (DSSS) breaks up the data signal into sequences and transmits to the receiver, which reassembles the sequences into the data signal.

HomeRF

While 802.11b and 802.11g were designed primarily for the corporate environment, HomeRF was developed from the start to meet the unique needs of the consumer in-home networking applications. In addition to offering solid technical features, simplicity, security, and ease of use, HomeRF networks are designed to be more affordable to home users than other wireless technologies.

HomeRF networks provide a range of up to 150 feet, sufficient to cover the typical home, garage, and yard. A recent ruling by the FCC allowed increased data rates for HomeRF networks, and products based on the second generation HomeRF 2.0 specification will take advantage of this ruling. HomeRF 2.0 enables the introduction of new types of devices, applications, and services including high-speed applications such as whole-house CD-quality audio distribution to wireless speakers and streaming video.

In addition to the expected benefits of home networking, such as shared Internet access, PCs, data files, and printers, HomeRF also supports high-quality voice and data. New information appliances and other devices will emerge that take advantage of the integrated voice and data capabilities. As services and devices evolve, HomeRF networks will support home entertainment, home automation, and even telemedicine applications.

An integrated network offers a simple, low-cost solution and makes it easy to add voice or data devices to the network when needed. Additionally, it improves access to and utilization of both traditional telephone lines and high-speed Internet connections.

The HomeRF specification incorporates the digital enhanced cordless telephony (DECT) standard. This standard supports the telephony features consumers expect plus a full range of enhanced features useful in both home and small-business applications. Cordless handsets using HomeRF will offer even more convenience than traditional cordless telephones, including more flexible phone placement. Today, cordless phones must be connected to telephone jacks in locations that may not be ideal to the end user; and because many homes include only two or three phone jacks, traditional cordless phones are confined to certain rooms.

With HomeRF there is no need for multiple base units to be tied to individual phone jacks. Only one connection to a single telephone jack is required, and additional cordless handsets can be purchased and placed wherever convenient. The ability to expand the voice network by simply adding handsets offers savings over purchasing multiple cordless phones.

Users also benefit from the convenience of sharing the same rich set of telephone features. Multiple handsets can also be used to enable users to place both external and intercom calls at the same time.

HomeRF technology also provides superior voice quality and security. Using 2.4 GHz technology, the HomeRF avoids interference from other cordless phones, remote controls, and baby monitors. Using FHSS technology, the voice channel changes fifty times every second, ensuring conversations are not overheard.

While today's home networks typically connect multiple PCs to enable Internet and printer sharing, home networks of tomorrow will enable sharing of unified voice, data, and video services. Users will realize new benefits and conveniences by including everything from new digital entertainment devices to traditional household appliances in the home network. Further, HomeRF support of both voice and data will allow speech-enabled applications for increased user convenience.

Following are examples of what users will be able to do with the availability of products that adhere to the HomeRF specification.

- Set up a wireless home network to share voice and data between PCs, peripherals, PC-enhanced cordless phones, and new devices such as portable remote displays or Web pads.
- Access the Internet from anywhere in and around the home from portable display devices.
- Share a single ISP connection between PCs and other new devices.
- Intelligently forward incoming telephone calls to multiple cordless handsets, fax machines, and voice mailboxes.
- Review incoming voice, fax, and e-mail messages from a small PC-enhanced cordless telephone handset.
- Activate other home electronics systems by simply speaking a command into a PC-enhanced cordless handset.

- Play multiplayer games, toys, and gaming consoles based on PC or Internet resources.
- Download MP3 and other audio files using audio streaming.
- Enjoy streaming audio and video from networked devices anywhere in the home.

■ SUMMARY

This chapter presents aspects of wireless LANs. For many years, wireless LANs were the one segment of the wireless market that lagged behind, technologically. Now, due to recent speed enhancements, wireless LANs are becoming more commonplace.

For further information, the following websites offer extensive coverage of wireless LANs.

www.wlana.org
www.solveIT.com
www.proxim.com
www.hydra.carleton.ca/info/wlan.html
www.orinocowireless.com
www.nokia.com/corporate/wlan/
www.wirelesslan.com
www.nortelnetworks.com
www.cisco.com
www.3com.com

In the next chapter we will discuss satellite communications.

REVIEW QUESTIONS

1. True/False Wireless LANs frequently augment rather than replace wired LAN networks.

2. In a typical WLAN configuration, a transmitter/receiver (transceiver) device is called the _____.
   a. terminal
   b. network interface card
   c. access point
   d. antenna

3. End users access the WLAN through _____.
   a. wireless LAN adapters
   b. PDAs
   c. JavaScript code
   d. wireless routers

4. WLAN data rates range up to _____ Mbps.
   a. 10
   b. 1.6
   c. 11
   d. 100

5. In the _____ mode, mobile terminals build the network by themselves.
   a. WMAN
   b. ad hoc
   c. point-to-point

6. The frequency range used currently for most enterprise-class unlicensed transmissions, including 802.11b, is the _____ industrial, scientific, and medical (ISM) band.

a. 900 MHz

b. 2.4 GHz

c. 800 MHz

7. Microwave LAN systems can connect remote distances up to _____ miles.

a. 2

b. 5

c. 20

d. 100

8. _____ is the most prevalent atmospheric factor that degrades microwave LAN performance.

a. rain

b. heat

c. light

9. Devices utilizing 802.11a are required to support speeds of _____, _____, and _____ Mbps.

a. 6, 12, 24

b. 12, 24, 36

c. 10, 100, 1,000

d. 1, 1.6, 11

10. The speed of an IEEE 802.11b WLAN is _____ Mbps.

a. 54

b. 100

c. 4

d. none of the above

11. The latest standard on WLANs revolves around the _____ WLAN standard.

a. IEEE 802.11b

b. IEEE 802.11a

c. IEEE 802.11g

d. none of the above

12. HomeRF is a _____ standard.

a. proprietary

b. de facto

c. de jure

d. TCP/IP

e. OSI

13. Movement of a wireless node between two micro-cells is referred to as _____.

a. roaming

b. traveling

c. transmission

d. error consensus

HANDS-ON EXERCISES

1. Using your browser, go to the wireless LAN consortium's website at www.solveIT.org. Research the various customer solutions found there and prepare a one-page report on at least three solutions.

2. Using your browser and a search engine such as Google (www.google.com), search for industry wireless LAN solutions. Prepare a report describing at least one solution.

3. Make some telephone calls and try to locate wireless LAN implementations in the area where you live. Prepare a slide show presentation of this implementation.

4. Using Microsoft Excel or a similar product, create a table comparing 802.11b and HomeRF.

5. Conduct research and produce a table listing the characteristics, advantages, and disadvantages of IEEE 802.11a, 802.11b, and 802.11g.

6. Conduct research and produce a chart listing at least five products based upon the HomeRF standard, along with a brief statement on the product, its cost, and its availability.

9 Satellite Communications

OBJECTIVES

After reading this chapter and completing the exercises, you will be able to:

- Learn the various types of satellite-based systems
- Learn the various uses for satellite-based systems
- Learn a few basics about orbits and how satellites get into orbit
- Learn examples of satellite systems in use today and those planned for the near future

THE INITIAL JOINT POLAR SYSTEM

For some two decades, the United States has provided continuous operational data from its National Oceanic and Atmospheric Administration (NOAA) series of polar orbiting meteorological satellites. Russia (formerly the USSR) has also maintained a series of meteor satellites in polar orbit. The United States currently maintains two satellites in polar orbit, one with an equator crossing time in the morning, the other in the afternoon. In 1998, EUMETSAT decided to join with the United States in providing a joint system of operational meteorological satellites in polar orbit. This Initial Joint Polar System (IJPS) will come into effect when EUMETSAT takes over operational responsibility for the so-called morning orbit with its new series of Metop satellites, while the United States continues to maintain a series of NOAA satellites in the afternoon orbit. In addition, the United States has operated the satellites of its Defense Meteorological Satellite Program (DMSP). Responsibility for operations in polar orbit will be shared from 2005, when the EUMETSAT Polar System becomes available.

The Initial Joint Polar System will include an NOAA satellite from the United States and a Metop satellite from Europe, in complementary orbits designed to ensure complete global data coverage at intervals of no more than 6 hours.

The IJPS will comprise the continuation of the current NOAA satellite series with NOAA-N and -N', with the new EUMETSAT satellite series Metop-1, -2, -3, the first of which being due for launch in 2005. The satellites will be produced independently by the United States and Europe, respectively, but will carry a core set of nearly identical instruments to ensure operational data continuity and coherence of the key meteorological observations. The core set of instruments on both satellite series includes the latest versions of AVHRR, HIRS, and AMSU-A, supplied by the United States; and MHS, provided by EUMETSAT. The plans include provision for the flight of AVHRR and AMSU-A on the first two Metop satellites. Beyond that, upgraded versions will be considered for flight on Metop-3 and later satellites.

In addition to the core set of common instruments, both satellite series will carry additional instrumentation for specific purposes. The NOAA satellites will carry the latest version of the SBUV instrument, while Metop will carry IASI, GOME, GRAS, and ASCAT, as detailed in the following table.

The United States is also working to converge its present parallel systems of civilian (NOAA) and defense (DMSP) meteorological satellites. These will come together when those two systems experience a further major upgrade, in the 2008 time frame, to become the National Polar Operational Environmental Satellite System (NPOESS). From that time there will be three polar satellites in the joint operational system, with overpass times evenly spaced throughout the day to ensure availability of global data at intervals of no longer than 4 hours.

In addition to these coordinated European/U.S. plans. Russia plans to provide a follow-on to its Meteor-2 and Meteor-3 series of polar meteorological satellites. The People's Republic of China has also launched experimental polar meteorological satellites with an imaging capability similar to that of AVHRR.

Instrument Payloads for the Initial Joint Polar System

Instrument on Metop-1, -2, -3	Instrument on NOAA-N -N'	Full Name	Primary Function
AVHRR/3*	AVHRR/3	Advanced Very High Resolution Radiometer	Global imagery of clouds, the ocean, and land surface
HIRS/4	HIRS/3	High Resolution Infrared Radiation Sounder	Temperature and humidity of the global atmosphere in cloud-free conditions
AMSU-A*	AMSU-A	Advanced Microwave Sounding Unit-A	Temperature of the global atmosphere in all weather conditions
MHS	MHS	Microwave Humidity Sounder	Humidity of the global atmosphere
IASI		Infrared Atmospheric Sounding Interferometer	Enhanced atmospheric soundings
GRAS		Global Navigation Satellite System Receiver for Atmospheric Sounding	Temperature of the upper troposphere and in the stratosphere with high vertical resolution
ASCAT		Advanced Scatterometer	Near-surface wind speeds over the global oceans
	SBUV	Solar Backscattered Ultraviolet ozone probe	Total atmospheric ozone
GOME-2*		Global Ozone Experiment-2	Monitoring profiles of ozone and other atmospheric constituents

Note: *Updated instruments considered for Metop-3

Source: www.noaa.gov.

SATELLITE COMMUNICATIONS—STRATOS GLOBAL CORPORATION

Stratos Global Corporation is a company that provides mobile and fixed remote communications for clients who do not have access to more traditional methods of communication because of extenuating geographical circumstances. Stratos is based in Toronto, Canada, and has annual revenue of $176.3 million. Stratos's communications are operated by using its own LEO satellites and digital satellite communications system. Stratos provides service to such sectors as oil and gas platforms, energy, mining, commercial fishing, humanitarian disaster relief, shipping, and the military. Stratos utilizes very small, lightweight mobile or fixed communications devices to connect these isolated industries with the outside world. Iridium is one of Stratos's most popular communications devices. The Iridium phones weigh less than 1 pound and can send and receive calls anywhere in the world. The Iridium phones provide up to 10 Kbps data and 2.4 Kbps voice transmission over its system. Another communications system developed by Stratos is the Inmarsat-M4 satellite terminal. This terminal combines high-speed data transmission capabilities with PSTN quality voice transmission. This system provides a connection point for analog devices, computers, and faxes. Stratos is definitely a world leader in communications equipment because of its use of satellite communications.

//www.stratosglobal.com/

STARBAND

A joint venture of Gilat, EchoStar, Microsoft, and ING Furman Selz Investments, StarBand was launched in 2000 using Gilat's Internet-via-satellite solution:

- Very small aperture terminal (VSAT) satellite communications equipment and services
- Patented Gilat software and advanced performance-enhancing technologies, including a quality of service solution that enables Internet traffic to be controlled by user, application, and time of day
- Advanced technical support
- Single-hop architecture, centralized network, and redundant operations, independent from terrestrial networks
- A national, scalable solution that can be expanded quickly and easily to households across the United States

StarBand is delivering on the promise of the high-speed Internet revolution to tens of thousands of American consumers and small office/home office (SOHO) users across the country. Now, anyone with a clear view of the southern sky—from the most rural locations to urban areas—can be part of the high-speed Internet revolution.

"We are excited to bring the reality of high-speed internet service to Americans across the country who have been denied access because their homes are located in areas not served by DSL or cable. Thanks to the innovative technology provided to us from Gilat Satellite Networks, consumers no longer have to worry about how far away they live from the telephone company's central office or if their cable company has upgraded the cable in their neighborhood. All they have to do is look up. If they can see the southern sky, they can get StarBand service," according to Zur Feldman, co-chairman and chief executive officer at StarBand Communications.

Gilat Delivers High-Speed Internet to U.S. Consumers

The Internet revolution is here, but access to the bandwidth that powers it is still out of reach for millions. Even in the United States, it is estimated that over 50 million households do not have access to cable modem or DSL technology. But 90 percent of U.S. single-family homes have satellite line-of-sight, which means help is on the way.

Gilat serves as the technology supplier, wholesale provider of VSAT satellite communications equipment, and provider of operations support and advanced technical services. With the power, reach, and convenience of satellite technology, the StarBand service offers broadband internet access to selected end users as well as several key advantages.

- American consumers and SOHO users who use the service can now connect to the Internet at browsing speeds up to 10 times higher than dial-up modem speeds: up to 500 Kbps downstream speeds. They also enjoy an always-on connection that saves time when connecting to the Internet and eliminates the need for a second telephone line.
- Another advantage is convenience: a one-stop hardware and services solution, complete with installation. No more multiple infrastructures, additional phone lines, coordination between service providers or need for a land-based ISP. The small satellite dish antenna, just 24 × 36 inches, is easily mounted on a roof, chimney, or on a pole in the yard. The antenna is then connected to the StarBand stand-alone satellite modem, which attaches to the user's computer via its USB or Ethernet port.
- Finally, StarBand gets the benefits of Gilat's patented software and advanced features, and was the first to offer a single antennal solution for both Internet and satellite television services through the partnership with EchoStar Communications Corp.

Gilat Consumer Internet Access Partnerships Worldwide

Gilat is continuing its trailblazing tradition by providing technology and support to ISPs for first-ever consumer satellite Internet services in Latin America, Asia, and Europe. More recently, Gilat has adopted a wholesale strategy, announcing consumer/SOHO projects with Star One and Universo Online in Brazil, Tiscali in Italy, Optus in Australia, Bharti Broadband in India, and Jingxin Hero in China.

For additional information about the StarBand system, go to www.echostar.com or www.msn.com.

BACKGROUND

Satellites are generally categorized by the altitude or geometry of their orbit around earth. The types of services these satellites provide are the determining factors in choosing what type of earth orbit each of the satellite systems will use. The different orbits allow for multiple paths and better use of the space around earth, as well as facilitate the specific needs of different satellite systems.

satellites

Following are the basic types of satellite systems and their ranges:

- LEO—low earth orbit (500–1,500 km)
- MEO—medium earth orbit (5,000–15,000 km)
- GEO—geostationary earth orbit (35,786 km)
- Polar orbits (variable heights)
- HEO—highly elliptical earth orbit (variable generally apogee = GEO altitude)

Each of these systems has their own advantages and disadvantages, and each is configured to make the best use of its associated attributes while minimizing detriments. Our discussion will begin with a brief history of satellites and then address some general concepts on satellites. Lastly, we will discuss the LEO satellite system configurations. Discussions on the other types will follow. An elementary discussion can be found in Chapter 1.

HISTORY

Sputnik

On October 4, 1957, the first artificial satellite, named **Sputnik,** was successfully launched and achieved orbit.[1] This Russian satellite was designed to measure and transmit information about the density of the upper atmosphere as it circled the earth. Before then, all atmospheric information that meteorologists had was based on "looking up" instead of "looking down." Since Sputnik, the science of satellites has advanced significantly. Specialized satellites now orbit the earth for telecommunications, global broadcasting, oceanic observations, scientific research, and military purposes, as well as for atmospheric observation. It is amazing to consider the rapid progress and evolution of satellite technology.

The first Soviet satellite program consisted of four Sputnik, or fellow traveler, satellites. As shown in Figure 9–1, Sputnik had a mass of 83.6 kg. As mentioned, it was designed to determine the density of the upper atmosphere and return data about the earth's ionosphere. However, its two radio transmitters only returned signals to earth for 21 days.

Meteorological satellites themselves have changed considerably since Sputnik. Weather observation and forecasting have become much more reliable today, thanks to the sophisticated satellites currently orbiting the earth.

In 1947, an unmanned American rocket carried a camera into outer space and recorded pictures of Earth from space.[2] The pictures showed how the atmosphere appeared from above. Cloud formations were clearly visible. These pictures gave scientists proof that weather observations could be made from space on a regular basis.

In 1960, the first weather satellite was launched into orbit around the earth. Called the television infrared observational satellite (Tiros), it carried a video camera to make regular observations of the atmosphere below. For the first time, meteorologists were able to compare their localized ground-based weather observations with broader pictures of the weather system. Weather forecasting took a dramatic leap forward. Figure 9–2 depicts the Tiros-1 satellite.[3] After the initial successes, nine more Tiros satellites were put into orbit during the 1960s.

In 1966, the United States placed its first weather satellite in high, geostationary orbit. Called the applications technology satellite (ATS), it traveled at the same speed the earth rotates, appearing to remain stationary with re-

---

[1] See nauts.com, truly an informative website.
[2] Information courtesy of nauts.com.
[3] Courtesy of www.geo-orbit.org.

Figure 9–1
Sputnik
(Courtesy of www.nauts.com)

Figure 9–2
Tiros-1 Measures 42
Inches in Diameter by 19
Inches High
(Courtesy of NASA)

spect to the earth below. From this GEO location 22,300 miles above the equator, ATS took the first pictures showing a whole hemisphere of the earth at once. With ATS images, meteorologists saw how clouds moved and storms formed over wide regions. The development of satellite weather technology had an enormous impact on the field of meteorology. The "big pictures" came into focus, and weather forecasting became more accurate.

COMMUNICATING WITH A SATELLITE

Transmission to and from a satellite are performed in select frequency bands. These bands are:

Frequency Band	Downlink	Uplink
C	3,700–4,200 MHz	5,925–6,425 MHz
Ku	11.7–12.2 GHz	14.0–14.5 GHz
Ka	17.7–21.2 GHz	27.5–31.0 GHz

LAUNCHING A SATELLITE

Let us discuss how satellites are launched into orbit. Satellites are launched by being carried on a rocket into space. The rocket is simply a huge canister or tube filled with either liquid or solid rocket fuel. If not carried on a rocket, satellites can sometimes be carried by the space shuttle (Figure 9–3).

After a rocket is launched with the satellite onboard, the rocket control mechanism uses an inertial guidance system to calculate any necessary modifications to the rocket's nozzles to tilt the rocket to the course described in the

Figure 9–3
Space Shuttle
(Courtesy of NASA)

approved flight plan. In most cases, the flight plan calls for the rocket to head east because the earth rotates to the east, giving the launch vehicle a free boost. The strength of this boost depends on the rotational velocity of the earth at the launch location. The boost is greatest at the equator where the distance around the earth is greatest, and is therefore rotating at the fastest rate. Just how big is the boost from an equatorial launch? To make a rough estimate, we can determine earth's circumference by multiplying its diameter times pi (3.1416). The diameter of earth is approximately 7,926 miles (12,753 km). Multiplying times pi yields a circumference of about 24,900 miles (40,065 km). To travel around that circumference in 24 hours, a point on the earth's surface has to move at 1,038 mph (1,669 km/h). A launch from Cape Canaveral, Florida, does not get as big a boost from the earth's rotational speed as it would on the equator. The Kennedy Space Center's Launch Complex 39-A, one of its launch facilities, is located at 28 degrees 36 minutes 29.7014 seconds north latitude. There the earth's rotational speed is about 894 mph (1,440 km/h). The difference in the earth's surface speed between the equator and Kennedy Space Center, then, is about 144 mph (229 km/h).

This 144 mph is truly significant. The reason the lower speed is important is the weight of the rocket. As an example, the space shuttle can weigh over 4.5 million pounds (2 million kg), depending on its payload (its load, such as a satellite, equipment, or space station component). This requires a great deal of energy to accelerate such a mass to 144 mph, and therefore a significant amount of fuel; thus, launching from the equator makes a significant difference. This issue is one that most people rarely think about.

ESCAPE AND ORBITAL VELOCITY

Whether a rocket wants to simply orbit the earth or escape its gravitational pull is determined by its velocity. To completely escape the earth's gravitational pull, and therefore enter space, a rocket must accelerate to at least 25,039 mph (40,320 km/h). This happens when NASA sends probes to the outer reaches of the solar system or when missions such as the Apollo moon mission transpire. Figure 9–4 shows the latest Titan rocket, the Titan 4. This rocket is used to send many of the latest satellites into orbit.

However, in the case of orbital velocity, or how much velocity is required to simply orbit the earth, it is a different issue, because earth's escape velocity is much greater than what is required to place an earth satellite into orbit. With satellites, the object is not to escape earth's gravity, but to balance it. This centers around orbital velocity. **Orbital velocity** is the velocity needed to achieve balance between gravity's pull on the satellite and the inertia of the satellite's motion—the satellite's tendency to keep going. This tendency to keep moving is due to the fact that space is a vacuum and an object, and once it is going in a direction, will continue in that direction indefinitely. This differs from an atmosphere in which an object will slow down until it finally stops. As far as orbital velocity, this is approximately 17,000 mph at an altitude

orbital velocity

Figure 9–4
Titan-4
(Courtesy of NASA)

of 150 miles (242 km). Gravity also enters the picture. Without gravity, the satellite's inertia would carry it off into space. Even with gravity, if the intended satellite goes too fast, it will eventually fly away. However, if the satellite goes too slowly, gravity will pull it back to earth. At the correct orbital velocity, gravity exactly balances the satellite's inertia, pulling down toward earth's center just enough to keep the path of the satellite curving like earth's curved surface, rather than flying off in a straight line.

The issue of orbital velocity is a little more complicated than it may seem on the surface. The orbital velocity of a satellite depends on its altitude above earth. The nearer the satellite is to the earth, the faster the *required* orbital velocity must be. At an altitude of 124 miles (200 km), the required orbital velocity is just over 17,000 mph (27,400 km/h). To maintain an orbit of 22,223 miles (35,786 km) above earth, the satellite must orbit at a speed of about 7,000 mph (11,300 km/h). That specific orbital speed and distance permits the satellite to make one revolution in 24 hours. Because the earth also rotates once in 24 hours, a satellite at 22,223 miles (35,786 km) altitude will stay in a fixed position relative to a point on the earth's surface. Because the satellite stays directly over the same spot all the time, this kind of orbit is called geostationary. Geostationary orbits are ideal for weather satellites and communications satellites.

In general, the rule is *the higher the orbit, the longer the satellite can stay in orbit*. At lower altitudes, a satellite runs into traces of the earth's atmosphere, which creates drag on the satellite. This drag causes the orbit of the satellite to decay until the satellite falls back into the atmosphere and burns up. At higher altitudes, where the vacuum of space is nearly complete, there is almost no drag and a satellite can stay in orbit almost indefinitely.

LEOs are actually in the two outer layers of the atmosphere. The outermost layer is called the exosphere. The exosphere is the most distant atmospheric region from the earth's surface, as shown in Figure 9–5. The upper

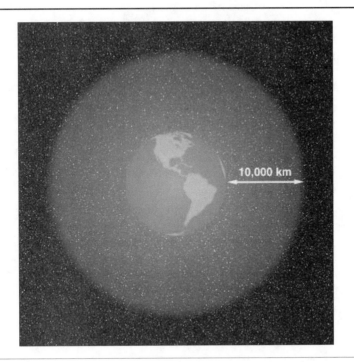

Figure 9–5
The Exosphere
(Courtesy of www.noaa.gov)

boundary of the layer extends to heights of perhaps 960 to 1,000 km and is relatively undefined. The exosphere is a transitional zone between the earth's atmosphere and interplanetary space. The next lower layer is the thermosphere (literally "heat sphere"), as shown in Figure 9–6. Most LEOs orbit in the thermosphere. Figure 9–7 depicts graphically the earth's atmosphere.

DIFFERENT TYPES OF SATELLITES

Of the several types of satellites, each is customized toward its mission.[4]

- *Weather satellites* pinpoint weather conditions and provide data for meteorologists to determine future weather conditions. Typical weather satellites include the Tiros, COSMOS, and GOES satellites. These satellites usually contain cameras that can return photos of the earth's weather, either from fixed geostationary positions or from polar orbits, as well as sensitive barometric and environmental sensors.
- *Broadcast satellites* broadcast television signals from a ground station to another ground station. These include a cable TV antenna and a home DirecTV/DishTV antenna.

---

[4] See www.howstuffworks.com, an excellent site for more information.

Figure 9–6
The Thermosphere
(Courtesy of www.noaa.gov)

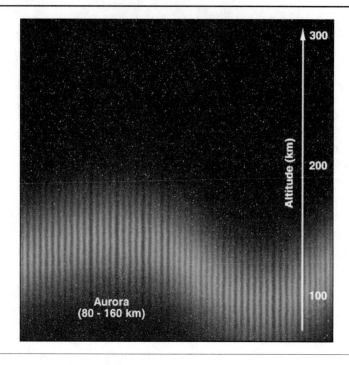

- *Communications satellites* allow telephone and data conversations to be relayed through the satellite from one point on the earth's surface to another point on the surface. Typical satellites include Iridium, Telstar, and Intelsat. Communications satellites use transponders. A **transponder** is a radio that receives a conversation at one frequency and then amplifies it and retransmits it back to earth on another frequency.
- *Scientific satellites* perform a variety of scientific missions. The Hubble space telescope is the most famous scientific satellite, but there are many others looking at everything from sun spots to gamma rays.
- *Navigational satellites* help ships and planes navigate. The most famous are the GPS Navstar satellites.
- *Rescue satellites* respond to radio distress signals.
- *Earth observation satellites* observe the planet for changes in everything from temperature to forestation to ice sheet coverage. The most famous is the Landsat series.
- *Military satellites* perform a wide range of surveillance activities. Intelligence-gathering possibilities using high-tech electronic and sophisticated photographic equipment reconnaissance are endless. Applications may include:
    Relaying encrypted communications
    Nuclear monitoring
    Observing enemy movements

transponder

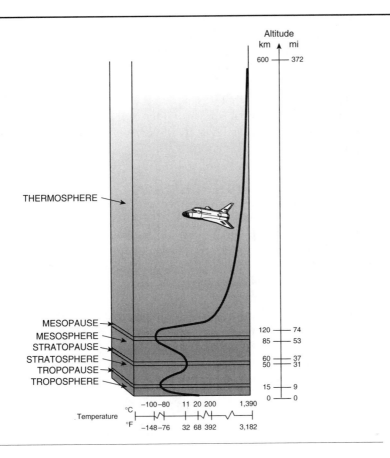

Figure 9–7
Atmospheric Layers of
Earth
(Courtesy of www.noaa.gov)

Early warning of missile launches
Eavesdropping on terrestrial radio links
Radar imaging
- *Photography satellites* use what are essentially large telescopes to take
pictures of militarily interesting areas.[5]

All satellites, regardless of their mission, have several basic components.
First, they have a metal or composite frame and body. This is usually very
thin, because structural integrity is not as much of an issue in gravity-less
space. Second, satellites have a source of power (usually solar cells) and batteries for storage. Arrays of solar cells provide power to charge rechargeable
batteries. Newer designs include using fuel cells. Power on most satellites is
precious and very limited, because the technology relating to solar cells is not
very advanced (i.e., solar cells cannot accumulate very much power). Nuclear

[5] See www.howstuffworks.com/framed.htm?parent=satellite.htm&url=http://spacelink.
nasa.gov/Instructional.Materials/Multimedia/Satellite.Tracking/.index.html to see what
satellites will pass over your location at any given time.

power has also been used on space probes to other planets. Nuclear-powered spacecraft do not suffer this power shortage, but, due to environmental concerns, nuclear power is only used for spacecraft that will *not* be circling the earth. Third, satellites have an onboard computer to control and monitor the different systems. This onboard computer system can be more or less advanced, depending on what it is expected to control. Fourth, satellites hold a radio system and antenna. Some satellites contain multiple antennas of varying strengths. Finally, they have an attitude control system to keep the satellite pointed in the right direction.

ORBITAL MECHANICS

Think about what happens when you throw a ball. Imagine that you are standing in a big field and throw a baseball as hard as you can. The ball may go 100 feet (30 m) and then hit the ground, but you put the ball in orbit. It's just that a ball's orbit is very short! Now imagine that you shot a rifle straight and level instead of throwing a ball. The bullet might travel 1 mile (1.6 km) before succumbing to gravity and hitting the ground. Now imagine that you shoot a very large cannon that is able to give its shell an extremely high initial velocity. Also imagine that our world is completely covered in water to remove any worries about hills, and that the cannon is shot straight and level. Its path might look like this:

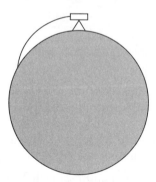

In this diagram you can see that the shell is going far enough to actually follow the curve of the earth for a period of time before hitting the ground.

One thing that needs to be considered is air resistance, or drag. Imagine that you took the cannon to the moon and mounted it on top of the highest mountain. The moon has no atmosphere and is completely surrounded by the vacuum of space. If you adjusted the speed of the shell just right and shot the cannon, the shell would follow the curve of the moon perfectly. The shell would fall at exactly the same rate that the curve of the moon falls away from it, so it would never hit the ground. Eventually the shell would curve all the way around the moon and ram right into the back of the cannon! On the moon you could actually have satellites in extremely low orbits like that—

just a couple of miles off the ground to avoid the mountains. In addition, satellites could conceivably be launched from cannons.

On the earth it is not so easy, because satellites have to get up above the atmosphere and into the vacuum of space to orbit for any length of time. The minimum to avoid atmospheric interference is about 200 miles (320 km). The Hubble space telescope orbits at an altitude of about 380 miles (600 km). The principle, however, is exactly the same. The speed of the satellite is adjusted so that it falls to earth at the same rate that the curve of the earth falls away from the satellite. The satellite is perpetually falling, but it never hits the ground.

LEO SATELLITES

LEO satellites are generally used for communications services or surveillance activities. In the communications arena, these services are satellite telephony; in the surveillance arena, LEOs are used primarily by the military to determine the activities of surface-based entities such as ships and troop formations. LEO satellites have certain advantages when compared with the more widely used GEO satellites. The LEO satellites' nearer proximity to the earth allow them the ability to communicate with earth stations using much less power than GEO satellites are required to use. This is because the satellite does not require as much power to send messages back and forth since the distance is not great. To transmit data over longer distances, much more powerful transmitters are required. The exception to this is the NASA space probes. These probes utilize low-power transmitters because the timeliness of the transmission is not as much of an issue as opposed to a businessperson talking to colleagues on a satellite telephone network. Therefore, a LEO satellite can be built much smaller than a GEO satellite that supplies the same type of service. These satellites range from over 1,000 kg (2,200 lb) to less than 10 kg (22 lb). Due to their smaller size, LEOs can be launched at around $50 million, as opposed to some GEO launches that costs in the neighborhood of $400 million in 1998 and over $500 million in 2002 (Goldin 1998). This cost also does not account for the price of the payload about the rocket. For example, hurricane watch satellites can cost $300 million, and missile warning satellites can cost $700 million, as in the case of the Defense Support Program (DSP) missile warning satellite launched in 2000.[6] The DSP satellite is a heat-sensing telescope that, from its vantage point high above the planet, looks down on earth in a constant vigil to detect missile launches or nuclear detonations.

Typically, a LEO satellite uses 20 dB (100 times) less power than a GEO satellite. The propagation delay for the LEO is much less than either GEO or

---

[6] For updates on the cost of satellites, go to www.howstuffworks.com/satellite7.htm. or the official NASA govt site: www.nasa.gov.

MEO. **Propagation delay** is the amount of time it takes for a signal to go from one point to another. In fact, the propagation delay is about 13 microseconds (ms) for a round-trip, two-way conversation. How does this compare to a GEO? The propagation delay for a LEO is about 13 ms, while this same trip takes 0.25 seconds and almost 0.5 seconds for GEO satellites.

Now that we have discussed the advantages of LEOs, there are a few disadvantages. A LEO, at an altitude of 1,000 km on an equatorial orbit, will circle the earth in about 100 minutes. This speed makes it difficult to locate, track, and communicate with. This, in turn, results in the need for highly sophisticated communications equipment, because it is a moving target and it is much more difficult to track and communicate with a moving target than a stationary one like a GEO. Another disadvantage of LEOs is that more satellites are needed to achieve the desired coverage of the earth's surface. The footprint of the LEO signal coverage is much less than satellites with higher orbits. For complete earth coverage, it requires forty to sixty (depending on exact orbital height) LEOs working together, while all-over earth coverage from GEOs requires only three to six satellites due to their higher altitude from the earth's surface. Finally, there exists a serious drawback concerning LEOs that is directly related to their orbital height. LEOs, because of their low orbital height, are actually in the upper ranges of the earth's atmosphere. This produces a drag effect and their orbits slowly decay. As their orbit decays, they fall into denser atmospheric layers and the decay increases. Eventually, the LEO satellite will crash. To forestall this eventuality as long as possible, LEOs typically use rocket motors which attempt to counter the orbital decay scenario. This works to delay the crash until the fuel runs out.

LEO Systems

Most LEOs are used as communications satellites. A key factor in the success of many LEO systems is personal communications services. To succeed, LEOs have to compete with terrestrial cellular phone systems such as PCS. To see a comparison of terrestrial-based versus satellite-based cellular calls, see Figure 9–8. Chapter 6 also describes terrestrial cellular telephony systems.

The rapidly increasing quality of terrestrial PCS puts a huge burden on satellite systems to perform better. To compete with current terrestrial communications systems, a satellite system cannot offer substandard services, especially at the high prices of $1.50 to $7.00 per minute. Satellite systems have the problem of overcoming technological barriers, such as the interference caused by storms on the earth's surface, solar flare activity, and the sheer distance over which they must transmit information from people's telephone conversations to messaging data as in the case of WAP telephony (see Chapter 4 for more information on WAP).

History and current events also play a key role in understanding the unique environment of LEOs. When Iridium and Globalstar were first being conceived in the late 1980s and early 1990s, worldwide cellular phone

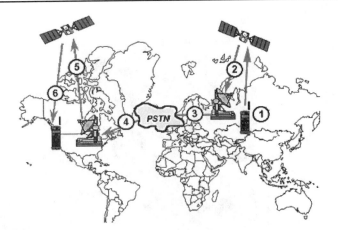

Figure 9–8
Satellite versus
Terrestrial Cellular
Calling
(Courtesy of www.globalstar.
com)

Making a Call via Satellite versus Cellular

1. A subscriber in Russia calls a friend in San Francisco on a Globalstar satellite phone. The signal is handled by a passing satellite.

2. The satellite relays the call to a gateway in its footprint.

3–4. The gateway converts the signal to work with the local PSTN and passes on the call. Depending on the distance between the callers, a Globalstar satellite call might pass through several gateways and PSTNs before locating the receiving phone. The PSTN uses the call's routing information to connect to another gateway that knows where the receiving phone is located.

5. The gateway located closest to the receiving phone converts the signal to Globalstar format and uplinks it to a satellite. This gateway knows that the receiving phone is in its contact area, because an earlier satellite relayed that phone's power-on registration message to the gateway. This information was stored in the gateway's visitor location register (VLR).

6. The call is relayed to the receiving phone and the call linkage is complete.

coverage was sparse. Gaining the necessary international government approvals to launch dozens of satellites in space and to establish on-ground base stations in hundreds of nations took years. By then, wireless services widely proliferated; prices for cellular telephone calls fell dramatically, and the handsets became smaller and easier to use. By the time the satellites were ready (meaning launched, set into orbit, established connectivity between the satellites and ground stations, and worked out all anomalies), cell phones had captured most of the market Globalstar and Iridium were targeting. In addition, initial satellite phones were large and bulky with sizable antennas which made them awkward to use. They were also the size of a medium suitcase. Now they are smaller, just a little larger than a typical cellular phone, as shown in Figure 9–9. These phones were costly, too, with first generation Iridium handsets going for as much as $3,000. The

Figure 9–9
Typical Iridium Phone
(Courtesy of www.iridium.
com)

Figure 9–9
Typical Iridium Phone
(Courtesy of www.iridium.
com)

original Iridium services cost the user as much as $5 to $7 per minute, quickly relegating satellite phones to the wealthy international business traveler and government agencies that needed communications in remote locations.

The mass-market consumer had hardly heard of Iridium or Globalstar and was certainly unwilling to pay their costly per-minute fees. These costs are now about $1.50 per minute. Some analysts say satellite phone systems amounted to overkill for most consumers, limiting them to a small target audience (Charney and Grice 2001). If you are part of that limited audience, however, then a satellite phone is a must.

IRIDIUM

The first and probably the most recognized name in LEO satellite constellations is Iridium. Motorola first proposed this system in 1987 and applied for an FCC license in 1990. These Iridium satellites, when set into orbit, can handle over 2,000 simultaneous full-duplex connections, or over 140,000 connections system wide (see Figure 9–10).

Originally when launched in the early 1990s, this system was famous for being on the cutting edge of communications technology. The idea of using LEO satellites for cellular telephony, anywhere in the world, was a bold initiative. Advantages to a satellite-based telephony system include having no cellular towers every 5 miles and no building permits.

With this advantage, why has Iridium failed to capture a profitable market share? Basically, it is an issue of sheer economics on the consumer side. For example, handsets were once priced around $3,000 (in 2002, they are $800).

Figure 9–10
Typical Iridium Satellites
(Courtesy of www.iridium.
com)

These were the size of a large briefcase and included a foldout antenna. In addition, the user would pay about $7 per minute for airtime use, which is also more than most nonglobal wireless services at $0.10 per minute (Grice 2000).

Iridium's design also posed a problem. There were sixty-six Iridium satellites launched so that the entire planet could be covered; but most of the users of wireless telephony do not particularly need global coverage, especially for the costs they would incur. There is little commercial need to provide coverage over the oceans or in deep forests or other inaccessible locations. Governments and scientists may need such coverage, but that target audience is too small to provide any positive return on investments for a system this large. The initial start-up costs for Iridium 66 are estimated at $7 billion.

The market situation of terrestrial-based cellular telephony systems has improved dramatically. The growth of the market for wireless communications has been staggering, and the subscriber base has increased as well. Government involvement by subsidizing the expansion of cellular telephony systems, particularly in Europe, has also been a factor.

The Iridium system has had an interesting history. It began in January 1999 and with billions of dollars from Motorola, its primary backer, Iridium has struggled toward profitability. A firm called Iridium Satellite[7] is now in charge of the Iridium system and has found some success (Grice 2000). In December 2000, the U.S. Defense Department signed a $72 million, 2-year deal to use Iridium Satellite's space-based phone network to provide secure voice communications for the navy and other government users (Grice 2000). The contract calls for Iridium Satellite to provide unlimited use for up to 20,000 government workers. Expected rates for nondefense users are in the $1.50 per

---

[7] Information on the new Iridium system can be found at www.iridium.com.

minute range. This is comparable with Globalstar, a major competitor (see below). One reason the cost per minute is lower than the initial Iridium system backed by Motorola is that the new firm purchased the Iridium system from the bankruptcy court at $25 million. This compares with the $5 billion Motorola put into the original Iridium system to get it going.

Despite the rosy market projections that have yet to pan out, the original (and now the new) Iridium's service has found a niche among industrial, personal, and military users: mining operators, offshore drilling rigs, Iditarod dog sled racers in Alaska, and CNN international correspondents. The ability to deliver service nearly anywhere on the globe has made satellite phones an attractive alternative to spotty cellular service for those willing to pay a hefty premium (Grice 2000). Appendix E provides information about a typical Iridium Satellite phone.

GLOBALSTAR

The Globalstar[8] satellite system began in January 2000. It is another LEO-based satellite system attempting to be commercially successful in the satellite-based telephony market. Unlike the original Iridium, which ceased operations with 63,000 customers (though this is difficult to ascertain), the Globalstar system has 75,000 subscribers as of August 2002. (Globalstar Reports 2001).

The Globalstar system is similar to Iridium. Unlike the primarily voice-oriented Iridium, Globalstar offers voice, fax, and paging services worldwide using a constellation of LEO satellites. Globalstar's market is mainly remote areas. The company hopes to create a profitable niche market by offering services to remote areas around the world.

Globalstar handsets are priced at $750 with per minute rates at 17¢/minute as of August 2002 (www.globalstar.com). Globalstar reduces set phone call rates. The implementation costs are $2.5 billion, which is just about half as much as Iridium, but is configured to manage 3 to 4 times the call volume. This increased volume is mainly due to the use of the advanced digital compression technique CDMA, which also allows for far higher voice quality. CDMA is discussed in more detail in Chapter 3.

Globalstar is a consortium of leading international telecommunications companies originally established in 1991 to deliver satellite telephony services through a network of exclusive service providers. The consortium is a partnership of the world's leading telecommunications service providers and equipment manufacturers, including cofounders Loral Space & Communications, Qualcomm Incorporated, Alenia, China Telecom (HK), DACOM, DaimlerChrysler Aerospace, Elsacom (a Finmeccanica Company), Hyundai, TE.SA.M (a France Telecom/Alcatel company), Space Systems/Loral; and Vodafone Group PLC.

---

[8] For more information on Globalstar, go to www.globalstar.com.

Specifically, the Globalstar system is designed to provide high-quality satellite-based telephony services to a broad range of users.

- Cellular users who roam outside of conventional coverage areas
- People who work in remote areas where terrestrial systems do not exist
- Residents of underserved markets who can use Globalstar's fixed-site phones to satisfy their needs for basic telephony
- International travelers who need to keep in constant touch

The Globalstar system offers the following services to users.

- One phone for both cellular and satellite calls
- Voice calling
- Short messaging service
- Global roaming
- Facsimile
- Data transmission

Globalstar's constellation of LEO satellites transmits calls from a Globalstar wireless phone or fixed phone station to a terrestrial gateway, where the calls are passed on to existing fixed and cellular telephone networks in more than 100 countries on six continents.

Gateways

Gateways are an integral part of the Globalstar ground segment, which also includes ground operations control centers (GOCCs), satellite operations control centers (SOCCs), and the Globalstar Data Network (GDN).

Each gateway, which is owned and managed by the service provider for the region (i.e., one gateway may serve several countries) in which the gateway is located, receives transmissions from orbiting satellites, processes calls, and switches them to the appropriate ground network. An individual gateway may service more than one country. Gateways consist of three or four dish antennas, a switching station, and remote operating controls. Because all of the switches and complex hardware are located on the ground, it is easier for Globalstar to maintain and upgrade its system than it is for systems that handle switching in orbit.

Gateways offer seamless integration with local and regional telephony and wireless networks. They utilize a standard high-speed, or T-1, interface to the existing PSTN. Encryption ensures voice and signaling security for individual transmissions. Chapter 1 discusses T-1 lines and other high performance solutions in more detail.

The GOCCs are responsible for planning and controlling the use of the satellites by the gateway terminals and for coordinating with the SOCC. GOCCs plan the communications schedules for the gateways and control the allocation of satellite resources to each gateway. Figure 9–11 shows typical Globalstar satellites.

The SOCC manages the Globalstar satellite constellation. It tracks satellites, controls their orbits, and provides telemetry and command (T&C) services for

Figure 9–11
Typical Globalstar
satellites
(Courtesy of www.globalstar.
com)

the constellation. Globalstar satellites continuously transmit spacecraft teleme-
try data that provide onboard diagnostic and status reports for the satellites. The
SOCC also oversees each satellite launch and any deployment activities. The
SOCC and GOCC facilities remain in constant contact through the GDN.

The GDN is the connective network that provides wide-area intercom-
munications facilities for the gateways, GOCCs, and SOCCs.

As a wholesaler, Globalstar sells access to its system to regional and local telecom service providers around the world. These Globalstar partners,
such as Qualcomm Inc., in turn form alliances with additional providers. This
lets Globalstar leverage the marketing, operating, and technical capabilities
of the world's premier communications service providers.

The Globalstar constellation consists of forty-eight LEO satellites, plus
an additional four satellites in orbit as spares. Each consists of an antenna, a
trapezoidal body, two solar arrays, and a magnetometer, and operates at an
altitude of 876 miles (1,414 km).

The satellites are placed in such an orbit that the polar regions—including
most of Greenland, small parts of Alaska, Canada, Scandinavia, Siberia, and
regions in the Southern Hemisphere, including Antarctica and parts of South
America—are not covered.[9] This is entirely acceptable because these areas
are not considered to be profitable. The Globalstar constellation of satellites
can then pick up signals from over 80 percent of the earth's surface. Several
satellites pick up a call, and this "path diversity" ensures that the call does
not get dropped even if one of the phones moves out of sight of one of the
satellites. If buildings or terrain block the phone signal, a "soft handover"
takes place, and the call's transmission is switched to an alternate satellite

---

[9] Eight orbital planes of six satellites each, inclined at 52 degrees to provide service on
earth from 70 degrees North latitude to 70 degrees South.

Figure 9–12
Typical Globalstar
satellite phone
(Courtesy of www.globalstar.
com)

with no interruption. This satellite now maintains transmission of the original signal to one of several terrestrial Globalstar gateways. Because all the switches and complex hardware are located on the ground in the gateways, Globalstar satellites are relatively simplistic, leading to dramatically increased system reliability. In short, they are simple relays. A typical Globalstar satellite phone can be seen in Figure 9–12.

Telit SAT-550

Telit's SAT-550 is a dual-mode Globalstar/GSM phone that provides callers with a familiar interface and user-friendly features in either the GSM or Globalstar mode. Its robust construction makes it an ideal working partner for remote area construction teams, oil and exploration workers or adventurers who keep safety in mind.

An easy-to-use data kit also gives the Telit SAT-550 complete asynchronous data capability, allowing users to connect PCs or Personal Digital Assistants (PDAs) to the Internet and to private networks at speeds up to 9.6 Kbps.

Advantages:

- Compatibility with existing GSM 900 cellular networks
- Globalstar mode when you're not within cellular coverage areas
- Superior digital voice quality in both Globalstar and GSM 900 modes
- Secure and reliable Globalstar data services[*]
- Reliable service virtually anywhere in the world

Specifications:

Dimensions: 224(H) 65.5(W) 50(D) mm
Weight: 425 g
Globalstar battery time (standard):   1.6 talk time hours/6 to 24 standby hours
GSM battery time (standard):   3.9 talk time hours/83 standby hours
Display: Graphical, 4 lines of 12 characters + one line of icons
Power source: Lithium - Ion battery
Special features: Clock and alarm, personalized ring audio function

Phone Features:

High contrast illuminated LCD
Dual-mode GSM 900/Globalstar
Automatic mode select on power up
Time clock with alarm
Dedicated keys for Satellite, emu functions, SMS and Phone book
Customizable ring tones
Programmable security level
Short Messaging Services (SMS) (caller can send and receive messages in all modes)
SMS and voicemail message waiting indicators
Support of SIM application tool kit

Services Supported:

Calling line ID
Automatic mode select on power up
Call forwarding
Call waiting/hold
Call barring
Short Message Service (SMS) (the user can send and receive SMS messages in all modes)

Source: www.globalstar.com/telit_sat-550.html.

Satellite Footprint

The satellite coverage beams for satellite phone communication links are large, as shown in Figure 9–13. Try to imagine the scale for the coverage of a typical 3,600-mile-diameter beam footprint.

Each footprint moves rapidly across the earth's surface. A satellite that passes directly overhead of a specific point on the earth's surface is visible to that spot for only about 15 minutes.

When a user places a Globalstar satellite call, the nearest satellite picks up the signal. Globalstar satellite phones can operate with a single satellite in view, though typically two to four satellites will be overhead. This simulta-

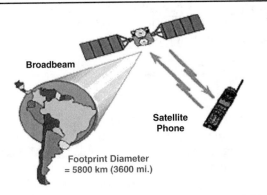

Figure 9–13
Footprint of a Satellite
(Courtesy of www.globalstar.
com)

neous coverage by multiple satellites is called path diversity, which enhances the quality of the Globalstar satellite call.

Path Diversity

Path diversity is a method of signal reception that combines multiple signals of varying power strengths into a single coherent signal, as shown in Figure 9–14. Satellite phones communicate with as many as three satellites simultaneously, combining those signals into a single, static-free signal. Globalstar satellite phones also alter power levels to compensate for shadowing and interference as needed.

As satellites move in and out of view, they will be seamlessly added to and removed from the calls in progress, reducing call interruption. This enables the Globalstar system to provide superior service to a wide variety of locations with less potential for signal blockage from buildings, terrain, or other natural features.[10]

Figure 9–14
Path Diversity
(Courtesy of www.globalstar.
com)

[10] Excellent information on Globalstar can be found at http://www.qualcomm.com/globalstar/about/satellites.html.

Globalstar, which is 38 percent owned by Loral Space & Communications filed for Chapter 11 Bankruptcy protection in February 2002. Globalstar, as of August 2002, is still struggling to be profitable.

TELEDESIC

Teledesic represents the vision of telecommunications pioneer Craig McCaw, the company's chairman. Teledesic's primary investors are McCaw, Bill Gates, Motorola, Saudi Prince Alwaleed Bin Talal, the Abu Dhabi Investment Company, and the Boeing Company. Service is targeted to begin in 2005. Founded in 1990, Teledesic is a private company based in Bellevue, Washington, a suburb of Seattle.[11]

Whereas Iridium focused on worldwide personal communications and Globalstar focused on remote services, Teledesic will provide personal broadband communication services. The Teledesic system consists of 288 satellites orbiting the earth at an altitude of 696 km. Connection speeds ranging from 16 Kbps to 1.244 Gbps will be capable of fulfilling almost any customer requirements. Teledesic claims that the system will be capable of supporting a peak capacity of 1 million full-duplex E-1 connections (2.048 Mbps) and maintaining millions of simultaneous users.

Using a constellation of LEO satellites, Teledesic and its international partners are creating a network to provide affordable, worldwide, fiberlike access to telecommunications services such as computer networking, broadband Internet access, high-quality voice, and other digital data needs. By 2005, Teledesic will provide guaranteed end-to-end QoS to meet the broadband needs of enterprises, businesses, schools, and individuals around the world. Figure 9–15 shows a typical Teledesic satellite.

The Teledesic Network

The Teledesic Network will enable service providers in countries worldwide to extend their networks, both in terms of geographic scope and in the kinds of services they can offer. Ground-based gateways will enable service providers to offer seamless links to other wireline and wireless networks, such as the Internet.

The Teledesic Network will consist of 288 operational satellites, divided into twelve planes, each with twenty-four satellites. To make efficient use of the radio spectrum, frequencies are allocated dynamically and will be reused many times within each satellite footprint. Within any circular area of 100 km radius, the Teledesic Network will support more than 500 Mbps of data to and from user terminals. The Teledesic Network will support bandwidth on demand, which will allow a user to request and release capacity as needed.

---

[11] Information on Teledesic can be found at www.teledesic.com.

Figure 9–15
Typical Teledesic Satellite
(Courtesy of www.teledesic.
com)

This will enable users to pay only for the capacity they actually use, and also for the network to support a much higher number of users.

Teledesic will operate in a portion of the high-frequency Ka band (28.6 to 29.1 GHz uplink and 18.8 to 19.3 GHz downlink). The Teledesic Network's low orbit eliminates the long signal delay experienced in communications through traditional geostationary satellites and will enable the use of small, low-power terminals and antennas. The compact terminals will mount on a rooftop and then be connected to a computer network or a PC inside the building.

The Teledesic Network is designed to support millions of simultaneous users. Most users will have two-way connections that will provide up to 64 Mbps on the downlink and up to 2 Mbps on the uplink. Broadband terminals will offer 64 Mbps of two-way capacity. This represents access speeds up to 2,000 times faster than today's standard analog modems. For example, transmitting a set of x-rays may take 4 hours over one of today's standard modems. The same images can be sent over the Teledesic Network in 7 seconds.

The price for this broadband connectivity is not set yet. End-user rates will be set by service providers, but Teledesic expects rates to be comparable with those of future urban wireline rates for broadband connectivity. Teledesic has had a short and interesting timeline (see Table 9–1).

The Teledesic Network will consist of a ground segment to include terminals, network gateways, network operations and control systems, and a space segment (the satellite-based switch network that provides the communication links among terminals).

See Figure 9–16 for a graphical illustration of this network.

Terminals will provide the interface both between the satellite network and the terrestrial end users and networks. They perform the translation

Table 9–1
Teledesic Timeline

1990	Company founded
1994	Initial system design completed; FCC application filed
1997	FCC license granted; World Radio Conference designates necessary international spectrum for service
1998	Motorola joins effort to build the Teledesic Network
1999	Teledesic signs major launch contract with Lockheed Martin
2005	Service targeted to begin

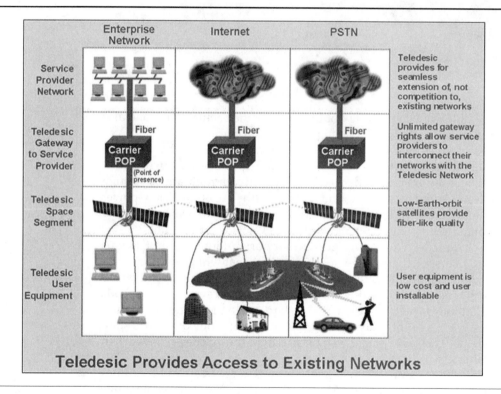

Teledesic Provides Access to Existing Networks

Figure 9–16
Teledesic Access to Other Networks
(Courtesy of www.teledesic.com)

between the Teledesic Network's internal protocols and the standard protocols of the terrestrial world, thus isolating the satellite-based core network from complexity and change.

Teledesic terminals communicate directly with the satellite network and support a wide range of data rates. The terminals also will interface with a wide range of standard network protocols, including IP, ISDN, and ATM. Although optimized for service to fixed-site terminals, the Teledesic Network is able to serve transportable and mobile terminals, such as those for maritime and aviation applications.

Most users will have two-way connections that provide up to 64 Mbps on the downlink and up to 2 Mbps on the uplink. Broadband terminals will offer 64 Mbps of two-way capacity. This represents access speeds up to 2,000 times faster than today's standard analog modems.

Terminals also will provide the interconnection points for the Teledesic Network's constellation operations control centers (COCCs) and network operations control centers (NOCCs). COCCs coordinate initial deployment of the satellites, replenishment of spares, fault diagnosis, repair, and de-orbiting. NOCCs include a variety of distributed network administration and control functions such as network databases, feature processors, network management, and billing systems.

Fast-Packet Switching

Teledesic's space-based network will use fast-packet switching. Communications are treated within the network as streams of short, fixed-length packets. Each packet contains a header that includes the destination address and sequence information, an error-control section used to verify the integrity of the header, and a payload section that carries the digitally encoded user data (voice, video, data, and so on). Conversion to and from the packet format takes place in the terminals at the edge of the network.

The topology of a LEO-based network is dynamic. The network must continually adapt to these changing conditions to achieve the optimal (least-delay) connections between terminals. The Teledesic Network uses a combination of destination-based packet addressing and a distributed, adaptive packet routing algorithm to achieve low delay and efficient throughput.

The Satellite Constellation

Each satellite will be a node in the fast-packet switch network and will have intersatellite communications links with other satellites in the same and adjacent orbital planes. This interconnection arrangement will form a full-bodied nonhierarchical mesh, or "geodesic," network that is tolerant to faults and local congestion.

To achieve high system capacity and channel density, each satellite is able to concentrate a large amount of capacity in its relatively small coverage area. Overlapping coverage areas plus the use of on-orbit spares permit the rapid repair of the network whenever a satellite failure results in a coverage gap. In essence, the system reliability is built into the constellation as a whole rather than being vulnerable to the failure of a single satellite.

The lowest frequency band with sufficient spectrum to meet Teledesic's broadband service, quality, and capacity objectives is the Ka band. The terminal-satellite communications links operate within the portion of the Ka frequency band that has been identified internationally for nongeostationary fixed satellite service and, in the United States, licensed for use by Teledesic. Downlinks operate between 18.8 GHz and 19.3 GHz, and uplinks operate between 28.6 GHz and 29.1 GHz. Communications links at these frequencies are degraded by rain and blocked by obstacles in the line of sight. To avoid obstacles and limit the portion of the path exposed to rain requires that the satellite serving a terminal be at a high elevation angle above the horizon. The Teledesic constellation ensures a minimum elevation angle (mask angle) of 40 degrees within its entire service area. Using this design, the Teledesic Network is able to achieve availability of 99.9 percent or greater.

Latency is a critical parameter of communications service quality, particularly for interactive communications and for many standard data protocols. To be compatible with the latency requirements of protocols developed for the terrestrial broadband infrastructure, Teledesic satellites operate at a low altitude, less than 1,400 km. The combination of a high mask angle and low earth orbit results in a relatively small satellite coverage zone, or footprint, that enables efficient spectrum reuse but requires a large number of satellites to serve the entire earth.

Network Capacity

To make efficient use of the radio spectrum, frequencies will be allocated dynamically and reused many times within each satellite footprint. The Teledesic Network will support bandwidth on demand, allowing a user to request and release capacity as needed. This enables users to pay only for the capacity they actually use, and for the network to support a much higher number of users.

SURREY SATELLITE TECHNOLOGY LTD.

Surrey Satellite Technology Ltd. (SSTL) has developed a unique application for LEOs. Iridium, Globalstar, and Teledesic systems are formed around a large-scale satellite network. SSTL plans to use as few as just one microsatellite to provide basic and inexpensive store-and-forward data communications services such as e-mail, Internet, and communications for remote areas (e.g., the Antarctic).

Potential users include the following:

- E-mail networks
- Embassies and foreign offices
- Military and national security
- Banks and multinationals
- Disaster relief teams
- Medics in remote areas

This system will not provide real-time voice communications services. SSTL will use an equatorial orbital altitude of 968.2 km. This altitude produces a ground track that repeats every 4 days. SSTL will use eight satellites for complete circuitous coverage.

ORBCOMM

ORBCOMM advertises itself as the world's first commercial provider of global LEO satellite data communication services.[12] The ORBCOMM system enables businesses to track remote and mobile assets such as trailers, railcars, locomotives, and heavy equipment; to monitor remote utility meters and oil and gas storage tanks, wells, and pipelines; and to stay in touch with remote workers anywhere on the globe.

Frequency Allocation

The ORBCOMM system uses 137 to 138 MHz and 400 MHz frequencies for transmissions down to mobile or fixed data communications devices, and 148 to 150 MHz frequencies for transmissions up to the satellites. These frequencies, approved for use by LEO satellite systems at the World Administrative Radio Conference in February 1992, were allocated by the FCC to little LEO mobile satellite services in January 1993. The FCC granted ORBCOMM a U.S. commercial license in October 1994.

Description

The ORBCOMM system uses LEO satellites instead of terrestrial fixed-site relay repeaters to provide worldwide geographic coverage. The system is capable of sending and receiving two-way alphanumeric packets, similar to two-way paging or e-mail.[13] The three main components of the ORBCOMM system are (1) the **space segment,** the constellation of satellites; (2) the **ground segment,** gateways that include the gateway control centers (GCCs) and gateway earth stations (GESs) and the network control center (NCC) located in the United States; and (3) **subscriber communicators** (SCs), handheld devices for personal messaging as well as fixed and mobile units for remote monitoring and tracking.

space segment

ground segment

subscriber communicators

ORBCOMM System Architecture

Vital messages generated by a variety of applications are collected and transmitted by an appropriate SC to a satellite in the ORBCOMM constellation.

---

[12] Information courtesy of ORBCOMM and can be found at www.orbcomm.com.
[13] See www.orbcomm.com.

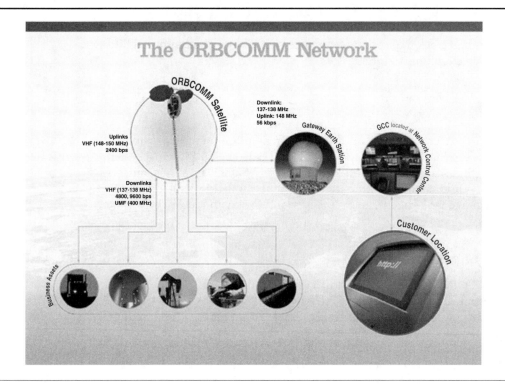

Figure 9–17
ORBCOMM Architecture
(Courtesy of www.orbcomm.com)

The satellite relays these messages to an ORBCOMM GES. The message is sent through a GCC to its destination, through the Internet or other terrestrial networks, to a personal computer, to a subscriber communicator, or to a pager (Figure 9–17).

Messages and data sent to a remote SC can be initiated from any computer using common e-mail systems, including the Internet, mail, and X 400. The NCC or GCC then transmits the information using ORBCOMM's global telecommunications network.

ORBCOMM Satellites

ORBCOMM's satellites are set up in constellations, strategically spaced so that every part of the world is covered by at least one satellite. ORBCOMM satellites are identical in function and overlapping in coverage, providing significant built-in redundancy, availability, and high assurance of uninterrupted service. Specifications in Table 9–2 apply to the ORBCOMM satellites, but more importantly they provide an example of the specifications for a typical LEO satellite.

Table 9–2
ORBCOMM System
Specifications and
Characteristics
(Courtesy of www.orbcomm.
com)

Deployed Length: 170"

Deployed Width *(at solar arrays)*: 88"

Electrical Power:
- 2 deployable sun tracking solar arrays
- 14-volt power system
- 5-volt regulated bus for digital electronics

Telemetry, Tracking, & Command:
- Spacecraft commanding and telemetry via VHF at 57.6 Kbps

Attitude Control:
- Nadir Pointed + /− 5 degrees using active magnetic controls, reaction wheel and gravity gradient
- New generation lightweight earth sensors and magnetometer

Stackable Design Allows Multiple Satellite Launches
- Launch Vehicles: Orbital Sciences' Taurus and Pegasus XL

Data Rate:
- 2,400 bps subscriber uplink
- 4,800 bps subscriber downlink
- 9,600 bps subscriber downlink (future)

Frequencies:
- Uplink: 148.00–150.05 MHz
- Downlink: 137.00–138.00 MHz
- Addressing: Internet and X .400
- Message Size: 6–250 bytes (typical)

Space Segment *(as licensed)*:
- 16 spacecraft-near polar 70° or 108°
- Up to 32 spacecraft-inclined 45°
- Attitude/Orbit: 825 km circular

Subscriber Communicator *(typical)*:
- Power Output: 5 watts
- Weight: 1–2 pounds
- Antenna: Whip

Similar to the other LEO-based systems, ORBCOMM has experienced financial difficulties. In fact, it changed hands in mid-2000 when International Licensees purchased the firm in bankruptcy court. These new owners are believed to have bought the assets of the firm for a fraction of the $810 million that investors sank into it (the company did not disclose the purchase price) (Reuters 2001). This should help ORBCOMM because it will allow the company to offer lower rates to its customers.

LEO satellites have had a turbulent time. After all, this frontier was a new one and has required a great deal of technological advance and patience. Globalstar and other ventures will find their place in orbit. It is expected that new and imaginative ideas will allow LEOs to become a valuable piece of the overall telecommunications pie.

MEOs

MEO systems operate at between 1,500 and 6,500 miles (10,000 km) above the earth, which is lower than GEO and higher than most LEOs. The MEO orbit is a compromise between the LEO and the GEO. Compared with LEOs, MEOs with their more distant orbit require fewer satellites to provide all-over global coverage than LEOs, because each satellite may be in view of any particular location for several hours. Compared with GEOs, MEOs can operate effectively with smaller mobile equipment and with less latency (signal delay). Their signal takes from 50 to 150 ms to make the round trip, compared with 1 to 2 seconds for a GEO. MEO satellites cover more earth area than LEO satellites but have a higher latency due to their higher orbit.

MEO satellites are in view of any point on the earth for a longer period of time than LEOs; therefore, they can be effectively used for communications, but they must be reachable. To combat the reachability problem, MEO systems often feature significant footprint coverage overlap from satellite to satellite, which in turn requires more sophisticated tracking and switching schemes than GEOs (which are stationary). Typically, MEO constellations have ten to seventeen satellites distributed over two or three orbital planes. MEO systems plan to include AstroLink, KaStar, and CyberStar.

Several systems in particular have helped form the use and development of MEO satellites. These are briefly discussed.

Telstar 1

The Telstar 1 was one of the first and most famous experimental satellites orbited in MEO.[14] Figure 9–18 depicts this satellite. On July 10, 1962, NASA launched the Telstar satellite. It was the first satellite to give television signals across the Atlantic. Its orbit was such that it was visible to both Europe and the United States simultaneously during one part of its orbit and both the United States and Japan during another orbit.

The Telstar satellite was in an orbit quite a distance from the earth, outside the Van Allen belts, which protect satellites and space shuttles in low orbits from most of the harmful effects of cosmic rays and geomagnetic storms. The satellite was also in a low orbit and thus carried the same disadvantage as before, only for a few minutes each hour.

Telstar was a 170-pound experimental satellite, created by AT&T's Bell Labs and launched into elliptical orbit by NASA. It transmitted international phone calls, television programs, radio signals, and newspaper stories. The first transatlantic transmission occurred on July 11, 1962, from a twin station in Andover, Maine, via the Telstar satellite. The success of Telstar and the earth stations, the first built for active satellite communications, illustrated the potential of a future worldwide satellite system to provide communications between continents.

Telstar was considered an unqualified success for AT&T; but then, without warning, the satellite fell silent. Broadcast from the satellite ended February 1963

---

[14] Information courtesy of Loral at www.loralskynet.com.

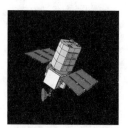

Figure 9–18
Telstar 1
(Courtesy of www.telstar.com)

after radiation damaged its transistors. Radiation from a nuclear test on earth destroyed Telstar's delicate circuitry. The satellite that captured the world's imagination was suddenly nothing more than an expensive piece of space junk.

Since then, more than ten Telnet satellites have been launched. As of 2001, Skynet, the term for the current fleet of Telstar satellites, reached over 85 percent of the populated world with leading-edge services in the communications arena. Table 9–3 shows the timeline for the Telstar satellites.

Currently, Skynet has seven Telstar satellites operating in orbit that provide the following services.

- Television broadcasting
- Cable and direct to home
- Sports and events
- Data transmission
- Internet applications
- Satellite news gathering
- Business television
- Distance learning
- Video conferencing

Telstar provides many services. For example, Telstar provides TV relay and broadcasting services for over 100 TV channels including:

- HBO-HDTV
- A&E Television Network
- Time Warner Cable
- AthenaTV
- InDemand
- OlympuSAT
- Playboy Television

Table 9–3
Telstar Timeline
(Courtesy of www.telstar.com)

1960s

The Skynet story began in August 1960, with the launch of Echo, an aluminum-coated balloon used to reflect microwave radio signals between Holmdel, New Jersey, and Goldstone, California. A joint experiment between Skynet and NASA, Echo provided the basis for all future engineering calculations involving satellite transmissions. Project Telstar grew out of the Echo experiment and was also a joint NASA-Skynet effort. Telstar 1, launched on July 10, 1962, was the world's first active communications satellite. These early developments demonstrated the feasibility of using satellites to transmit multiple simultaneous telephone calls, high-speed data, and domestic and international video in both black and white and color.

1970s

In the 1970s, Skynet, through a joint venture with GTE, leased four Comstar satellites from Comsat Corp. The Comstar fleet provided voice communications to the United States through downlinks to seven large earth stations strategically located across the country.

1980s

When the Comstar fleet went out of service in the 1980s, the Telstar series was reestablished. Launched in 1983, the Telstar 3 satellites provided exceptional service to the broadcast industry by adding capacity for short and periodic feeds, as well as telephony, point-to-point data, and restoration applications.

1990s–2003

Skynet launched Skynet 4, which provided new services such as VSAT, digital television, and data transport. Skynet 4 has more transponders than any satellite (over 200), making it one of the most powerful satellites serving the United States and its territories. Skynet was acquired by LORAL Space & Communications Ltd. in 1997. After four decades of excellence, Skynet has become part of one of the world's most resourceful satellite-based operations.

In addition, Telstar provides a satellite news gathering (SNG) service that includes expansive transponder capacity, a state-of-the art inventory and reservation system, a dedicated conference bridge, and a service management center responsible for supervising and managing interference-free access to transponders.

Telstar also provides a satellite communications solution that is complementary to terrestrial infrastructure for voice, data, and business video applications. The solution offers unique rate plans designed to bill only for the capacity used, making it affordable and cost effective. The variety of applications served are as follows:

- Restoration services for enterprises and ISPs
- Video conferencing
- Telemedicine
- File transfer and work collaboration
- Special/short-term event support
- SNG supports disaster recovery
- Database access and updates

TRW's Odyssey

TRW launched its $2.5 billion satellite network of twelve satellites and pro-
vided voice telephone data and short message services in late 2000 in a project called Odyssey. The Odyssey satellite network provides global voice, fax,
and message communications services to users equipped with portable
handsets. The system was developed as a joint commercial venture by TRW
and Teleglobe Canada. This system is called Odyssey Worldwide Services.

TRW started operations at the end of 1998. TRW estimated that it would
cost $1.8 billion to build and launch the twelve satellites (with two ground
spares), install ground stations in the United States, and operate the system
for a year. This amount rapidly expanded to at least $2.5 billion. Costs for
consumers are $250 to $300 for the cellular phone handset, $24 for the
monthly charge, and $0.65/minute use charges and $0.10/minute access
fees (under the best business use plan). TRW received their FCC license in
early 1995. Six satellites giving limited service to selected regions provided
initial operations. The full constellation of twelve satellites has now been
launched.

The satellites connect the user to a Ka band ground station, where the call
is routed into the normal telephone system. No satellite crosslinks are used in
this constellation. Each region has one ground station equipped with four
16.4 foot antennas. Ten stations are presently being utilized. Channel capacity of each satellite is 3,000 digital voice circuits at 4.8 Kbps. The total beam
width (how wide the signal is broadcast to the earth's surface) is 40 degrees,
which will provide coverage to regions 7,400 km in diameter.

Landsat

The Landsat Program is the longest running enterprise for acqusition of imagery of the earth from space. The first Landsat satellite was launched in
1972; the most recent, Landsat 7, was launched on April 15, 1999. Landsat
satellites have gradually grown in size as more capabilities have been added,
with the latest, Landsat 7, weighing approximately 4,800 pounds (2,200 kg).
The spacecraft is about 14 feet long (4.3 m) and 9 feet (2.8 m) in diameter.
Figure 9–19 depicts this satellite.

The instruments on the Landsat satellites have acquired millions of images of points on the earth's surface. The images, archived in the United
States and at Landsat receiving stations around the world, are a unique resource for global change research and applications in agriculture, geology,
forestry, regional planning, education, and national security.

The value of the Landsat Program was recognized by Congress in October 1992, when it passed the Land Remote Sensing Policy Act (Public Law
102-555), authorizing the procurement of Landsat 7 and assuring the continued availabilty of Landsat digital data and images, at the lowest possible cost,
to traditional and new users of the data. Table 9–4 provides a summary of the
Landsat satellites.

Figure 9–19
Landsat 7 Satellite
(Courtesy of NASA)

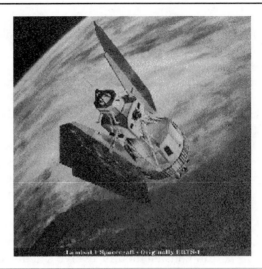

Figure 9–19
Landsat 7 Satellite
(Courtesy of NASA)

Other MEO satellite systems are planned, but none are in orbit at this time. As of 2001, there are several LEOs in operation as well as many GEOs. The MEO market is yet to meet its potential.

GEOs

GEOs are the earliest and the most prevalent type of satellite system to date. GEOs have been launched and placed into orbit since the mid-1970s. They are circular orbits orientated in the plane of the earth's equator. In a geosynchronous orbit, the satellite appears stationary (i.e., in a fixed position) to an observer on earth. More technically, a geosynchronous orbit is a circular prograde orbit in the equatorial plane with an orbital period equal to that of the earth; this is achieved with an orbital radius of 6.6107 (equatorial) earth radii, or an orbital height of 35,786 km. A satellite in a geosynchronous orbit will appear fixed above the surface of the earth (i.e., at a fixed latitude and longitude).

The footprint, or service area, of a geosynchronous satellite covers almost one-third of the earth's surface (from about 75 degrees south to about 75 degrees north latitude), so that near-global coverage can be achieved with a minimum of three to four satellites in orbit.

By placing the satellite at an altitude where its orbital period exactly matches the rotation of the earth (approximately 35,800 km), the satellite appears to hover over one spot on the earth's equator and thus appears to stay stationary over the same point.

A geosynchronous satellite completes one orbit revolution in circular orbit, around the earth, every 24 hours. If the orbit is in the equatorial plane, and

System	Launch (End of service)
Landsat 1	7/23/72
	(1/6/78)
Landsat 2	1/22/75
	(2/25/82)
Landsat 3	3/5/78
	(3/31/83)
Landsat 4*	7/16/82
Landsat 5	3/1/84
Landsat 6	10/5/93
	(10/5/93)
Landsat 7	4/99

Table 9–4
Landsat Program -
System Summary
(Courtesy of NASA)

if rotation is in the same direction as the earth (rotating at the same angular velocity as the Earth) and it overlies the same point on the globe permanently, then the satellite is termed geosynchronous.

Geosynchronous means that a satellite makes one orbit every 24 hours so that it is synchronized with the rotation period of the earth. As stated, this will happen when a satellite is in a circular orbit at a rough distance of 36,000 km above the surface of the earth, or roughly 42,000 km from the center of the earth. The orbital location of geosynchronous satellites is called the Clarke Belt, in honor of Arthur C. Clarke who first published the theory of locating geosynchronous satellites in earth's equatorial plane for use in fixed communications purposes. For brevity, the term *geosynchronous satellite* is often shortened to *geo satellite.*

Earth stations transmit a signal to a satellite in orbit; this act is called an **uplink.** Geosynchronous satellites receive the uplinked signal, amplify it, shift it to a lower frequency, and then couple the outgoing signal to the transmitting array of onboard satellite antenna where the signal is focused into a narrow beam and sent back to earth. The act of sending the signal back to earth is called a **downlink.** The onboard satellite electronics, which receive the uplinked signal, amplify it, and shift the frequencies, is called a transponder. Figure 9–20 depicts this process.

uplink

downlink

In the transponder layout (i.e., configuration) of a satellite, the channels are arranged with a band of frequencies between each one, typically 4 MHz. This is known as a guard band and its purpose is to provide isolation between each channel. Note also that there are usually at least two sets of frequencies per transponder and they are at opposite polarities; this is known as frequency reuse. Frequency reuse is a technique in which two transponders share the same frequency; however, by handling signals at opposite polarities, the two signals will not interfere with each other, so each frequency is

Figure 9–20
Uplink and Downlink
Process

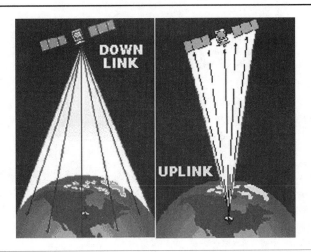

reused. This fact is especially important because satellites cost from between $150 million to $1 billion to build and launch (this includes the cost to launch plus the satellite itself). Without finding ways to reuse frequencies in each satellite, the orbits would quickly fill up, and too many resources would be pumped into less efficient programs.

To further avoid co-channel interference, each set of polarities is typically offset from each other by an amount equal to one-half of their bandwidth (i.e., the band ends of one set of polarities fall on the center frequencies of the other set of polarities). In the current age of digital compression, frequency reuse is further obtained by digitally compressing multiple channels per transponder (i.e., more use of each frequency is obtained by using compression techniques to cram more information into each uplink/downlink signal).

On newer satellites, some individual transponders can be reconfigured from earth control stations to be combined to provide a customer with more bandwidth or can be divided to lessen the bandwidth to customer specifications (for instance, a 72 MHz transponder can be broken into two 36 MHz bandwidth units).

Whereas a low altitude orbiting satellite is hidden for part of its orbit around the earth, at least one GEO satellite is always directly visible from the same earth geographical area, day or night. The field of view of a satellite in GEO is fixed. A GEO is essential for a permanent satellite communications system between two points and is ideal for repeat observations to a fixed geographical global area. Examples of GEO satellite systems follow.

These two satellite-TV providers are in the midst of a merger announced in Oct 2001. Regulatory approval is still pending.

Direct Broadcast Services

There are two providers of direct broadcast services (DBS)—DirecTV and DishTV.

DirecTV

DirecTV Inc., provider of the nation's leading digital satellite television service, uses seven DirecTV-owned satellites that deliver the high-power DirecTV programming service to millions of customers in the United States. DirecTV has launched five satellites and obtained in 2000 the two Primestar satellites from that defunct system.

DirecTV currently has a license for forty-six high-power frequencies at three orbital slots, all of which provide full coverage of the continental United States. DirecTV is the nation's leading provider of digital television entertainment service with more than 8.7 million customers.

Along with broadcasting more than 100 television channels, DirecTV has broadened out into the Internet services arena by providing Internet access to home or business-based users. DirecTV claims a subscriber base of 9.8 customers.

Internet Access via DirectPC.   Using their own satellites, access to the Internet is provided to home and business-based subscribers. The DirectPC configuration transmits and receives signals directly via the satellite without needing an additional dedicated phone line to support the high on-time connection for the upstream piece of the broadband service.

DishTV

DirecTV is not the only satellite-based provider of Internet services to home and business-based subscribers. The other provider is DishTV, provided by EchoStar Communications. DishTV reports 6 million customers.[15] Internet connectivity on the DishTV satellites is provided via its sister service, StarBand.

StarBand is headquartered in McLean, Virginia. It advertises itself as America's first consumer two-way, always-on, high-speed satellite Internet service provider (see the case study at the start of this chapter). StarBand began operations in April 2000, with the assistance of strategic partners such as Gilat Satellite Networks, Microsoft Corporation, and EchoStar Communications.

Identical to DirecTV, the StarBand system consists of a 24 × 36 inch satellite dish mounted on or near the home (1.2 m satellite dish in Alaska). The dish must have a clear unobstructed view of the southern sky. Two standard coaxial cables connect the satellite dish antenna to a PC or a StarBand satellite modem.

---

[15] Information courtesy of DirecTV at www.directv.com.

Figure 9–21
Accessing the Internet
via DBS
(Courtesy of www.directv.
com)

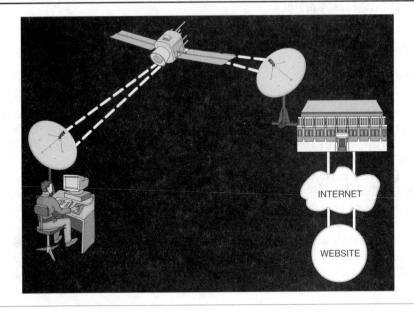

The antenna sends requests to the Internet and receives Internet content via the satellite, which is in orbit approximately 22,300 miles above the equator. The satellite communicates with the StarBand hub facility, which has a direct connection to the Internet.

In addition to the two-way satellite Internet service, the StarBand system can also be configured to receive satellite television service from EchoStar's Dish Network.

Inside the home, the StarBand system consists of a satellite modem that attaches to a home PC through a USB or Ethernet port. No telephone connection, dial-up account, or other Internet service providers are necessary. StarBand does it all. Figure 9–21 shows this process, regardless of whether the DirectPC or StarBand system is used.

Intelsat

Intelsat (Figure 9–22) has a fleet of twenty satellites. In contrast to the direct broadcast services, Intelsat offers Internet, broadcast, telephony, and corporate network services to leading companies in more than 200 countries and territories worldwide. In an expansion mode, Intelsat is launching nine more satellites between 2001 and 2003. All are in geostationary orbit at a distance of 22,300 miles (36,000 km). Figure 9–23 shows that the coverage of the earth's surface by the Intelsat fleet of satellites is extensive.

Intelsat's customer base includes hundreds of Internet service providers, telecommunications companies, broadcasters, and corporate network service

Figure 9–22
Intelsat Satellite
(Courtesy of www.intelsat.
com)

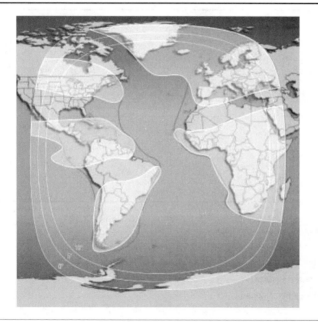

Figure 9–23
Intelsat Coverage of the
Earth's Surface
(Courtesy of www.intelsat.
com)

providers worldwide. As compared with systems such as Iridium, which are primarily U.S. based, Intelsat is international in scope. The headquarters are in Washington, D.C., but other major offices are in Australia, France, Germany, Hong Kong (PRC), India, Norway, South Africa, Singapore, and the United Kingdom. Table 9–5 shows the revenues generated by region.

Revenues generated by various industry segments, not just cellular telephony as in Iridium, are in Table 9–6.

Even Intelsat's ownership is international in scope. Intelsat has more than 200 investor companies in 145 countries (see a sampling in Table 9–7).

Table 9–5
Intelsat Regional
Revenues
(Courtesy of www.intelsat.
com)

Europe	28%
North America & Caribbean	23%
Asia & Pacific	19%
Latin America	14%
Middle East & North Africa	8%
Sub-Saharan Africa	8%

Table 9–6
Revenues by Industry
Segment
(Courtesy of www.intelsat.
com)

Carriers	42%
Internet	14%
Corporate network	25%
Broadcast	19%

Table 9–7
Intelsat Investor
Companies
(Courtesy of www.intelsat.
com)

| Lockheed Martin Global Telecommunications |
| Videsh Sanchar Nigam Limited |
| FRANCE TELECOM |
| Telenor Broadband Services AS |
| British Telecommunications PLC |
| Teleglobe Inc. |
| Deutsche Telekom AG |
| Telecom Italia S.p.A. |
| Telstra Corporation Limited |
| EMBRATEL |
| Cable & Wireless PLC |
| CHINA TELECOM |
| T RK TELEKOM NIKASYON A.S. |
| TELECOM COLOMBIA |
| Telecommunication Company of Iran |

Inmarsat

Inmarsat was the world's first global mobile satellite communications operator and advertises itself as the only one to offer a mature range of modern communications services to maritime, land mobile, aeronautical, and other users.[16]

- Formed as a maritime-focused intergovernmental organization over 20 years ago, Inmarsat has been a limited company since 1999, serving a broad range of markets. Starting with a user base of 900 ships in the early 1980s, it now supports links for phone, fax, and data communications at up to 64 Kbps to more than 210,000 ship, vehicle, aircraft, and portable terminals.
- Inmarsat Ltd. is a subsidiary of the Inmarsat Ventures PLC holding company. It operates a constellation of geostationary satellites designed to extend phone, fax, and data communications worldwide. The constellation comprises five third-generation satellites backed by four earlier spacecraft. These satellites are controlled from Inmarsat's headquarters in London, which is also home to Inmarsat Ventures, as well as the small IGO created to supervise the company's public service duties for the maritime community (Global Maritime Distress and Safety System) and aviation (air traffic control communications). Inmarsat has regional offices in Dubai, Singapore, and India.

  Users of the Inmarsat system of satellites include ship owners and managers, journalists and broadcasters, health and disaster relief workers, land transport fleet operators, airlines, airline passengers and air traffic controllers, government workers, national emergency and civil defense agencies, and peacekeeping forces.
- The central aspect of the strategy is the new Inmarsat I-4 satellite system, which, beginning in 2004, will support the Inmarsat Broadband Global Area Network (B-GAN). This system will be capable of handling mobile data communications at up to 432 Kbps for Internet access, mobile multimedia, and many other advanced applications.

The Satellites

Inmarsat's primary satellite constellation consists of four Inmarsat-3 satellites in geostationary orbit. Between them, the main (global) beams of the satellites provide overlapping coverage of the whole surface of the earth apart from the poles. So, thanks to Inmarsat, it has become possible to extend the reach of terrestrial wired and cellular networks to almost anywhere on earth (see Figure 9–24).

A geostationary satellite follows a circular orbit in the plane of the equator at a height of 35,600 km, so that it appears to hover over a chosen point on

---

[16] Information courtesy of INMARSAT at http://217.204.152.210.

Figure 9–24
A Typical Inmarsat
Satellite
(Courtesy of www.inmarsat.
com)

the earth's surface. Three such satellites are enough to cover most of the globe, and mobile users rarely have to switch from one satellite to another. Other mobile satellite systems use larger numbers of satellites in lower, nongeostationary orbits. From the user's point of view, they move across the sky at a comparatively high speed, often requiring a switch from one satellite to another in mid-communication and risking the possibility of an interrupted call.

A call from an Inmarsat mobile terminal goes directly to the satellite overhead, which routes it back down to a gateway on the ground called a land earth station (LES). From there the call is passed into the public phone network.

The Inmarsat-3 satellites are backed up by a fifth Inmarsat-3 and four previous generation Inmarsat-2s, also in geostationary orbit.

The Network

To deliver its services, Inmarsat operates a worldwide network of ground stations in addition to the satellite constellation.

- The satellites are controlled from the satellite control center (SCC) at Inmarsat HQ in London. The control teams there are responsible for keeping the satellites in position above the equator, and for ensuring that the onboard systems are fully functional at all times.
- Data on the status of the nine Inmarsat satellites are supplied to the SCC by four tracking, telemetry, and control (TT&C) stations located at Fucino, Italy; Beijing, China; Lake Cowichan, western Canada; and Pennant Point, eastern Canada. There is also a backup station at Eik, Norway.
- Data from a user terminal are transmitted to a satellite and then down to a LES, which acts as a gateway into the terrestrial telecom networks. There are about forty LESs, located in thirty countries.

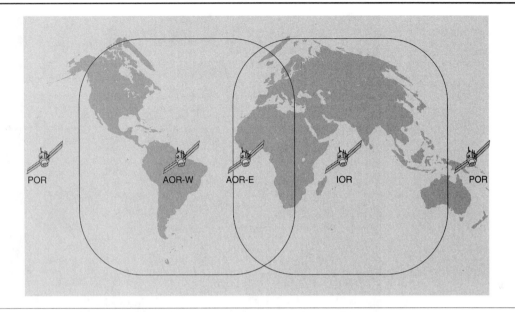

Figure 9–25
Inmarsat Global Coverage
(Courtesy of www.inmarsat.com)

- The flow of communications traffic through the Inmarsat network is monitored and managed by the network operations center (NOC) at Inmarsat HQ in London.
- The NOC is supported by network coordination stations (NCSs). Their primary role is to help set up each call by assigning a channel to the MES and the appropriate LES. There is one NCS for each ocean region and for each Inmarsat system (Inmarsat A, B, C, and so on). Each NCS communicates with all the land earth station operators in its ocean region, the other NCS, and the NOC, making it possible to distribute operational information throughout the system. Figure 9–25 shows Inmarsat global coverage.

Geostationary Operational Environmental Satellite

Another GEO system used extensively is the geostationary operational environmental satellite (GOES) system. Like many weather satellites, GOES was developed and launched by NASA. Since 1968, there have been eleven GOES satellites launched; nine are still operational. Figure 9–26 depicts the latest GOES satellite—GOES 10. Once it became operational, GOES was turned over to NOAA for day-to-day administration. The GOES mission is

Figure 9–26
Typical GOES Satellite
(Courtesy of NASA)

to provide timely global weather information. GOES imagery is commonly featured on many TV weather reports across the United States and the world. For example, the hurricane images during hurricane season are provided via the GOES system. The GOES Program maintains two satellites operating in conjunction to provide observational coverage of 60 percent of the earth. The GOES satellite system has remained an essential cornerstone of weather observations and forecasting for 25 years.

The GOES satellites carry a space environment monitor (SEM), which investigates solar particle emissions and helps study the effect of solar activity on the earth's telecommunications systems. The SEM detects solar protons, alpha particles, solar electrons, solar x-rays, and magnetic fields. In addition to observations, the GOES platform (the satellite stationed over the Pacific Ocean) has been used to create and operate PEACESAT (Pan-Pacific Educational and Cultural Experiments by Satellite). PEACESAT provides satellite telecommunications to serve the educational, economic development, medical, and cultural needs of many Pacific island nations and territories.

Table 9–8 provides a timeline of GOES development.

GOES observations have proven helpful in monitoring dust storms, volcano eruptions, and even the spread of forest fires (see Figure 9–27).

A data collection system added to GOES allows it to receive and relay environmental data. These data could have been collected by widely dispersed

| **Table 9–8** |
| GOES Timeline |
| (Courtesy of NASA) |

1978–1987

GOES-4 made the first vertical temperature and moisture measurements from synchronous orbit. From these cross-sections, the altitudes and temperatures of clouds were determined and a three-dimensional picture of their distribution drawn for more accurate weather prediction. Using GOES imagery, meteorologists were able to measure the frame-to-frame movement of selected clouds at different altitudes and to obtain their wind direction and speed in order to better understand atmospheric circulation patterns.

GOES-7 had the added capability of transponding signals at 406 MHz from emergency locator transmitters on ships and planes in distress; this would greatly aid in search-and-rescue efforts. GOES-7 was able to participate in an experiment to show the effectiveness of satellites for use in the international search and rescue satellite aided tracking system (SARSAT). SARSAT became operational with the launch of GOES-8 in 1988.

1988–

The GOES Program continues to improve; new technological innovations and sensors (such as Doppler radar, popular with weather stations across the country) were added. The modern GOES satellite assists with the observation and, when possible, the prediction of fairly local weather events including thunderstorms, tornadoes, flash floods, and even snow squalls.

surface platforms such as river and rain gauges, seismometers, tide gauges, buoys, ships, and automatic weather stations. Platforms transmit sensor data to the satellite at regular intervals, upon request by the satellite, or in an emergency alarm mode when a sensor receives information exceeding a preset level (such as when river water levels rise).

PEACESAT

PEACESAT is best known as a public service satellite telecommunications network that links educational institutions, regional organizations, and governments in the Pacific Islands. The capabilities of PEACESAT are provided via the GOES satellite system.

Public Service Telecommunications

Public service telecommunications may be defined as "the use of telecommunications and information technologies by government, educational, and nonprofit organizations for education, medical and health services, emergency and disaster management, environment and resource management, and economic development."[17]

---

[17] Read the entire PEACESAT Mission at www.obake.peacesat.hawaii.edu/info/2-miss.htm.

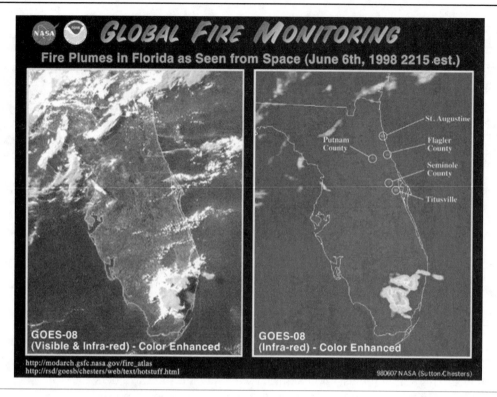

Figure 9–27
Fires Seen from GOES Satellite Images: NASA-GSFC, Data from NOAA GOES
(Courtesy of NASA)

The capability of PEACESAT, via GOES, to provide telecommunications services to this region is crucial. In more developed economies, telecommunications is far less of a cost and technology barrier than it is in developing or emerging economies. For example, it is possible for residents in Hawaii to call New York at $0.15 a minute, but the same call would cost many of the Pacific island economies over $3 per minute. In addition, while access to information via the Internet is provided at a very small cost in Hawaii and in other developed economies, the same service for a user spending 10 hours per month in other Pacific islands could cost over $100 per month.

PEACESAT also provides information to these islands in the areas of education, health and medical emergencies, emergency management, and technology transfer. Some of the projects that PEACESAT is working on to support these program areas include the development of digital voice, data, and compressed video teleconferencing capabilities within the region; building collaborative relationships to support these levels of technology; developing an electronic mail system for health and emergency operations; and developing a network interface among emergency agencies in Hawaii.

POLAR ORBITS

Because geosynchronous satellites cannot cover the poles of the planet, another method for providing satellite coverage is required. Polar synchronous satellites are a solution to the challenge, as shown in Figure 9–28.

Polar orbits are a type of LEO orbit. They are much closer to the earth than GEOs (GEOs are at 35,800 km altitude; polar orbiting satellites are at 850 km altitude). By definition, a polar satellite has an inclination of 90 degrees to the equator. By placing a polar orbit satellite at an altitude of about 850 km above the earth's surface, a polar orbit of roughly 100 minutes can be achieved. This will allow the earth to rotate beneath the satellite so that one polar satellite is used per application. For more continuous coverage, more than one polar orbiting satellite may be used.

Polar orbit satellites can be used for one of two applications or uses—to view only the poles (to fill in gaps of GEO coverage) or to view the same place on earth at the same time each 24-hour day. The critical design goal is to place a polar satellite in an orbit that is low enough to allow a relatively short orbital period, but at the same time keep its orbit altitude sufficient to permit observation of a sufficiently wide path. This would enable the satellite instrumentation time to scan the earth below, because the earth will be rotating more slowly in relation to the satellite.

With the two applications of polar satellites come many uses. Polar satellites can carry sensors that are sensitive to both visible light and infrared radiation. They can measure the temperature and humidity in the Earth's atmosphere. They can sense and record surface ground and surface seawater temperatures, and monitor cloud cover and water or ice boundaries. When *Discovery* airs programs about the theory of global warming, the pictures shown are from polar satellites. Polar satellites are also able to receive, measure, process, and retransmit data from various scientific platforms such as balloons, buoys, and remote automatic stations. Polar satellites can carry emergency search-and-rescue transponders to help locate downed airplanes or ships in distress.

Usually a polar orbit is fixed in space, and the earth rotates underneath, so it travels from the North Pole to the South Pole. A typical polar satellite can cover the entire globe every 14 days.

Sun-Synchronous Orbits

A special polar orbit that crosses the equator and each latitude at the same time each day is called a sun-synchronous orbit. Sun-synchronous orbits are also called helio-synchronous orbits, and the timeliness of their crossing a certain latitude at almost the same time every day can come in handy when wanting to do comparative studies or provide a regular interval of communication each day. The timing of the orbital period is caused by the centripetal force of the spinning earth.

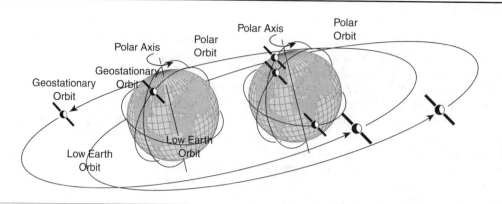

Figure 9–28
Polar versus Geostationary Orbit
(Courtesy of NASA)

One of the reasons to use this type of orbit, also called a dawn-to-dusk orbit, is that it enables the satellite to always have its solar panels in the sun. An example of a dawn-to-dusk satellite orbit is the one occupied by the "Radarsat" satellite. Radarsat is only 798 km above the earth, and circles the globe from the North Pole to the South Pole. Because Radarsat can keep its solar panels pointed toward the sun almost constantly, it does not have to rely much on batteries. Radarsat takes pictures that are useful for agriculture, oceanography, forestry, hydrology, geology, cartography, and meteorology.

HEOs

Highly elliptical orbits are very different from the other orbits because of their shape. Instead of orbiting the earth at a relatively constant distance, they dip and dive, then shoot out further than most satellites. The top of an orbit is called the apogee, and the lower end is the perigee. The apogees and perigees of HEOs range from 50,000 km to 500 km above the earth's surface, respectively.

HEOs for earth applications were initially used by the Russians to provide communications to their northern regions not covered by GEO satellite networks.

The Russians, for example, used the Molnya satellite system over Siberia. The area of interest would be one with a major population center or an arena otherwise necessary to cover. After the apogee period of orbit, a handoff must occur between the satellite about to descend and the rising satellite after it. Without the handoff, any communication being transmitted will be lost.

Some examples of HEOs are the Russian Molnya system, which employs three satellites in three 12-hour orbits around the earth. Each satellite is separated from the next by 120 degrees, with the **apogee** distance at 39,354 km above the earth's surface and the **perigee** at 1,000 km. The timing of the satellites in their orbits (especially having more than one satellite in the same orbit) is crucial, and if for some reason the timing is shot, there could be loss of service for any given amount of time. Another example of an HEO is the Russian Tundra satellite system. Unlike the Molnya, the Tundra only employs two satellites. Each satellite is separated by 180 degrees around the earth, and is responsible for its own 24-hour orbit. The Tundra satellites have their perigees and apogees at 17,951 km and 53,622 km, respectively.

apogee

perigee

OTHER OPTIONS

With the many different satellite technologies and improvements becoming more readily available, corporations and governments are finding more innovative ways of using satellites to their advantage. Two ways in particular are the very small aperture terminal (VSAT) and a system called the HALO Network, for high altitude long operation (HALO) aircraft.

VSAT

A VSAT is a remote terminal in a satellite communications network. The term VSAT refers to the small size of the antenna dish, normally 2.4 m or smaller in diameter. This size is small when compared with larger earth station antennas, which can run 11 m or more in diameter. The satellite is essentially a repeater (acts like a mirror) that reflects transmissions from the terminal back to earth. This enables one terminal in the network to communicate with another terminal hundreds of miles away by bouncing signals off of a satellite in the sky. Another application of the VSAT is that it enables broadcasting information from one terminal to many others, as the retransmission of the signal from the satellite back to earth normally covers a large geographical area.

The VSAT network architecture can be star, full mesh, or hybrid. Star means that one central site communicates with all the remotes. This is the architecture of satellite TV broadcasting, in which one large central antenna transmits to many small remote antennas; however, VSATs can offer two-way communication, with the remotes also transmitting and receiving from the central site, like in an Ethernet network. If two remotes communicate, the sending remote transmits to the central site. The central site retransmits the message to the intended remote. This means two hops to the satellite. Full mesh means that any terminal in the network can communicate with any other terminal, with only one trip to the satellite. There is no intermediate stop at a central site. Hybrid means the network architecture employs both star and full mesh. An example is a network that uses star for data transmissions to a central site and also provides voice communications between all remote terminals (full mesh).

Applications for VSATs include all kinds of communications, including wireless Internet and telephony. Satellite can fill in pockets where other technologies cannot reach. It provides global coverage in a single network, and is unequaled for broadcast applications, such as the delivery of the same information to a large group of users. In addition to providing fast access to the Internet for the individual user, satellites can also provide reliable two-way Internet backbone connectivity, entirely bypassing local infrastructure. For regions of the world that lack affordable high-bandwidth terrestrial links, Internet via satellite can be the fastest economical means for an ISP to gain high-bandwidth access to the Internet.

The HALO Network

The HALO Network uses a data communications hub aboard a specially designed HALO aircraft, flying at altitudes high above commercial air traffic and adverse weather, to provide customers with access to broadband services, regardless of their location.

Pioneered by Angel Technologies and their partners, the HALO Network is designed to provide broadband communications services over a footprint 50 to 75 miles in diameter to users in metropolitan areas. The idea is that super-high-flying aircraft are cheaper to launch than rockets, especially because the atmosphere need not be broken. The regular landing and refueling of aircraft will make maintenance of the craft and the transmitting of equipment easy. No more will space shuttle journeys be necessary to recover a satellite or make repairs.

The abilities of the network are impressive. The HALO Network will have an initial transactional capacity greater than 16 Gbps (i.e., comparable to more than 5,000 fully loaded, symmetric DS-1 connections). After the program takes off, Angel says they will increase capabilities according to market demand. Access to this system will be easy, as long as users are within the footprint and have purchased the necessary equipment (e.g., a network interface card and a small external auto tracking antennae). Any information destined for nonsubscribers or recipients out of the HALO Network area will be switched over to the terrestrial communications systems.

Advantages of the HALO system are many. Each aircraft will be flying at a much higher altitude than any tower could reach. The increased area of service lowers the number of hubs and other transmission equipment needed to serve a particular area. The HALO flights can also be used to supplement satellite networks and fill the gaps during nonsatellite coverage. Because the planes may go up or come down as needed, financing can be incremental, and Angel plans on implementing their service on a city-by-city basis. Unlike Iridium, HALO has a way out if the profits are not turned.

■ SUMMARY

With the advances in satellite technology, we find ourselves with the same struggle as always: unlimited wants and needs, and limited resources. Thus far we have been able to find different orbits in which to launch satellites and calculate distances and forces in the atmosphere and beyond. We are able to measure more accurately, from outer space, aspects of our planet that would have remained unknown to us had we not traveled there. We have come a long way since Sputnik, and we are ever finding more ways to use our knowledge and technology. One thing throughout is certain: We continue to find new limits and we continue to break them.

In the next chapter we will discuss the global positioning system, which is a unique and increasingly useful aspect of satellite communications.

REVIEW QUESTIONS

1. MEO stands for _____.
   a. middle earth orbit
   b. mid earth orbit
   c. medium earth orbit
   d. elliptical earth orbit

2. LEOs are typically used for _____.
   a. weather forecasting
   b. scientific experiments
   c. cellular communications

3. Odyssey is a _____ based system of satellites.
   a. LEO
   b. MEO
   c. GEO

4. The _____ satellites orbit at 5,000 to 15,000 km.
   a. MEO
   b. GEO
   c. LEO
   d. none of the above

5. Solar activity primarily affects _____.
   a. GEOs
   b. MEOs
   c. LEOs

6. Landsat is an example of a _____ based satellite system.
   a. LEO
   b. GEO
   c. MEO

7. The _____ satellite systems appear stationary to those on the ground.
   a. LEO
   b. MEO
   c. GEO

8. Geosynchronous means _____.
   a. these satellites always appear stationary to those on the ground
   b. orbits that are circular
   c. orbits around the equator
   d. all of the above

9. A transponder is a _____.
   a. transmitter
   b. receiver
   c. solar energy cell for satellites
   d. transmitter and receiver

10. LEOs have a deteriorating orbit due to _____.
   a. their unique orbital attitude
   b. atmospheric effects
   c. solar activity
   d. a and b only
   e. none of the above

11. The _____ satellite system requires a fewer number of satellites to fully cover the earth's surface.
   a. LEO
   b. MEO
   c. GEO

12. The first satellite-based cellular telephony system was _____.
   a. Odyssey
   b. Iridium
   c. GEOS

13. Inmarsat is an example of a _____ based satellite system.
   a. LEO
   b. MEO
   c. GEO

14. The orbital velocity of the satellite depends upon _____.
   a. the altitude from the earth
   b. the initial launch speed
   c. how many solar cells the satellite utilizes each 24-hour period

15. A _____ orbit is excellent for mapping the earth's surface.
   a. higher
   b. equatorial
   c. polar

16. A _____ view must be available at all times for proper reception of DirecTV or DishTV.
   a. northern
   b. southern
   c. easterly
   d. does not matter

HANDS-ON EXERCISES

1. Using your browser, go to www.un.org, the website for the United Nations. Look for information on satellites. What role do they play in the satellite arena?

2. Using a graphics program such as Paint or Adobe Photoshop, create a picture of the earth showing typical LEO, MEO, and GEO configurations.

3. Conduct research using online and other sources and produce a table in Excel (or a similar program) comparing both StarBand and DirectPC. Include items such as initial installation costs, monthly costs, and services provided.

4. The GOES performs many services for a wide range of customers. Using your browser, conduct an Internet search and make a list of the capabilities of GOES and who will benefit.

5. Research the important SARSAT system. Provide a list of its benefits, and who benefits from these capabilities.

10 Global Positioning System

OBJECTIVES

After reading this chapter and completing the exercises, you will be able to:

- Learn how the global positioning system evolved
- Learn how the global positioning system operates
- Learn the applications of the global positioning system
- Learn the future of the global positioning system

In the previous chapter we discussed the broader aspect of satellite technology. In this chapter, we discuss a somewhat unique, and increasingly useful, aspect of satellite communications, namely the global positioning system.

GPS IN THE PITS
DIFFERENTIAL GPS APPLICATIONS AT THE MORENCI COPPER MINE

In the past 6 years, Phelps Dodge has integrated GPS position and navigation functions into all operations at its Morenci copper mine. The resulting increases in efficiency, productivity, and safety have made this investment worthwhile.

Phelps Dodge Morenci

Phelps Dodge's Morenci copper mine in Southeastern Arizona is one of the world's largest, producing more than a billion pounds of copper per year. At Morenci, Phelps Dodge (PD) uses GPS technology not only for surveying and monitoring the position of heavy equipment, but also to precisely navigate electric drills to designed blasthole patterns and exact target depths. Mine management can track material and vehicles, and engineering can calculate surface-to-surface volume, navigate vehicles, and design and construct roads using information from a real-time kinematic GPS (RTK-GPS) system.

Open-Pit Mining

An open-pit copper mine looks like a giant sports stadium, with horizontal surfaces, called *benches,* that descend like steps into the pit. Vehicles pass from bench to bench along ramps and roads, which are repositioned as the benches are mined away. To excavate the ore, automated drills make a pattern of blast holes in a bench, and explosive charges are loaded into the holes. After the blast, shovels load the broken rock into trucks. The trucks take the rock to be crushed, then dump it in stockpiles. An acid solution sprayed on the stockpiles leaches the copper out of the rock as it trickles through the pile. After the solution is recovered, the copper is extracted from it.

GPS technology first came to Morenci in 1995, when PD began to use GPS for surveying. GPS was first used for machine guidance in 1997, when GPS was installed on the first electric shovel at Morenci. Since then, GPS technology has found many applications at the mine.

One of six GPS-equipped drills at Morenci. The distance from the top of the mast to the ground is about 115 feet.

Morenci's GPS System

Morenci's RF network consists of more than 3,000 voice and data radios. Our GPS system is composed of multiple open-architecture RF networks. Two overlapping RF networks using 900-MHz radios provide position correction data to the mobile units. These base stations sit at the top of the pit's 3,000-foot walls. The first network (R1) uses a base station and three mobile repeaters to cover the electric shovels and drills. The second network (R2) is an

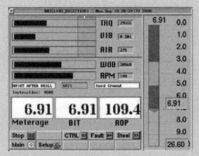

The display that the drill operator sees, showing rock hardness and depth.

overlay of the same network with a base station at a different location. This allows seamless coverage of the mine with rapid transmission of data.

We have a separate survey network that transmits on three 450-MHz UHF frequencies. This network covers a larger area so that mineral exploration can go on.

Rotary Drills

The rocks from a good blast fall where they are meant to, not into an area where they endanger workers or block a road. A good blast results in an even surface below the shot so that the shovel can remove the material efficiently and the next set of drill holes will be drilled to a uniform depth and with the correct spacing behind the last pattern.

To maximize the likelihood that a blast will meet these criteria, holes for blasting must be drilled in a pattern optimal for the location and the type of rock (geology) involved. Large rocks or boulders create waste, and have to be rehandled, reblasted, and crushed for proper recovery.

Before the first drill at Morenci was outfitted with a GPS system, laying out drill patterns and planning the utilization for seventeen drills was the responsibility of one drill supervisor. But when PD adopted GPS-controlled drilling in 1998, our surveyors, who were looking for new things to do because GPS had already made their work more efficient, became pattern layout and design engineers. Now instead of personnel going out to the field, surveying the new pattern area, returning to the office to design the pattern, and returning once again to the field to lay out a staked pattern, the surveyors design the patterns in the office, then send them via radio to the drills.

Each drill is equipped with two GPS receivers, which enable the drill operator to navigate from hole to hole and pattern to pattern with great accuracy. The dual GPS receivers allow the operators and management to always know the true direction the drill is operating without having the drill move around to "find" its heading. This is necessary for all single-receiver type GPS applications. This solution saves valuable time for immediate navigation. At over $300/hour to run an electric drill, any additional time needed to determine the drill's correct position and direction is very costly.

As the blast hole is drilled, the control software collects information about rock hardness and drill productivity. Engineers and supervisors download or view production information (X, Y, and Z coordinates) in near real time via our radio network without interrupting productivity in the field. Before GPS equipped drills at PD, surveyors spent an average of six man-hours daily locating blasthole patterns; now, with GPS on the drills, we spend less than two man-hours daily. This is a saving of more than $50,000 annually in survey time alone.

The drill navigation system provides more than productivity statistics. The information that engineering uses is really only the tip of the drill mast, so to speak.

The system automatically downloads diagnostic information from the drills and uploads it into Microsoft Access and SQL databases. We can monitor GPS coverage and availability, drill productivity, and operator accuracy and efficiency and track the use of consumables (such as drill bits, steel, and stabilizers).

Another important bit of information we gather from our drills is something we call the *Blastability Index.* The GPS system records information as it drills down the hole in 1-inch increments. It indicates the type and hardness of rock. Information from the drill control system is correlated with geologic zones and data gathered by our geologists. We create contoured drawings of hardness and define areas where we have had good fragmentation results. Then, our fragmentation personnel can replicate these results in other areas. This information is used not only for blasthole loading (explosives) but for more efficient utilization of the drill fleet.

The navigation system is used to precisely drill to the designed hole and target depth. Before GPS was installed on the drills, this depth was usually the operator's best guess and cost many hours annually of rework, redrilling to achieve correct depth or backfilling holes that were drilled too deep. The accuracy of the GPS drills now gives us more control and the ability to achieve consistent fragmentation results.

Electric Shovels

After the blast, electric-powered shovels scoop up the broken rock. The shovels are equipped with onboard computers, software, a GPS receiver, and data radios and receivers.

Engineers transmit planning designs to the machine's onboard computer. The operator sees a display showing the boundaries of the ore deposit, powerlines, miscellaneous obstacles, and mining limits. Next to the map, an elevation gauge tells the operator how much material needs to be cut or filled to achieve the desired new elevation for the bench. The elevation is displayed in 0.1-foot increments, but the shovels are not sufficiently maneuverable to move material with that kind of precision. A shovel with a 54-cubic-yard bucket can maintain grade within a foot or two, and commonly within − 0.5 feet. The system records the machine's progress, updates the operator's display, and transmits the same information back to the engineering office for analysis and documentation. There, the record of the shovel's movements is used in volume-to-volume calculations to ascertain exactly how much material was mined and precisely where it was mined. Daily production reports go to geologists, mine planners, shift supervisors, and shovel operators.

Savings and other benefits. Since the GPS systems were installed on the shovels, savings due to reduced error in material routing have been substantial. Still more money has been saved on materials and worker hours needed for targeting. Because the shovel operator can see locations of power lines, hazards, and such features on the display, the likelihood of hitting them is much reduced. Also, the shovels can go to work right after a blast, instead of having to wait for workers to set targets. Finally, work can proceed in darkness and inclement weather.

User screen inside track dozer, showing the operator that the dozer is below grade, and needs to fill.

Haul Trucks

The electric shovels load the ore into haul trucks, which were outfitted with GPS in 1997 to track haulage routes, destinations, and travel times.

Dispatching software routes the trucks on the best path. A truck's position is updated every 30 seconds. The system records information about road speeds, congestion points, and cycle times, and later analyzes these data to help identify traffic problems. The locations where the truck was loaded and dumped are recorded and linked with data from the shovels so that the characteristics of each truckload can be stored for future stockpile modeling.

Haul truck operators can record points on the fly to identify any problem spots on the roads or ramps, such as rough spots, excessively steep grades, insufficient turning clearance, drainage problems, or insufficient banking on turns. This information goes to the surveyors, who are responsible for scheduling repair or redesign.

Track Dozers

Track-type bulldozers build at least one new ramp a week and also build and maintain roads in the mine. Like the onboard systems in the electric shovels, the track dozers' onboard systems track each vehicle's position and elevation and advise the operator of what remains to be cut and filled.

The dozers' navigation system can use display or elevation files. The display files can show the operator elevation in plan, profile, digital, or gauge views. Design limits for steepness, banking, crowning, and turning clearance displayed with text direct the dozer operators. The system converts these files from DXF files generated at the office. Elevation files can be generated from digital terrain models.

The system helps the dozer operators improve road quality, making road surfaces smoother and the ramp grades more uniform, without requiring a lot of surveying time. Slight variations in ramp grade make truck drivers change gears frequently, which causes equipment wear. At Morenci, reducing truck maintenance costs by 1 percent saves $400,000 per year, so even seemingly small improvements in the roads pay off.

The display that the dozer operator sees.
Source: Courtesy of Caterpillar Inc.

GPS also controls the track dozers that maintain and build new ore stockpiles. Volume calculations must be accurate and grade control precise in these areas, because the ore must be spread in even layers to be leached efficiently.

GPS-equipped track dozers also work on the benches. After each blast, dozers level the surface that the shovel will work on and replace the safety berms that keep the shovels from going over the edge of the bench.

As with the other equipment, the GPS system enables the dozers to work safely at night.

The GPS Librarian

All operations at the mine, including drilling, blasting, loading, truck dispatching, road maintenance, and leach stockpiling, use GPS extensively. However, not all equipment is in use at once, and we have fewer GPS units than we have vehicles. Every vehicle carries wiring and brackets for a GPS unit, and the units can be hot-swapped between vehicles in about an hour and a half. So that GPS units are not sitting idle, a GPS librarian, on call 24 hours a day, controls which vehicles receive GPS units for which tasks.

The mine's managers have ranked GPS applications at the mine based on how much time and money they save and how much they reduce the potential for accidents. Each application has a ranking from 1 to 5, where 1 is the highest ranking. For example, applying GPS to dozers that are building ramps and roads rates a 1 for safety, a 3 for time savings, and a 2 for cost savings. Receiving requests for GPS units, and looking at the rankings for each application, the librarian allocates GPS units to vehicles in such a way as to maximize efficiency, safety, and cost savings— not just on a first-come, first-served basis. As a result, the electric shovels have GPS units installed at all times. Next in line come the track dozers, whose priority depends on their application. Dozers allocated to road maintenance and fragmentation operations share equal priority over leach stockpile dozers because they require a higher level of survey control than the stockpile dozers do.

Some Difficulties

Backup plans for outages. Systems are in place at the mine to cope with partial or full satellite outages. For example, we learned in advance from a GPS almanac Web site that Satellite Vehicle Number 13 would be taken out of service for four hours of health testing. This would create the potential for some systems to lose positioning capability for that period. In four hours, a large shovel could theoretically misroute as much as $15,000 worth of material. In preparation for the satellite outage, the engineers lowered the system's precision and quality levels so that it could function with fewer satellites in view. If all else failed, the surveying staff was prepared to place targets in the field by conventional means, but this was not necessary.

Since then we have developed a standard operating procedure for satellite outages. When we anticipate an outage, we place vehicles in areas where they will encounter only one type of material, so that the accuracy with which they are positioned is less critical that it normally is.

Multipath interference. The high walls of the pit and the large vehicles we use both create multipath interference. Although our equipment uses algorithms that compensate for this interference, we take this phenomenon into account when planning our operations.

Deep trouble. Another problem with our GPS system is that as the pit deepens, satellite visibility within it deteriorates. Computer projections based on future mine plans show that in less than two years, we will begin to experience six-hour windows during which fewer than three satellites will be visible. We are testing pseudolites in the mine to solve this problem.

Beyond the Napkin

At Morenci, we have come a long way from paper-napkin designs and hood-top directions. Our GPS navigation and positioning capability has improved operations in every area. Clearly, systems using GPS will operate equipment with less and less human interaction. Autonomous vehicles will revolutionize the mining business by reducing labor costs while maximizing the efficiency of the equipment. Most important, machine guidance and automation will increase worker safety. The fewer the workers who must walk around near the equipment, the better.

As we upgrade our software and add more GPS units, we expect that the benefits will keep accruing.

The Morenci Mine at a Glance

Size 18 square miles, 3,000 feet deep—the largest copper mine in North America

Production 840,000 tons of rock removed each day, 1 billion pounds of copper extracted each year

Transportation More than 30 miles of roads; average haul 5.5 miles

Personnel 2,500 employees, 2,500 contractors

GPS-Controlled Equipment

- 14 electric shovels
- 2 loaders
- 88 haul trucks
- 11 track dozers (of 31)
- 5 drills (of 17)

Manufacturers

Each GPS vehicle has an RTK-GPS receiver (Trimble 7400), and a radio (TrimComm 900 MHz) for data transmission, all from **Trimble Navigation** (Sunnyvale, California). The computers on board the electric shovels (**P&H Mining Equipment**, Milwaukee, Wisconsin) and loaders (Caterpillar 994, **Caterpillar,** Concord, California) run Caterpillar's Computer Aided Earthmoving System (CAES), as do the Caterpillar 930E, 793B, and 793C and Komatsu 930E (**Komatsu,** Vernon Hills, Illinois) haul trucks and the Caterpillar D10 and D11 track dozers. The 49R drills from **Bucyrus International** (South Milwaukee, Wisconsin) are equipped with Drill Manager (DM) from **Aquila Mining Systems** (Montreal, Quebec, Canada).

CAES and DM use Caterpillar's METSManager software to manage wireless file transfer between the office and the equipment.

We are currently testing IP networks with Caterpillar using TCIP 900 MHz. We also have a separate network using Freewave technology from **Leica Geosystems Inc.** (Torrance, California), which we are using to test Leica's Dozer 2000 machine guidance software and surveying applications.

Scott Shields is senior surveyor, mine engineering, at Phelps Dodge Morenci. **Janet A. Flinn,** formerly a mining engineer at Phelps Dodge Morenci, is now a Web engineer at Fitchburg Mutual Insurance Company, Fitchburg, Massachusetts. **Andres Obregen** is a senior engineering technician at Phelps Dodge Morenci.

Source: www.aquilamsl.com.

GPS—MAPTECH INC.

Global positioning system (GPS) is a navigation and location technology that was developed by the U.S. Department of Defense and reached full operating capability on July 17, 1995. GPS operates by using a system of orbiting satellites in space and a control station in Colorado Springs on earth. The earth control station is supplemented by five monitoring stations and three antennas spread across the earth. GPS can provide the exact latitude, longitude, altitude, and velocity of a person or object equipped with a GPS receiver on earth. This position is calculated by measuring the distance from the group of satellites, which provide exact reference points in space. There are two types of GPS, the Standard Positioning Service (SPS), which is what nonmilitary private citizens or companies may use, and the Precise Positioning Service (PPS), which is what the military uses. The signal on the SPS is intentionally off by a few degrees to protect the national security interests of the United States.

Maptech Inc. is a surveying and mapping company in Jackson, Mississippi. Maptech offers such services as topographic mapping, hydrographic surveys, construction layout, 3D modeling, flood studies, pipeline surveys, oil field surveys, and GPS services. Although GPS services are a particular specialization of Maptech, they utilize GPS in all of their other surveying efforts. The use of GPS by Maptech allows them to greatly improve the accuracy of their surveys.

Some of Maptech's clients include the U.S. Army Corps of Engineers (levy elevations on the Mississippi River), the U.S. Fish and Wildlife Service (flood plain survey on the Louisiana Delta), and the Tennessee Valley Authority. Through the use of GPS, Maptech has been able to break into new areas of surveying and complete these surveys to the highest degree of accuracy and professionalism.

Source: http://www.maptech.com.

BATTELLE GPS

Battelle is a leading transportation company that does many case studies on GPS, which is a data collection device. The California Air Resources Board (CARB) brought about a data collection effort to produce results and data needed to improve the process of heavy-duty truck emission estimations. The GPSs were used to record vehicle activity for the heavy-duty trucks in normal and over-the-road use.

The purpose of this project was to establish truck travel patterns in rural and urban areas for a number of vehicle classes and also to analyze characteristics of heavy-duty truck travel by producing, among other summaries, speed profiles, trip patterns, and start patterns. They also collected data in an effort to support congestion management activities. Battelle's solution was to outfit 140 heavy-duty trucks with GPS devices that collected data such as vehicle type, body type, and fuel type. It also included information such as speed and length of driving. GPS and Battelle have allowed California to study its emissions so that the company may better regulate dangerous and nondangerous emissions.

Source: www.battelle.org/transportation/case-studies.stm.

HISTORICAL PRECEDENTS OF GPS

Throughout history, one of the most critical issues for travelers has been how to get from location to location without becoming lost. Use of landmarks, compasses, dead reckoning, and celestial navigation has been common in the past. In the recent past, more accurate means of navigation have been used such as radar, sonar, and satellites. The development of a computer with the ability to make rapid complex computations enabled the U.S. Department of Defense (DOD) to launch a system of twenty-four satellites. From this system, we can calculate the exact position of a person, ship, or other object anywhere in the world. The uses of this system are broad, ranging from military to civilian services. GPS is easily used by experts and novices on both land and sea.

The beginnings of GPS, however, began many centuries ago. To ancient travelers on land and sea, landmarks and crude estimations of the location of stars and other celestial bodies were the main ways of navigation. The main problems with landmarks were they were only usable by the local population, and they were subject to destruction. One of the most famous landmarks in the ancient world was the Lighthouse of Alexandria. Built around 280 B.C., the lighthouse served as a guide to ships coming to harbor on an otherwise unmarked coast. It was destroyed in the fourteenth century after being damaged by several earthquakes.

The development of the compass freed navigators from landmarks. Although the phenomenon of magnetism was known to the ancient Greeks, Romans, and Chinese, it was not until the thirteenth century that people discovered a free-floating magnet aligns itself with the earth's magnetic field. This discovery led to the invention of the compass, which revolutionized travel by reducing the reliance on landmarks and stars for navigation.

A few centuries later, Columbus used a navigation method called deduced (or dead) reckoning to chart his way to the New World. Dead reckoning works by starting from a known position, such as a port, and calculating the distance traveled in a day and marking that spot on a chart. Each day's ending position would be the starting point for the next day's measurements. Speed was measured by throwing a piece of flotsam over the side of the ship and counting out how long the flotsam took to pass preset points on the ship. Use of a compass and this process allowed the navigator to keep a course and measure the distance traveled. This method was tedious and difficult, but until the Renaissance it was the best method available.

Developments during the fifteenth, sixteenth, and seventeenth centuries in the fields of astronomy, geometry, trigonometry, and physics led to the development of celestial navigation, a method of navigation utilizing the stars, moon, and sun. To calculate one's position on earth by looking at various objects in the sky, however, a few facts are needed. One needed to know the exact time, which at sea meant knowing the longitude. In 1762, English clockmaker John Harrison invented a chronometer, the first accurate mechanical clock in existence. After one knew the local time and could reference it to another time such as Greenwich mean time, an almanac describing the positions of stars at certain times could be used at sea. Another necessary item discovered for celestial navigation was the **sextant,** a system that con- **sextant** sisted of two mirrors and a telescope mounted on a metal frame (see Figure 10–1). It was a way to measure the position and angle of a celestial body in relation to the apparent horizon. As a star or other body was brought into view, light was reflected by one mirror, then another, before entering the telescope. By lining up the star with the horizon, the altitude of the star could be measured by examining where the lower part of the sextant listed the angles from 0 to 90 degrees. After the angle was obtained from the arc on the

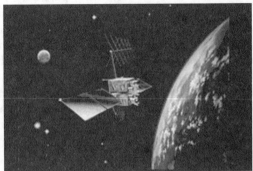

Figure 10–1
Typical GPS Satellites

sextant, a process was then performed to eliminate several errors. Once this occurred, the sextant had a true measurement, including an error in the sextant itself (sextants are not scientifically precise), the error caused by the inability to see the true horizon due to the height of the eye and the curve of the earth, atmospheric refraction error, parallax error (the difference between the image seen through the viewfinder and the image seen by the taking lens), and errors caused by the semidiameter of objects closer to the earth such as the moon and planets. After these corrections were made, travelers could then determine their position on the edge of a circle whose radius was defined by the altitude of the celestial body observed. This procedure must be performed at least one more, and preferably two more, times to accurately locate one's position on the globe, because of natural human error. Typically, a traveler would take these measurements several times and then average the results. This is why it generally took many hours to do this centuries ago. Unfortunately, the most accurate measurements taken in this manner can only pinpoint a position within 60 miles. Although this is good enough for crossing an ocean, it does not work for more accurate calculations needed for surveying and mapmaking.

DESIGN AND PRINCIPLES OF OPERATION

The U.S. DoD developed a system of satellites for finding an exact position anywhere on the globe, for such purposes as finding locations of nuclear submarines when they surface. The DoD has allowed civilian use of the GPS free of charge; however, they introduced an error called selective availability, which limits GPS to approximately 100 m of accuracy for reasons of national security. This selective availability was eliminated by President Bill Clinton in May 2000, providing a current accuracy of approximately 10 m. The DoD was given until 2006 to develop other ways of degrading performance in time of war; for example, they could conceivably (or legally in any case) switch SA back on. Figures 10–1a and 10–1b show typical GPS satellites. Figure 10–1a is the original GPS satellite manufactured by Rockwell. Figure 10–1b provides a look at the newest version (1995 and later) made by Lockheed Martin.

To provide global coverage, twenty-four satellites must orbit the earth every 12 hours. These Navstar satellites, made by Rockwell International, have an orbital plane 55 degrees to the equatorial plane. It would have been possible to place them in a geosynchronous orbit, the advantage being that it would take much fewer than twenty-four satellites to provide global coverage. This placement, however, would provide much less accuracy as one moved farther from the equator—a phenomenon known as dilution of precision. It should be noted that there are often more than twenty-four operational satellites as new ones are launched to replace older satellites. The satellite orbits repeat almost the same ground track (as the earth turns beneath them) once each day. The orbit altitude is such that the satellites repeat the same track and configuration over any point approximately each 24 hours (4 minutes earlier each day). There are six orbital planes (with nominally four SVs in each), equally spaced (60 degrees apart), and inclined at about 55 degrees with respect to the equatorial plane. This constellation provides the user with between five and eight SVs visible from any point on the earth (see Figures 10–2 and 10–3 for these orbits). Another issue is that if you were to reposition the satellites into a geosynchronous orbit (geosynchronous orbits are generally twice as high as the current GPS orbits), then another range of issues would arise such as signal strength. This would mean that a GPS receiver would have to be much more powerful than current models, to accurately contact a geosynchronous GPS satellite. This would increase price and potentially raise health issues and a range of other issues.

Although the calculations involved are similar in some ways, the complexity of calculating one's position using GPS would be extremely difficult without the use of the microprocessor. The satellites in the GPS form a set of reference points other than celestial bodies which receivers can use to locate a position on earth. They form one segment of the three parts that comprise the GPS system. The other two parts consist of the users and their receivers, and the control stations on the ground that monitor the satellites and update information about their position. The information these ground control stations collect is put into an almanac or catalog such as the *Yuma Almanac*. The information these ground control stations collect is also in the pseudorandom code sent out by the satellites. To find an exact position, receivers use the catalogs to discover the location of the satellite from which it is receiving. They also need to know the exact time, triangulate a position based on the information received from the satellites, and correct for any errors caused by the

Figure 10–2
GPS Orbits

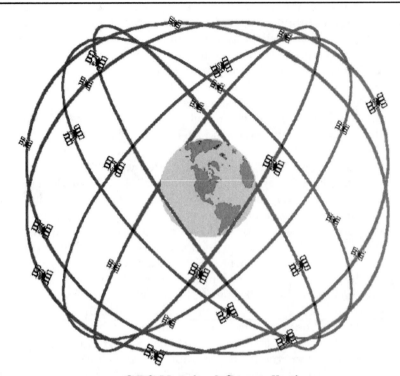

GPS Nominal Constellation
24 Satellites in 6 Orbital Planes
4 Satellites in each Plane
20,200 km Altitudes, 55 Degree Inclination

atmosphere. The satellites send out a signal, which is called a pseudorandom code. It is easy for the receivers to pick up due to the amplification of the signal using **information theory**.[1]

information theory

Information theory is a concept invented by Claude Shannon and presented to the world in 1948. He proposed a linear schematic model of a communications system. This was a novel idea. At the time, communication was

---

[1] This is a general discussion. To be specific, as Peter Dana of the University of Texas correctly points out: "We don't triangulate [no measured angles are used in the position solution]. You might say we quadra-laterate [four ranges], but really it is a hyperbolic solution using four [or more] relative ranges [the receiver clock bias is unknown]. From that we find the position that minimizes the residuals. The satellites do each send an Almanac, a catalog of positions for the constellation, but what we use in position determination is the Ephemeris Data, sent by each satellite only for itself. The Almanacs are far from exact."

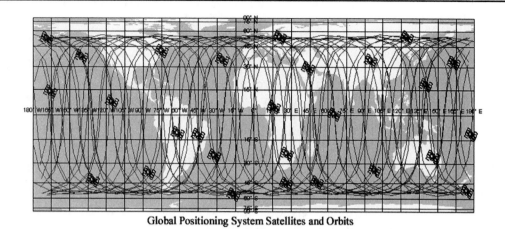

Global Positioning System Satellites and Orbits

Figure 10–3
Global GPS Orbits

then thought of as requiring electromagnetic waves to be transmitted down a wire. The idea that one could transmit pictures, words, sounds and so on by sending a stream of 1s and 0s down a wire was fundamentally new. In defining information, Shannon identified the critical relationships among the elements of a communication system: the power at the source of a signal, the bandwidth or frequency range of an information channel through which the signal travels, and the noise of the **channel,** such as unpredictable static on a radio, which will alter the signal by the time it reaches the last element of the system, the receiver, which must decode the signal. In telecommunications, a channel is the path over a wire or fiber, or in wireless systems, the slice of radiospectrum used to transmit the message through free space. In short, Shannon's equations told engineers how much information could be transmitted over the channels of an ideal system.[2]

channel

    This pseudorandom code is a highly complex, nonrepeating signal that is known to both the satellites and the receivers. Since the signal is essentially the same as the background noise, the receiver can measure where the signal boosts the background noise and pick up on the original signal instead of the background noise. Receivers pick up on this signal and calculate the distance from the receiver by multiplying the velocity of the signal, which is the speed of light by the time it took to get there. To calculate the time it took to travel that distance, the receiver must measure the difference between the code it expects and the code it is receiving. The difference in the timing of the code is the time it took the signal to travel from the satellite to the position where the receiver is located. Once the receiver does this with three separate satellites it

[2] For updates, see www.lucent.com/minds/infotheory/what5.html.

can combine these measurements to deduce its location down to two points, one of which can usually be thrown away. A fourth satellite measurement is also required to determine the exact time. The satellites have atomic clocks to measure the time precisely, which is necessary because the signals they are sending out are moving at the speed of light. Even a small time error here would cause a large error in distance. The receiver does not have an atomic clock due to the costs involved. A measurement from the fourth satellite allows the imperfect clock on the GPS receiver to discover exactly how different it is from the atomic clock. It does this using a simple application of algebra, solving for four variables using four equations. Besides the complicated math involved in solving for four variables, the receiver must take into account other errors caused by the atmosphere. This is done by applying a general mathematical model which estimates the error caused by the atmosphere. Another error that the receiver must deal with is caused by the GPS signal bouncing off other objects before coming to the receiver. Even with all these errors which can be corrected, the DoD introduces another error referred to as selective availability. This is done to make sure no terrorist or enemy groups can use GPS to make super accurate weapons. Differential GPS can overcome even this forced error, so the U.S. government is looking into discontinuing the policy of selective availability.

Differential GPS came into being as a result of techniques implemented to get around the errors caused by selective availability. Differential GPS has had a profound effect on GPS as a resource. Using differential GPS allows GPS to become much more than a system for navigation. It becomes a precise measuring tool for everyone. Differential GPS works using two receivers, one that is stationary and one that is mobile. Because the satellites are so far away, the stationary and mobile receivers have essentially the same errors. The idea behind differential GPS is to have one receiver measure the timing errors and then provide correction information to the other receivers that are roving around. In this way, nearly all errors are eliminated from the system, including selective availability error. With differential GPS, even more possibilities for the use of the technology become available.

The applications for GPS and differential GPS are as different as the people using them. GPS was developed for and by the military and as such still has a multitude of applications. It is used to find locations of units in the field so they can avoid friendly fire. Submarines that surface in the middle of the ocean can find their exact position quickly. These uses and others continue to allow the U.S. military to safely deploy and operate their mission throughout the world. GPS is not limited to the military world though. Many civilian application technologies have exploded onto the scene in recent years, with the promise of many more to come. One such idea is a technology known as free flight. This system would use GPS and other technologies to create and manage airplane flight paths. The pilot would be able to chart a course using GPS and information from other planes with GPS coordinated by stations on the ground to chart a shorter, more fuel-efficient flight plan. Another GPS application technology that would be useful for everyone would be an automatic vehicle location system. This system would use GPS to find a car's position and tie that information into the geographic information system, which is a database storing a map in a computer. These maps are much more complex than paper maps, as they show terrain in three dimensions. Using this software with a car would allow people to plot a course from city to city and within cities with ease. GPS technologies have found their way into many other fields as well, such as surveying, mining, and oil drilling. The applications possible for GPS are only limited by the imagination of the human mind. In the near future, GPS may be as vital to our lives as gasoline and electricity.

THE CONTROVERSIES

As mentioned, GPS was not accurate enough for many civilian applications, partly because of the DoD's encryption, and partly because of atmospheric conditions and other errors. Even after Clinton addressed this issue of encryption in 2000, controversies still abounded. As various commercial GPS applications developed, such as for use in surveying, investors remained skeptical about being able to rely on a system which was both encrypted and operated by the U.S. government. Clinton addressed this issue years earlier when, in 1996, he signed the Presidential Decision Directive on GPS, guaranteeing that GPS signals, although government controlled, would be permanently available for civilian use. Today, applications employing the knowledge of precision location are giving rise to new innovations, with promise for more in the future. Russia has developed its own satellite navigation system, **Glonass,** and **Glonass** Europe objects strongly to using the U.S. system (as a sole means system), preferring to create its own rather than being subject to the whims of the U.S. government. This situation is currently being negotiated.

While business is rushing to develop new products using the GPS-enabled chip sets, expected to sell in the near future for about $10, individuals are expressing growing concerns about their privacy. Most people do not want the government and businesses to have access to their whereabouts at any given moment. With a GPS card in consumers' computers, Internet advertising agencies can target their customers within a geographic area, say a distance of 3 miles. Knowing that government agencies and commercial businesses alike will know their every move, as is currently the case with commercial truck drivers, customers believe the system is an invasion of privacy. The government recently mandated that the FCC put a system in place which will accurately pinpoint the location of a cellular phone calling 911 for an emergency. This is currently being implemented. This would require that a GPS card be installed in all cell phones, but would also render older cell phones noncompliant (Borland 2000). This FCC regulation has raised some privacy concerns (Ross 2000). Some believe this is another attempt at "Big Brother" tracking private citizens. Further, the DoD must remove all the GPS

Figure 10–4
The Segments of GPS

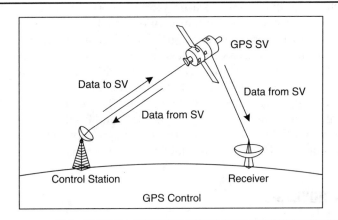

signal dithering and deliberate ephemeris errors, which were added for encryption purposes, for civilian access by 2006. All of this is made possible at no charge by the government to commercial interests and with no access fees to the end user.

The GPS consists of three segments—satellite, control, and user. We will now address each in some detail. Figure 10–4 graphically depicts this configuration.

SATELLITE SEGMENT

The Navstar GPS contains twenty-four satellites, maintaining an altitude of approximately 12,500 miles above the earth. These satellites are divided into six planes of four satellites each, spaced equally around the equator, and the ascending nodes of each has an inclination of 55¡. There are three types of satellites: Block I, launched from 1976 to 1982; Blocks II and IIA, launched from 1989 to 1993; and the newer Block IIR satellites, which have improved nuclear explosion detection and onboard computer memories. New Block IIR Navstar satellites replace the older models when they are no longer functional. The clocks on board each satellite are synchronized with an accuracy of three billionths of a second.

All of the GPS satellites are stabilized in orbit on three axes through the use of reaction wheels and magnetic torque. The satellites must deal with forces such as solar pressure; the gravitational fields of the sun, earth, and moon; and thermal radiation from the satellite itself. Orbit predictions must take these factors into consideration, in addition to the rotation rate and axis of the earth. The orbit maintained by each satellite allows it to repeat the same path once every 24 hours, completing it 4 minutes earlier every day. Each satellite orbits the earth with its solar panels facing the sun at right angles and its antennas facing earth. Each satellite broadcasts pseudorandom noise

(PRN) codes to modulate the signal, allowing receivers to calculate the signal arrival time compared with the receiving unit's local clock. The satellites each transmit their signal at two frequencies, the L1 (1575.42 MHz) and L2 (1227.60 MHz) signals. The PRN codes are superimposed on the data in the L1 signal. The precision (P) code and a data message are contained in the L2 signal. Although all the satellites are broadcasting on the same frequency, each satellite's PRN is unique, allowing a receiver to track each satellite separately. This process is known as code division multiple access (CDMA).

Other factors can interfere with the satellite-originated signal arriving on time to the receiver; therefore, the satellites broadcast over two frequencies to allow correction for them. Some of these factors include interference from the **troposphere** and the total electron content in the **ionosphere,** along the path the signal travels from the satellite to the receiver. The arrival time difference of the two signals can be computed by the receiver, allowing a more accurate prediction of its position. To understand this process, it is necessary to discuss the earth's atmosphere.[3]

troposphere

ionosphere

Layers of the Atmosphere

The atmosphere of the earth may be divided into several distinct layers, as shown in Figure 10–5.

Troposphere Layer

The troposphere is where all weather takes place; it is the region of rising and falling packets of air. The air pressure at the top of the troposphere is only 10 percent of that at sea level (0.1 atmosphere). A thin buffer zone lies between the troposphere and the next layer called the tropopause.

Stratosphere and Ozone Layers

Above the troposphere is the **stratosphere,** where airflow is mostly horizontal. The thin ozone layer in the upper stratosphere has a high concentration of ozone, a particularly reactive form of oxygen. This layer is primarily responsible for absorbing the ultraviolet radiation from the sun. The formation of this layer is a delicate matter, because only when oxygen is produced in the atmosphere can an ozone layer form and prevent an intense flux of ultraviolet radiation from reaching the surface, where it is quite hazardous to the evolution of life.

stratosphere

Mesosphere and Ionosphere Layers

Above the stratosphere is the **mesosphere** and above that is the ionosphere (or thermosphere), where many atoms are ionized (have gained or lost electrons so they have a net electrical charge). The ionosphere is very thin, but it

mesosphere

---

[3] For additional information on the earth's atmosphere, go to
www.csep10.phys.utk.edu/astr161/lect/earth/atmosphere.html.

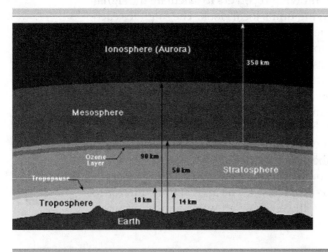

Figure 10–5
Layers of the Earth's Atmosphere

is where aurorae take place, and it is responsible for absorbing the most energetic photons from the sun and for reflecting radio waves, thereby making long-distance radio communications possible. The structure of the ionosphere is strongly influenced by the charged particle wind from the sun (solar wind), which is in turn governed by the level of solar activity.

CONTROL SEGMENT

The accuracy of the Navstar satellites is maintained by the master control station, located at Schriever Air Force Base in Colorado Springs, by five unmanned monitor stations and four antenna stations situated around the world. There are two stations in the Pacific, located in Hawaii and Kwajalein; one unmanned station in the Atlantic, on Ascension Island; and the other in the Indian Ocean at Diego Garcia. The monitor stations gather information from all the satellites and relay it to the master control station. It is at the master control station that the information is calculated for accuracy and relayed back to the satellites to update their navigation messages. A Kalman filter is updated every 15 minutes and updates the satellite clock frequency and phase, the position and velocity of the satellites, and the clock frequency and phase of the monitor stations. A Kalman filter is used to determine the best estimate of any variable, such as location, altitude, and velocity. All this is performed using a closed-loop control system, in which the monitor stations receive the L-band pseudorange signals from the satellites, calculate corrections by smoothing the PRN using the Kalman filter, predict the ephemeris (coordinates of an object in space), and generate a new message. The new message travels via the S-band ground antenna where it is uplinked to the satellites (SNS).

USER SEGMENT

A minimum of four GPS satellites are required for a receiver to calculate its three-dimensional position. The twenty-four-satellite constellation configuration ensures a minimum of six satellites within view of any receiver at any point on the globe at any given moment. It is possible for twelve satellites to be within range of a single receiver. The greater the number of satellites in range, the greater the accuracy of the position solution calculated by the GPS receiver. Hand-held GPS units are compact, typically the size of a cell phone or smaller. Major U.S. manufacturers include Trimble Navigation, Ltd., Ashtech-Magellan Corp., and Novatel. See Figures 10–6 through 10–8 for a list of these products.

First, the antenna on each GPS receiver converts the satellite's radio signals into an electric current which is then filtered, amplified, and processed according to the capabilities of the particular receiver used. The signal is filtered, amplified, and mixed to a lower intermediate frequency (IF). Newer model receivers will usually convert the signal to digital samples, then process the information with specialized digital signal processing components in addition to a microprocessor. A correlator then determines whether the PRN code is early or late and the tracking function adjusts the local code generator accordingly. The information about the clock and ephemeris in the translated data message is then used to compute the exact position of the satellite and its clock phase, completing the information the receiver needs to compute a position solution. The accuracy of the position is determined by the computations, number of satellites used, and the quality, type, and purpose of the receiver. Between atmospheric anomalies, encryption codes, and other interfering factors, the basic system is usually accurate to within 10 m or will be sometime between now and 2006 when DoD provides the accuracy the U.S. military has enjoyed in the past. Until recently, the hand-held units have been considered more of a specialized gadget or an executive novelty as the cost was significant (usually over $1,000). Now these devices are around $200.

Figure 10–6
A Sample List of GPS
Hand-held Receivers

Product
Garmin eMap
Mobile Type, 12 Channel Receiver, 1.7 2.2 Inch Backlit Mono Display, 500 Waypoints,
    PC Interface, (2) AA Battery
Lowest price: $159.00

Garmin eMap Deluxe
Mobile Type, 12 Channel Receiver, 1.7 2.2 Inch Backlit Mono Display, 500 Waypoints,
    PC Interface, (2) AA Battery
Lowest price: $239.02

Garmin eMap Deluxe w/Metroguide
Mobile Type, 12 Channel Receiver, 1.7 2.2 Inch Backlit Mono Display, 500 Waypoints,
    PC Interface, (2) AA Battery
Lowest price: $327.99

Garmin eTrex
Outdoor Type, 12 Channel Receiver, 2.1 1.1 Inch Backlit Mono Display, 500 Waypoints,
    (2) AA Battery, 22 Hour Battery Life
Lowest price: $109.99

Garmin eTrex Legend
Outdoor Type, 12 Channel Receiver, 2.1 1.1 Inch Display, 500 Waypoints, (2) AA
    Battery, 18 Hour Battery Life
Lowest price: $214.99

Garmin eTrex Summit
Outdoor Type, 12 Channel Receiver, 2.1 1.1 Inch Backlit Mono Display, 500 Waypoints,
    22 Hour Battery Life
Lowest price: $203.99

DIFFERENTIAL GPS

To correct the inaccuracies inherent in the system, an earth-based solution known as the differential system helps to improve GPS calculations. The differential system employs the use of ground-based receiver/transmitters situated on surveyed points. These provide accurate, reliable readings, which are used in differential receivers to compute a position solution much closer to the actual location. These locations read the satellite position, compare it with the ground-base known location, and broadcast an accurate signal for use in differential receivers, creating an accuracy level within a few meters. The lower price and the improved accuracy level of the GPS receivers are spurring new applications which, in turn, increase consumer demand.

Figure 10–7
A Sampling of GPS
Software for the PC

AGIS Mapping 1.62

Description

Plot your own geographic information or download and display data from a variety of sources on the Web. The app has a multidocument interface and can be used to add high-quality vector map displays to your documents such as MS Word. A number of data import formats are supported, including MapInfo, ARCInfo, and Garmin GPS. A built-in scripting language can be used to create animation; automate map displays using data from other sources, such as an MS Access database; and serve images to interactive map displays on Web pages. Features include full control over virtually every aspect of a map display, many layers of maps and data, thematic mapping, a number of map projections, simplified access to digital charts of the world, searching for named data points, and world base-map data. Version 1.62 corrects a problem running examples in the earlier release. This version has a 30-day trial. The cost to register this product is $49.

GPSdb 1.6

Description

GPSdb is an application designed to help maintain waypoints, tracks, routes, and so on, for Garmin GPS systems. It also logs and displays NMEA data. Features include real-time track recording, the ability to upload and download waypoints and routes, a distance calculator, and coordinate system support. The unregistered version is limited to 50 waypoints; however, all functionality, including routes, tracks, groups, import/export, upload/download, and NMEA, is unrestricted.

QuoVadis 1.62c

Description

QuoVadis is a GPS-based moving map software for bitmapped files, enabling you to prepare routes, tracks, and waypoints, using scanned maps, for Garmin GPS receivers. It can show your position on scanned maps in online driving mode. It also tracks, plots, and saves your trail. You can see where you have been by viewing different kinds of bitmapped maps, using several standard bitmap formats. Previous upgrades have included a new graphics engine, the ability to work with maps from CD-ROM, calibration-by-map-projection functionality, a new GUI, and French support.

The 1.62c upgrade includes support for Etrex/Emap, support for Top50, an Austrian Map, a Swiss Map, and 100-track replay.

The trial period lasts for 25 days, after which the GPS interface will be disabled. The cost of the software is $99.

Figure 10–8
GPS PC Software for
PalmPilots (Courtesy
CNET.com)

HandMap Deluxe 3.5

Description

HandMap Deluxe is a full-featured **electronic street directory.** You can pan and zoom in and out. You can **search for specifics, such as parks, street names,** and so on. You can also **create your own custom landmarks and travel routes.** HandMap Deluxe has the functionality of the *HandMap* standard version, but also supports the new layered map format. The Focus Street function quickly finds cross-streets. Version 2.7 includes the following improvements: more refined searches; support for new U.S. maps, with a new, cleaner look; support for new Populated Place Markers; and a quick-search shortcut that lets you start typing the name of a street or landmark.

Version 3.5 now supports beaming. The registered version accommodates viewing of third-party maps. Registration is $16.

QuoVadis for Palm 2.0

Description

QuoVadis for the Palm platform is a grayscale, real-time, street-level mapping system. This program has the following features: manual or GPS tracking; seamless integration of multiple maps; zoomable and searchable labeled maps; and multiple-instance searches for streets, intersections, shopping centers, airports, and other landmarks. All of these features are utilized from QuoVadis's intuitive interface. In addition, this program supports GPS systems and displays navigation and satellite information screens. Version 2.0 contains an updated map library, significant speed improvements, better compression of map data, and several other enhancements.

Note: The unregistered version of this program has only one map included and registration costs $44.95.

PlaceTrace 1.2

Description

PlaceTrace is a map display program for the PalmPilot that can track your travels using a GPS receiver. Your position is shown about every second and recorded once per minute. Maps are hotsync'd from the included Windows program, PlaceTraceBase, which contains street-level detail for all 50 states and U.S. possessions. Create your own labeled waypoints to aid navigation. This demo version is limited to street information in three counties, and PlaceTraceBase and printing are disabled. The PalmPilot program is fully functional.

Source: CNET.com.

GEODETIC DATUMS

Geodetic datum(s)[4] define the size and shape of the earth, as well as the origin and orientation of the coordinate systems used to map the earth. Geodetic means something relating to, or determined by, **geodesy**. Geodesy is quite simply defined as a branch of applied mathematics that determines by observation and measurement the exact positions of points and the figures and areas of large portions of the earth's surface, the shape and size of the earth, and the variations of the terrestrial gravity and magnetism. The word *datum* is simply the singular form of the more popular word *data*, and is something upon which an argument is derived or from which an intellectual system of any sort is decreased.

Hundreds of different datums have been used to frame position descriptions since the first estimates of the earth's size were made by the Greek philosopher Aristotle, who also was the teacher of Alexander the Great. The geodetic datum has evolved from those describing a spherical earth to ellipsoidal models, which were derived from years of satellite measurements. Modern geodetic datums range from flat models used for plane surveying to complex models used for international applications, which completely describe the size, shape, orientation, gravity field, and angular velocity of the earth. Various disciplines such as cartography, surveying, astronomy, and navigation make good use of geodetic datums.

Obviously, one of the more popular applications of the science of geodesy is GPS. While the science of geodesy dates back to Aristotle (and actually even further back to the first Babylonians of 3500 B.C.), its current principles are used in GPS which dates back to July 17, 1995, when it became fully operational.

Geodetic datum(s)

geodesy

Referencing Geodetic Datums

Referencing geodetic coordinates to the wrong datum can result in position errors of hundreds of meters, which could be a crucial error if it occurs during the navigational process of directing a plane, a ship, or a missile's target coordinates. Different nations and agencies use different datums as the basis to coordinate their systems that are used to identify positions in geographic information systems, precise positioning systems (PPSs), and navigation systems. The diversity of datums in use today and the technological advancements that have made global positioning measurements with submeter

---

[4] The plural of geodetic datum is pretty well known to be *datums*. The *Geodetic Glossary* explicitly defines this plural form (U.S. Department of Commerce, National Oceanic and Atmospheric Administration, National Geodetic Survey, 1986, Page 51). Even the *Random House Unabridged,* 2nd ed., 1987, give *datums* as the accepted plural of datum for surveying usage. This is according to the noted GPS expert, Peter Dana of the University of Texas. See www.pdana.com.

accuracy require a careful datum selection and careful conversion between coordinates in different datums.

GEOMETRIC EARTH MODELS

Geodetic datums and their coordinate reference systems were developed to describe the geographic positions for surveying, mapping, and navigating the globe. Throughout history, the shape of the earth was thought to be flat; then spherical models were sufficiently accurate to allow for global exploration, navigation, and mapping. True geodetic datums were employed only after the late 1700s, when measurements showed that the earth was ellipsoidal in shape. Early ideas of the earth's shape resulted in descriptions of the earth as an oyster, a rectangular box, a circular disk, a cylindrical column, a spherical ball, and even a very round pear shape by none other than Columbus in the last years of his life. Flat earth models are still used today for plane surveying over distances short enough for the curvature of earth to be insignificant (less than 10 km). Spherical models represent the shape of the earth having a specified radius. These models are often used for short-range navigation and for global distance approximations. Spherical models fail to show the exact shape of the earth, as the slight flattening of the earth at the poles results in a significant difference between the average spherical radius and the actual measured polar radius of the earth. Ellipsoidal earth models are required for accurate range and bearing calculations over long distances.

An ellipsoid is defined as a surface of all plane sections, which are ellipses or circles, while ellipsoidal is defined as resembling an ellipsoid. GPS navigation receivers use ellipsoidal earth models to compute position and waypoint information. Ellipsoidal models define an ellipsoid with equatorial radius and a polar radius, and the best of these models can represent the shape of the earth over the smoothed, averaged seasurface to within about 100 m. These reference ellipsoids are usually defined by semimajor and flattening. The semimajor represents the equatorial radius, and the flattening represents the relationship between the equatorial and polar radii. Other reference ellipsoid parameters such as semiminor axis and eccentricity can be computed from these terms. The semiminor axis is more commonly referred to as the polar radius.

EARTH SURFACES

The earth has a highly irregular and constantly changing surface in that it consists of mountainous terrain, flat plains, oceans, and terrace plateaus that are all moving as dictated by magnetic poles and tectonic activity. Models of the surface of the earth are used in navigation, surveying, and mapping. Topographic and sea-level models attempt to model the physical variations of the earth's surface, while gravity models and geoids are used to represent local variations in the gravity that change the local definition of a level surface. A geoid is defined as the surface within or around the earth that is everywhere normal to the direction of gravity, and coincides with mean sea level in the oceans, and approximates to the shape of an ellipsoid of revolution. The topographical surface of the earth is the actual surface of the land and sea at some moment in time. Aircraft navigators have special interest in maintaining a positive height vector above this surface. Sea level is the average surface of the oceans. Believe it or not, tidal forces and gravity differences from location to location can cause even this smoothed surface to vary by hundreds of meters over the globe. Gravity models attempt to describe the variations in the gravity field in specific detail. Plane and geodetic surveying uses the idea of a plane perpendicular to the gravity surface of the earth, the direction perpendicular to a plumb bob pointing toward the center of mass of the earth. Local variations in gravity, caused by variations in the earth's core and surface materials, cause this gravity surface to be quite irregular.

Geoid models attempt to represent the surface of the entire earth over both land and ocean as though the surface resulted from gravity alone. Bomford describes this surface as that which would exist if the sea was admitted under the land portion of the planet by small frictionless channels (Dana). The World Geodetic System 84 (WGS-84), created in 1984, defines geoid heights for the entire earth, and the U.S. National Imagery and Mapping Agency (formerly the Defense Mapping Agency) publishes a 10 by 10 degree grid of geoid heights for the WGS-84 geoid. By using a four-point linear interpolation algorithm at the four closest grid points, the geoid height can be determined for any location.

DATUM(S) REVISITED

Datum types include horizontal, vertical, and complete. Hundreds of geodetic datums are in use around the world, but the GPS is based on the WGS-84. Parameters for simplistic XYZ conversion between many datums and WGS-84 are published by the U.S. National Imagery and Mapping Agency. There is also a GPS time (T) quantity. See the next figure, Figure 10–9, for a graphical representation of this concept.

Datum Shifts

Coordinate values resulting from interpreting latitude, longitude, and height values based on one datum, as though they were based in another datum, can cause position errors in three dimensions of up to 1 km. These errors are known as shifts, and are not to be taken lightly, because they affect the navigational instructions given to ships, planes, and so on, based on these datums. The result of a shift not corrected in time could result in serious disaster.

Figure 10–9
Measurements via the
GPS System

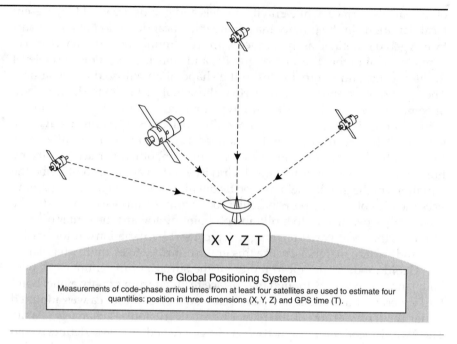

X Y Z T

The Global Positioning System
Measurements of code-phase arrival times from at least four satellites are used to estimate four
quantities: position in three dimensions (X, Y, Z) and GPS time (T).

Datum Conversions

Datum conversions are accomplished by various methods or formulas. A complete datum conversion is based on seven parameter transformations that include three translation parameters, three rotation parameters, and a scale parameter. A simple three-parameter conversion between latitude, longitude, and height in different datums can be accomplished by conversion through earth-centered, earth-fixed (ECEF) XYZ Cartesian coordinates in one reference datum, and three origin offsets that approximate differences in rotation, translation, and scale. The standard Molodensky formulas can be used to convert latitude, longitude, and ellipsoid height in one datum to another datum if the delta XYZ constants for that conversion are available and other coordinates (ECEF XYZ) are not required.

This brings up the second integral part of understanding the global positioning system, namely coordinate systems. To fully appreciate the GPS, we must explore two areas of science and the GPS itself. The first area was geodetic datums; the second is coordinate systems.

COORDINATE SYSTEMS

There are many different coordinate systems, based on a variety of geodetic datums, units, projections, and reference systems in use today. These systems are used to provide precise positioning for local and global systems used in navigation and geographic information systems for the location of points in space. There are many basic coordinate systems familiar to students of geometry and trigonometry, and these systems can represent points in two-dimensional or three-dimensional space. Rene Descartes (1596–1650) introduced systems of coordinates based on orthogonal (right angle) coordinates, and these two- and three-dimensional systems used in analytic geometry are often referred to as Cartesian systems. Similar systems based on angles from baselines are often referred to as polar systems.

Plane Coordinate Systems

Two-dimensional coordinate systems are defined with respect to a single plane. Three-dimensional coordinate systems can be defined with respect to two orthogonal planes. Orthogonal simply means two lines lying or intersecting at right angles, or two lines that are mutually perpendicular.

Global Coordinate Systems

The most commonly used coordinate system today is the latitude, longitude, and height system. The prime meridian and the equator are the reference planes used to define latitude and longitude. (Remember back to the fifth grade?) The geodetic longitude of a point is the angle between a reference plane and a plane passing through the point, both planes being perpendicular to the equatorial plane. The geodetic latitude (there are many other defined latitudes) of a point is the angle from the equatorial plane to the vertical direction of a line normal to the reference ellipsoid. The geodetic height at a point is the distance from the reference ellipsoid to the point in a direction normal to the ellipsoid.

ECEF X, Y, Z

ECEF is a type of Cartesian coordinate that is also used to define three-dimensional positions. These coordinates (X, Y, Z) define three-dimensional positions with respect to the center of mass of the reference ellipsoid. The x-axis is defined by the intersection of the plane as defined by the prime meridian and the equatorial plane. The z-axis points toward the North Pole. The y-axis completes a right-handed orthogonal system by a plane 90 degrees east of the x-axis and its intersection with the equator.

Universal Transverse Mercator

The universal transverse mercator (UTM) coordinates define two-dimensional, horizontal positions. UTM zone characters designate 8-degree zones extending north and south from the equator. UTM zone numbers designate 6-degree longitudinal strips extending from 80 degrees south latitude to 84 degrees north latitude. There are special UTM zones between 0 degrees and 36 degrees longitude above 72 degrees latitude, and a special zone 32 between 56 degrees and 64 degrees north latitude. Each zone has a central meridian. Zone 14, for example, has a central meridian of 99 degrees west longitude. Eastings are measured from the central meridian (with a 500 km false easting to ensure positive coordinates). Northings are measured from the equator (with a 10,000 km false northing for positions south of the equator).

Military Grid Reference System

The military grid reference system (MGRS) is an extension of the UTM system. UTM zone numbers and zone characters are used to identify an area 6 degrees in the east-west extent and 8 degrees in the north-south extent. UTM zone numbers and designators are followed by 100 km square easting and northing identifiers. The system uses a set of alphabetic characters for the 100 km grid squares. Starting at the 180 degree meridian, the characters A to Z (omitting I and O) are used for 18 degrees before starting over. Starting from the equator and going north, the characters A to V (again, omitting I and O) are used for 100 km squares, repeating every 2,000 km. Northing designators normally begin with A at the equator for odd-numbered UTM easting zones. For even-numbered easting zones, the northing designators are offset by five characters, starting at the equator with F. South of the equator, the characters continue the pattern set north of the equator. Complicating the system even further (as if this were possible), ellipsoid junctions (spheroid junctions in the terminology of MGRS) require a shift of ten characters in the northing 100 km grid square designators. Different geodetic datums using different reference ellipsoids use different starting-row offset numbers to accomplish this. OnStar® customers have absolutely no idea what all goes into their service!

World Geographic Reference System

The world geographic reference system (GEOREF) is used specifically for aircraft navigation, and is based on longitude and latitude. The globe is divided into twelve bands of latitude and twenty-four zones of longitude, each 15 degrees in extent. These 15-degree areas are further divided into 1-degree units identified by fifteen characters. Let us assume that the character designation method for the GEOREF is just as complicated as that of the MGRS, but only slightly different.

Many nations have defined grid systems based on coordinates that cover their territory. Some examples include Australia, Great Britain, Belgium, Ireland, and Italy. The United States uses the state plane system, which was developed in the 1930s, and was based on the North American Datum 1927 (NAD-27). NAD-27 coordinates are based on the unit of measurement known as the foot. While the NAD-27 state plane system has been superseded by the NAD-83 system, maps in the NAD-27 coordinates (in feet) are still in use. The state plane system 1983 is based on the North American Datum 1983 (NAD-83), and NAD-83 coordinates are based on meter.

THE FUTURE OF GPS

Although GPS and its technology have been available for some time now, the costs, inaccuracy, government ownership, and privacy issues have prevented this useful and practical service from being utilized fully. Future prospects are more promising and there is not another system of its type likely to take its place for personal, commercial, and military use.

One method through which GPS cards could be implemented for use in computers is to include the geographic location through the network layer of the open systems interconnect (OSI) protocol stack. Known as the GEO (geographic) system, this method would consist of three main components: GeoRouters, GeoHosts, and GeoNodes. The GeoRouter is essentially an IP router which is geographically aware, meaning it is able to track the geographic location of its service area. The GeoRouter would require a routing table which has been adapted to include the geographic location information. GeoHost application programming interface (API) software would be installed on all host computers, performing the GPS location updates and determining the local geographic address. A GeoNode would use scheduling algorithms to store incoming geographic messages for multicasting to subnets or wireless cells at a later time.

If GPS cards become commonplace technology, will consumers object to the government or to commercial enterprises knowing their exact location? If your computer's IP address includes the precise geographic location, it would be available to anyone on the Internet. Is this good business or is it an invasion of privacy? The courts will have to determine the legal issues involved with this technology. The lives of many are enhanced through GPS already. The safe and speedy navigation of ships and boats of all sizes, the ability to track a package from the distributor to the consumer, or pinpointing the yield effects of fertilizer application down to the square meter, thus reducing fertilizer pollution runoff, are all practical applications of today's GPS. Many new cars can combine the information from their GPS card, cell phone connections with a central assistance location, and onboard maps provided on CD-ROM to plot directions to the next location. If the car is stolen, the GPS locator can relay its location to assist the police in returning the stolen vehicle. Do you know precisely where you are? Whether you are ready or not, GPS can tell you precisely where you stand.

More GPS products are coming on the market as vendors see a viable market for these devices and software, for example, GPS receivers, GPS software for the PC, and even GPS software for the PalmPilot. (See Figures 10–5, 10-6, and 10-7 for a listing of these products as of mid-2001.)

Figure 10–10

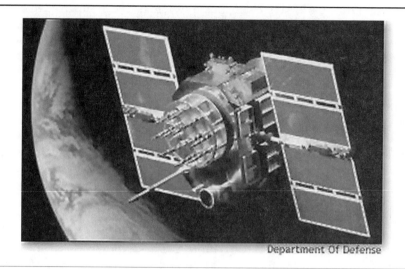

Department Of Defense

Recent governmental events surround GPS. It appears that the Pentagon wants to turn up the power on its network of satellites used to guide U.S. troops and the bombs they fire. The $200 million proposal is one of the military's first to fulfill Defense Secretary Donald H. Rumsfeld's plans to protect the United States from a space-based Pearl Harbor (Beefing up satellites www.msnbc.com/news/748505.asp?0dm=C1757"beefing upsatellites for space wars").

This $200 million would pay to upgrade the newest GPS satellites, which have yet to be launched (see Figure 10–10 for a pictorial of those satellites). That would allow the transmission to military receivers of signals that are eight times more powerful than those sent by the current generation of satellites (Beefing up satellites www.msnbc.com/news/748505.asp?0dm=C1757 "beefing upsatellites for space wars").

- These boosted signals would be powerful enough to burn through electronic jamming put up by an adversary. An important aspect is that American troops would not get lost and satellite-guided smart bombs would still find their targets.
- The ever more popular GPS devices used by civilians would not be affected or receive the boosted signal, defense officials say.
- By 2006, enough new satellites would be in orbit so that troops with GPS receivers should be able to receive a boosted signal anywhere on the earth's surface, according to the plan.
- The Bush administration is seeking about $50 million for the program in its proposed 2003 budget.
- The military uses the system for navigation and targeting. During the Gulf War, U.S. tanks relied on GPS directions to find their way around the desert in Iraq and Saudi Arabia.

- The increased power would only be transmitted on channels used exclusively by the military. Over-the-counter GPS locators used by foreign militaries, merchant ships, and expensive cars would continue to receive the low-power transmissions, leaving them more vulnerable to jamming.
- Just one boosted signal would make it easier for receivers to find other low-power satellites, even in an environment full of electronic noise thrown up to drown out the GPS signal, military officials say.
- The plan would allow the U.S. military to jam an adversary's over-the-counter GPS equipment on a battlefield, but still use its own.

■ SUMMARY

This chapter discusses various aspects of the global positioning system—its evolution, operations, applications, and its future.

The next chapter delves into the world of paging systems. We will find out who uses paging systems, and the protocols, components, and operations of the various paging systems.

REVIEW QUESTIONS

1. A predecessor to GPS in the field of navigation was the _____.
   a. swiss movement clock
   b. Teledyne
   c. radiophone
   d. sextant

2. Discovery of the _____ led to the discovery of the compass.
   a. Van Allen belts
   b. ozone layer
   c. free-floating magnet
   d. sextant

3. The orignator of the GPS was the _____.
   a. U.S. Department of Interior
   b. European Union
   c. U.S. Department of Defense
   d. NASA

4. Typically, there are _____ GPS satellites orbiting the earth at any given time.
   a. 16
   b. 24
   c. 3
   d. 66

5. GPS stands for _____.
   a. geodesic positioning service
   b. global positioning system
   c. general posit service
   d. global plane-surveying system

6. The satellites in the GPS form a set of _____.
   a. celestial bodies
   b. triangular points
   c. orbital position points
   d. reference points

7. Selective availability is _____.
   a. the ability to select a point on the earth and then find out where it is
   b. the ability to roughly gauge positions on the earth
   c. a limitation of GPS
   d. the limited availability of GPS products

8. GPSs use _____ signals.
   a. unidirectional
   b. spot beaming
   c. omnidirectional
   d. gamma ray

9. A minimum of _____ GPS satellites are needed to determine a location.
   a. three
   b. four
   c. two
   d. one

10. The first GPS satellite was launched in _____.
    a. 1996
    b. 1998
    c. 1990
    d. 1995

11. The service will require all new cell phones to have a GPS card installed in them.
    a. 911
    b. FEMA
    c. trunked radio
    d. DoD dispatch

12. There is always a minimum of _____ GPS satellites within view of a potential receiver at any time.
    a. three
    b. four
    c. six
    d. twenty-four

13. The GEOREF is a GPS system used for _____.
    a. aircraft navigation
    b. maritime navigation
    c. astronomy applications
    d. person location finding

14. GPS products can now be purchased for use on the following devices (circle all that apply):
    a. personal computers
    b. PalmPilots
    c. GPS receivers
    d. watches
    e. household appliances

HANDS-ON EXERCISES

1. Use your browser to compile a list of products that are GPS enabled. Likely sites include www.zdnet.com and www.cnet.com. Divide this list into products that operate on a PC, those that operate on GPS receivers, and those that operate on a PDA.

2. Go online and use your browser to conduct in-depth research into the history of navigation. Develop a five-page report on your findings.

3. Using a drawing package such as Photopaint, compose a picture of someone getting a location fix via the GPS.

4. Go online for pictures of GPS satellites. Using Excel or a similar package, develop a table listing each of the GPS satellites in orbit and when they were launched.

11 Paging Systems

In the previous chapter we discussed the global positioning system. This chapter deals with the intricacies involved in a paging system.

OBJECTIVES

After reading this chapter and completing the exercises, you will be able to:

- Learn the various parties who are involved in paging systems
- Learn the various protocols that operate in paging systems
- Learn the components and operation of paging systems

VODAFONE CALLS TIME ON THE TROUBLEMAKERS

Vodafone Launches the PubAlert Service

Licensees are increasingly banding together in local Pubwatch schemes to protect themselves, their families, and their customers against crime and rowdy behaviour on their premises. In a drive to help licensees stay one step ahead of any trouble, Vodafone, in partnership with an independent, government-approved body, the National Pubwatch Steering Committee, has announced the launch of its PubAlert pager service at the Pub Club and Leisure Show.

The National Pubwatch Steering Group (NPSG), which is sponsored by a number of prominent MPs and one of the major breweries, is calling for a national Pubwatch scheme to complement local initiatives. Vodafone is also a sponsor of the NPSG.

PubAlert is a specialist package designed by Vodafone Paging to meet the specific requirements of Pubwatch members. Paging is the best form of communication for this type of scheme, offering a fast, reliable, and cost-effective means of sending messages to a group of scheme members, at the same time. With the aid of a Group Call number, a member can notify other pubs and clubs of any incident or presence of known troublemakers in or near the premises within minutes, with just one call. As a local Pubwatch member, the police will find this information invaluable. Equally, the police can page licensees of any criminals coming in from outside of the area or scams being peddled, for example.

Some local authorities are supporting Pubwatch schemes by integrating CCTV monitoring of town centres with paging. Having a pager within the CCTV control room provides real-time information to the police which allows offenders to be tracked, and pubs and clubs to be notified of potential problems before they occur.

To encourage members to wear their pagers at all times, personal messages as well as football and The National Lottery results can be received. These additional services are included in the PubAlert package.

Commenting, John Smith, Managing Director of Vodafone Paging, said; "As one of the preferred suppliers to the National Pubwatch Steering Group, Vodafone Paging offers a comprehensive and proven PubAlert package backed by outstanding levels of customer support and advice on setting up initiatives. Pubwatch schemes have recorded notable successes in the fight against crime and anti-social behaviour generally, especially when integrated into a CCTV facility within towns. Newbury is one example of where the Pubwatch scheme, supported by the police and local authorities, has been an instant success from day one."

Raoul Devaux, Chairman of the National Pubwatch Steering Committee, said: "Controlled and implemented by their members and with the support of the police as well as professional communications support, local Pubwatches are making an acknowledged contribution to combating crime and disorder. In view of this success, the NPSG is pressing for a nationally co-ordinated scheme to support all licensees and to promote a safer drinking environment for their customers."

Existing scheme coordinators and police officers have endorsed the scheme, commenting that PubAlert has helped to catch and successfully prosecute local troublemakers, and reduce violence and the number of disturbances in their areas.

Source: http://www.vodafone.com.

WHO IS INVOLVED IN THE PAGING SYSTEMS INDUSTRY?

Paging Carriers

paging carriers

Paging carriers operate the infrastructure that delivers data to the subscriber. They are profit driven and closely scrutinized by the investment community. The services that carriers provide must cover the cost of doing business and generate a healthy profit. Therefore, they are cautious when it comes to initiating new services that may interfere with their existing revenue-generating services.

killer applications (apps)

Carriers are aware of the growing interest in the two-way wireless Internet market and are interested in hosting the **killer applications (apps)** that can increase their revenue stream. However, hosting smart two-way products on their networks is a new challenge, one that carriers are approaching with appropriate caution. Killer apps are software and/or hardware applications that can entice users to just that device or system, for that one special application alone. An example would be the Visicalc spreadsheet application, arguably the first killer app in the PC arena.

Capacity issues are of special concern to carriers. Carriers face numerous challenges in raising capital for new infrastructure, locating new transmitter and receiver sites, and acquiring additional RF spectrum, especially in this post dot.com era when venture capitalists are more cautious with their funds. In addition, it simply takes time to expand networks to bring additional capacity online, and an application that requires too much bandwidth will simply not be accepted by paging carriers.

Subscriber Device Suppliers

subscriber device suppliers

Subscriber device suppliers generate revenue by selling wireless devices and are anxious to find and host killer apps that increase demand for their products. Therefore, they have a vested interest in working with the other key players to remove barriers to success. Because the technology needed to design and build wireless devices is highly sophisticated, device suppliers tend to be large corporations.

Although device suppliers may have resources that can be applied to address problems related to wireless communications, they do not control all the pieces of the system. For example, device suppliers may be able to guide and influence the best advanced messaging configuration for the carrier's infrastructure, but they do not control these networks. Similarly, they can try to influence infrastructure suppliers to add support for technologically advanced features, but they cannot mandate or dictate the timing.

Subscriber device suppliers do not have domain knowledge in all industries. Therefore, they must rely on other companies to identify and develop applications that make sense in a wireless environment. In short, device suppliers must work together with the other key players to achieve mutual success.

Infrastructure Suppliers

Infrastructure suppliers are benefited by technologies such as advanced two-way messaging that create a demand for additional infrastructure; however, these advances in technology require continued product development. The development cycle for this equipment typically takes many months. Infrastructure equipment is a capital investment that carriers expect will serve their needs for a very long time. Enormous challenges face infrastructure suppliers in trying to bring products to market that meet a quickly developing market opportunity, are highly reliable, and are backward compatible to existing systems and protocols.

infrastructure suppliers

Infrastructure products are sophisticated. Therefore, the suppliers tend to be midsize to large companies. Paging infrastructure suppliers are not the only source of infrastructure components.

Some paging carriers have actually developed pieces of their own infrastructure, rather than depend on the product development timing and priorities of established suppliers.

Infrastructure suppliers, subscriber device suppliers, and carriers have traditionally worked closely together to establish standards and promote technologies that benefit all. The role of the independent developer is a new addition to the mix, and is yet to be well integrated into the symbiotic relationship.

Information Content Providers

Information is the fuel that drives the Internet revolution. **Information content providers** control this resource. These companies run the gamut from creators of information to suppliers or redirectors of information feeds. These companies can be huge organizations with staffs devoted to producing news, weather, sports, and other information. They can also be very small organizations or individuals who provide specialized information.

information content providers

Information such as news and weather may be intended for public distribution and consumption. Other information may simply exist in company databases that support the mission of the business, but could be very useful if accessed wirelessly. It is likely that some killer apps will need to use information controlled by information content providers. These applications may perhaps package and present the information in a new and improved manner.

Subscribers (Users)

Subscribers, or users, drive the market. They are the reason the other players exist. These consumers run the gamut from highly technical to very nontechnical, from very demanding to easily satisfied, from heavy users to occasional users. Successful products are those that are well designed for the intended user and that meet the needs of the majority of users. This means that easy to use is generally preferred over technically challenging, speedy execution and response is preferred over slow and sluggish response, simple is preferred over complex. The successful developer architects the killer app based on a thorough understanding of the users' needs and behaviors.

DESIGN AND PRINCIPLES OF OPERATION

The Paging Infrastructure— System Overview

A paging network is a collection of paging terminals, system controllers, transmitters, receivers, and data links. The network components must be engineered to make sure the paging system has optimal coverage and response capabilities with minimal interference. Network operators use sophisticated mapping tools to determine infrastructure site locations, to maximize coverage and minimize interference.

Paging systems are in some ways similar to cellular networks, yet in others are very different. Like cellular systems, virtually all paging networks use more than one transmitter. One-way paging networks rely on a simultaneous broadcast of messages to cover an area where a subscriber is likely to be located. Because one-way subscriber devices have no way of telling the system where they are, several transmitters must send the same message over a wide area using the same frequency. This method of message transmission differs greatly from the use of cells with towers that can locate users.

Radio signals can and do cause interference when different messages are transmitted using the same frequency at the same time. Therefore, it is important that paging systems use precisely controlled timing when transmitting messages.

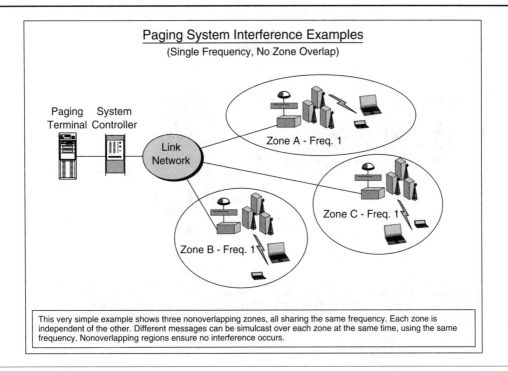

Figure 11–1
Nonoverlapping Paging Systems, One Frequency

Theoretically, two-way messaging networks could target delivery of messages much like cellular networks. In reality, however, paging carriers may use the same network to support both one-way and two-way subscribers. Furthermore, not all paging networks support single transmitter or cell-based targeted message delivery. Even if targeted messaging is available for two-way subscribers, there is still the issue of sending broadcast messages to many two-way subscribers. The following figures depict possibilities for paging system zone coverage and frequency use.

Figure 11–1 shows three nonoverlapping **zones** in a system that has a single frequency. The zones might represent widely separated areas such as Atlanta, Chicago, and Houston. Many transmitters cover each zone. The geographic dispersion and relatively low power of the paging transmitters ensures that the different messages can be simultaneously broadcast within each zone at the same time using the single frequency without interference.

zones

This very simple example shows three nonoverlapping zones, all sharing the same frequency. Each zone is independent of the other. Different messages can be simulcast over each zone at the same time, using the same frequency. Nonoverlapping regions ensure that no interference occurs.

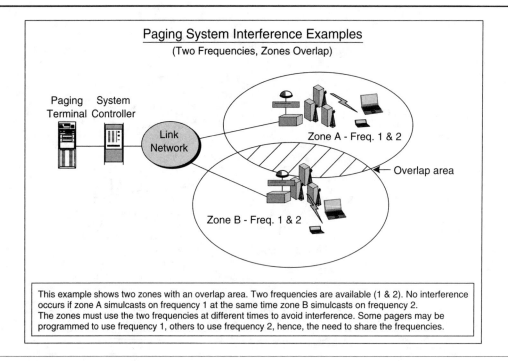

Figure 11–2
Overlapping Paging Systems, Two Frequencies

Figure 11–2 shows the case where two zones have an overlap region, but two separate frequencies are available to the paging operator. The example might be the Dallas and Ft. Worth markets. Dallas could be zone A; Ft. Worth could be zone B. Because of the close proximity of the two cities, we have an overlap area. Because the operator has two separate frequencies, it is possible to alternate the use of the frequencies. Frequency 1 can be used in zone A, while frequency 2 is used in zone B. Then the zones can switch. Frequency 2 can be used in zone A, while frequency 1 is used in zone B. Both frequencies are needed in both zones, because some pagers may be programmed for one specific frequency. Care must be taken to ensure frequency use is coordinated to prevent both zones from using the same frequency at the same time to send different messages.

This example shows two zones with an overlap area. Two frequencies are available (1 and 2). No interference occurs if zone A simulcasts on frequency 1 at the same time zone B simulcasts on frequency 2.

The zones must use the two frequencies at different times to avoid interference. Some pagers may be programmed to use frequency 1, others to use frequency 2, hence, the need to share the frequencies.

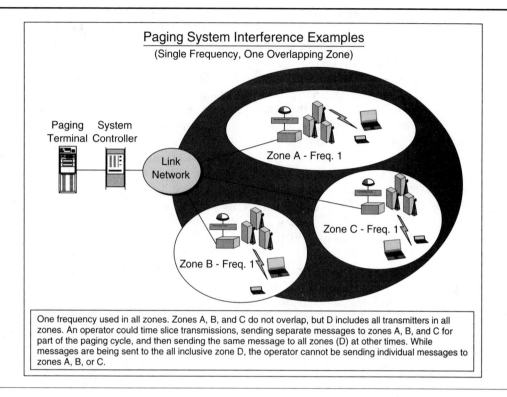

Paging System Interference Examples
(Single Frequency, One Overlapping Zone)

One frequency used in all zones. Zones A, B, and C do not overlap, but D includes all transmitters in all zones. An operator could time slice transmissions, sending separate messages to zones A, B, and C for part of the paging cycle, and then sending the same message to all zones (D) at other times. While messages are being sent to the all inclusive zone D, the operator cannot be sending individual messages to zones A, B, or C.

Figure 11–3
Overlapping Paging Systems, One Frequency

Figure 11–3 shows a system that is set up as three nonoverlapping zones (A, B, and C), and one super zone (D), which is actually the combination of all three zones. This could be the case in a region including several major cities such as New York, Philadelphia, and Washington. In this example, some subscribers pay for local service in their home city, while others pay for a wide area service that covers the whole region. In this example, the operator has a single frequency which must be shared. The operator can statically configure the scheduler to send separate messages to each of the three nonoverlapping zones during part of the paging cycle and then send messages to the wide area subscribers during the remainder of the paging cycle. During the time messages are sent to the whole region, all transmitters are simulcasting the same information. This solution permits the carrier to send a single message to all three zones, to be transmitted at the same time, to reach subscribers who may be in any of the three zones.

Note that this example is just one way a carrier could provide wide area coverage using a single frequency. The other obvious alternative is to simply treat the regions as three nonoverlapping zones, as shown in Figure 11–1. In this case three separate copies of a message intended for a roaming subscriber could be sent, one to each zone. There is no requirement that the messages be transmitted at the same time because the regions do not overlap. In fact, depending on the proportion of local to roaming subscribers or the balance of message traffic to each zone, this alternative may be preferable to the super zone solution, because it usually requires that only roaming messages be simulcast during the time periods reserved for wide area (zone D) coverage. If there are not enough messages to completely fill the (statically configured) reserved time slot for the super zone, the unused portion of the time period cannot be used to send local messages. Of course, if the system controller (the scheduler) is capable of dynamically adjusting its schedule to switch from wide area to local coverage when roaming message queues are empty, then this problem is alleviated. Many paging systems do not implement these sophisticated scheduling algorithms, however. In the event that multiple system controllers are used to handle different zones, the likelihood that a super zone could be used is further diminished, as few system controllers contain the external synchronization timing and logic needed to implement this.

One frequency is used in all zones. Zones A, B, and C do not overlap, but D includes all transmitters in all zones. An operator could time slice transmissions, sending separate messages to zones A, B, and C for part of the paging cycle, and then sending the same message to all zones (D) at other times. While messages are being sent to the all-inclusive zone D, the operator cannot be sending individual messages to zones A, B, or C.

Paging systems are store-and-forward systems. They accept messages for delivery to paging devices, and store them for a brief time before delivery. This storage is necessary because pages must be scheduled for delivery. The storage interval is typically a few seconds, but it depends on a number of factors including the over-the-air (OTA) protocols used, the system utilization at the time the message arrives, and the pager and system configuration settings. In the worst case, it could take several minutes for a message to be transmitted.

Paging systems are highly reliable, but it is possible for messages to be lost while in the paging network. This is rare but can occur due to system component failures or, in some systems, excessive network traffic. Excessive message traffic can have a strong impact on message delivery time. Users or applications that send excessive data through a paging network can cause significant delays for all messages that use the network. A sample paging system is shown in Figure 11–4.

Figure 11–4 provides an overview of what a simple paging system contains, although a few notes are warranted. First, not all paging systems have all the components shown. For example, one-way paging systems may not have the Internet access components and certainly will not have the receivers. Some systems have an external network management system; others handle the operations, maintenance, and control (OMC) functions directly within the paging terminal or controller. Some suppliers combine the functions of the paging terminal and system controller into a single box or subsystem. Al-

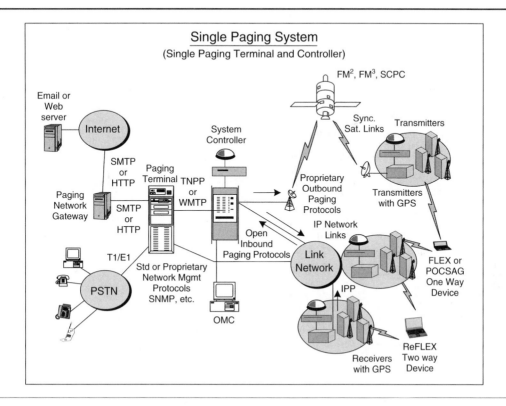

Figure 11–4
Single Paging System

though Figure 11–4 shows both satellite and link network connections to the transmitter site (typically IP), actual systems usually will have one type of distribution network, not both. Figure 11–4 also shows a collection of transmitters and receivers in small, localized areas. The illustration is misleading, because in real systems, transmitters and receivers are interspersed across the geographical area as needed to provide adequate inbound and outbound coverage.

Additional components needed to support the carrier's business have been omitted from Figure 11–4. For example, all paging systems have some provision or link to a billing system. Most systems need ancillary equipment such as operator terminals and printers, which are necessary for subscriber provisioning and system configuration.

Figure 11–4 illustrates a single paging system. This configuration might exist for very small paging operators who serve a small geographic area. However, larger systems are usually distributed and include many paging terminals and controllers. Each distributed system serves a portion of a carrier's total market. Figure 11–5 shows an example of two interconnected paging systems.

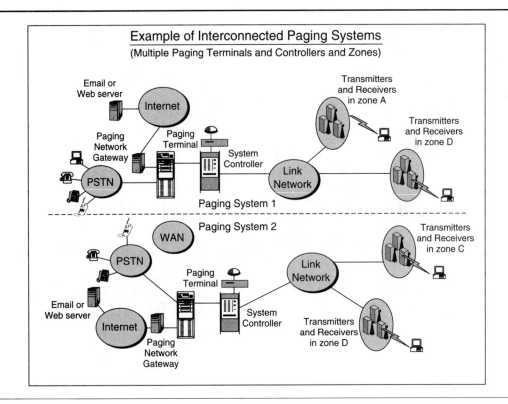

Figure 11–5
Interconnected Paging Systems

In the type of system depicted in Figure 11–5, interconnection is not limited to two systems. Some large carriers provide nationwide coverage that requires many terminals and transmitters. Other interconnect possibilities exist as well. For example, some network suppliers permit many paging terminals to send traffic to a single system controller, which then sends paging traffic to a wider collection of transmitters.

This section has served as a high-level introduction and view of a paging system. Now it is time to explain the functions of the various components.

Paging Terminals

The paging terminal is usually the entry point to the paging system. It connects callers to the paging system, accepts and validates messaging requests, performs other paging administration services, and forwards messages on to other subsystems in the paging system. It may offer features in addition to basic paging such as integrated voice mail services. It interfaces with billing systems, operator-assisted paging systems, and various internet gateways and servers.

In the past, the most common origination point for messages was the Public Switching Telephone Network (PSTN). With the introduction of alphanumeric and two-way messaging, sources for paging have multiplied and now include personal computers, specialized paging input devices, sites on the Internet, and other pagers or wireless devices. Regardless of source, messages usually pass through the paging terminal.

Paging terminals interface to the PSTN to permit subscribers to dial up and send pages from their telephones using dual tone multifrequency (DTMF). Messages entered from the PSTN are usually numeric, but various techniques have been used in some paging terminals to permit callers to send limited alpha messages from a telephone. Sending alpha messages from a telephone keypad is tedious, however, generally requiring the use of various key combinations to select alpha characters.

Many alphanumeric messaging terminals permit access to alphabetic entry devices including PCs or other specialized terminals that can compose alpha messages. These messages may be sent directly to the paging terminal through a dial-up modem connection or a direct connection via a dedicated circuit. Some vendors support access to the paging terminal through a separate gateway.

Most paging terminals do not directly support the mail and Internet protocols such as SMTP and HTTP used on the Internet. Thus, terminals cannot directly interact with e-mail or websites. Infrastructure suppliers and other vendors have developed gateways that translate between the Internet protocols and the paging protocols supported by the paging terminals, as shown in Figure 11–4.

Paging terminals contain information about paging subscribers. Generally the subscriber database is distributed across many paging terminals in a network. Each subscriber has a single home terminal that stores service information. Subscriber profiles include their personal identification numbers (PINs), information on the type of device, information regarding the subscribed services, service limitations, and subscriber configuration parameters. Typical service options may include voice mail and message storage. Limitations may include the maximum number of messages permitted in some time frame, maximum message lengths, and so on. Subscriber configuration options may include user-selected passwords, custom greetings, and the ability to divert or forward messages.

Paging terminals process requests for paging services regardless of the point of origin. The process generally includes a voice response script or other menu-driven option of services from which a caller makes a selection. The subscriber to whom the caller wishes to send a message is provided, either automatically by the telephone network based on the telephone number dialed, or PIN entered by the caller. The terminal uses this information to validate the called subscriber. If the called subscriber information resides in the paging terminal that answers the call, then the information is validated and used to determine the services and features the caller can request for the called subscriber.

If the called subscriber's information is not located in the paging terminal that answers the call, but is located in another home terminal, then the paging terminal may be designed to forward the request to the home terminal for processing. If the subscriber information is not found, the network rejects the call.

The home paging terminal translates a subscriber ID into the internal address (capcode) of the device used by the subscriber. Paging devices accept and decode messages addressed to their unique capcode. Paging devices do not understand subscriber IDs, which are assigned by service providers and may be either unique telephone numbers or PINs.

The paging terminal gathers all the necessary information from the caller including the message to be sent. It then sends this information along for further processing through the paging system. In some cases this means the information is sent to a separate system controller that schedules, batches, and encodes the information for delivery to transmitters.

Some paging terminals include the queuing, batching, scheduling, and encoding functions in the terminal itself. Other architectures use a separate system controller to perform some or all of these functions. The paging terminal generally holds the message until the system controller accepts it. The terminal may store the message for later retrieval and retransmission.

Paging terminals usually have an accounting function built in which records the call or message details for later billing. The information can be transferred to a separate billing system for processing. (The billing interface is not shown in Figures 11–4 or 11–5.)

One-way subscriber devices cannot tell the paging network their location because they do not have transmitters. For these subscriber devices, the paging carrier typically includes in the subscriber database record all the potential subscriber location zones or areas. The paging system instructs the system controller to simultaneously broadcast messages to all these zones.

Some one-way systems, particularly in Asia, support a roaming feature whereby a subscriber can notify the paging terminal of location during a defined period. This permits the paging system to provide message delivery to a smaller area.

Two-way messaging devices have the transmitter built in. These devices can notify the paging system or subscriber location; thus, messages can be sent to a particular area. The paging device actually informs the paging system of the strongest transmitter signal. The system then knows which signal should be used to send messages to the device. This is possible because each transmitter sends a unique code with the message that is detected by the paging device. Because the system does not need to activate all transmitters when sending a message to a subscriber, this capability in some situations can greatly increase network capacity. In other circumstances, however, the system may need to inquire of the location of subscriber devices before sending a message. These small "where are you" messages are generally simulcast over a wide area reducing RF channel savings from targeted delivery.

Paging terminals that support two-way devices must track subscriber locations to efficiently target message delivery. Two-way terminals generally use a request/response messaging protocol to match outbound commands and messages to replies from the subscriber device.

Two-way subscriber devices can confirm receipt of messages which helps ensure message delivery. The paging terminal will hold on to two-way messages for a system-configured time until it receives confirmation. If confirmation is not received within the configured time, which is typically several days, then the terminal discards the message.

System Control Switches

System controllers queue, batch, encode, and schedule messages received from paging terminals for delivery to transmitter sites. They also process two-way inbound messages that originate from receiver sites. The system controller distributes messages to the transmitters. In doing this, it must optimize the use of the distribution links and over-the-air RF spectrum. Some network suppliers combine the system control function in the paging terminal equipment. Others keep this function separate, as shown in Figure 11–4.

The system controller queues messages until they can be scheduled for delivery to the transmitters. The length of time the messages may remain in queue is a function of the supported OTA protocol and the message traffic destined for a particular area. The controller may have to send one-way messages to several dispersed geographic areas to reach the intended subscribers. For two-way messages, inbound messages help locate the subscriber, so the target area can be much smaller. Not all system controllers support both one-way and two-way messaging.

Paging systems typically simulcast a message over a wide area to reach the intended subscriber. To avoid interference in a geographical area covered by more than one transmitter, the controller must schedule the messages so that they are simulcast from all the transmitters at exactly the same time.

The message launch time is sent to each transmitter controller along with the message. The message must arrive at the transmitter controller just before the designated launch time, so the transmitter controller has time to process the message and send it to the transmitter's power amplifier at precisely the right time. If the message arrives too early, it may cause the transmitter controller's input queue to overflow. If it arrives too late, the message will not be transmitted.

Messages that arrive from the paging terminal are sent using one of several paging terminal protocols. To schedule and transmit them over the distribution network, they must be encoded using a different protocol by the system controller. The encoding into a different protocol provides some error detection and correction to protect messages transmitted OTA.

In two-way paging systems, such as those used for the advanced messaging devices, the system controller plays a key role in locating subscribers and processing inbound messages received from receivers. The two-way protocols must handle user-initiated inbound peer-to-peer messages. Scheduling inbound messages minimizes collisions and optimizes the use of the inbound RF channels. The two-way protocols define the messages necessary to request inbound transmissions and schedule the transmissions. The system controller must implement these complex message flows.

The system controller may also encode, transmit, and receive system management and control messages over the same distribution infrastructure used for basic messaging. These management messages include configuration information, diagnostic information, and new software for the transmitters and receivers. Different infrastructure systems handle the operations and maintenance functions differently. Some embed this functionality within the controller directly; others leave it to a specialized network management system. Depending on the network management protocols and the type of distribution network used, the system controller may schedule these messages.

The protocols used between the paging terminal and system controller and between the system controller and the transmitters/receivers generally determine the type of data networks used. Some protocols require serial synchronous circuit switched or direct connect networks; others require packet switched networks. The system controller must provide interfaces to the appropriate networks. Not all system controllers support all types of networks. Newer controllers tend to support IP-based networks, which are gaining favor because they can be used for text and voice messaging, as well as network administration.

System controllers generally come in either a redundant or nonredundant configuration. Redundant configurations better ensure that the paging system remains operational when system hardware or subsystems fail. The controller software manages the failure detection and switchover functions.

Transmitters

Transmitters transmit messages from the paging infrastructure to the subscriber devices. Paging transmitters range in power from less than 100 watts to around 300 watts. Typical commercial paging systems include hundreds of transmitters located so as to provide adequate coverage over the intended service area.

Transmitters operate in specific frequency ranges. Different frequencies are required in different regions of the world, because the governing bodies in each country set aside different frequencies for use by services such as paging, cellular telephone, and radio. A paging transmitter designed for use in the United States may not work in Europe due to different allocated paging frequencies.

Paging systems operate at relatively low data rates, typically ranging from 1,200 bps to 6,400 bps on each outbound RF channel. Effective data rates are even lower because the transmission protocols include overhead needed for assembling messages into batches and error detection and correction. Moving to higher data rates requires the paging carrier to install more transmitters with closer site spacing to achieve the same level of coverage as with lower data rates.

Paging protocols have traditionally delivered messages over the air using 25 kHz outbound RF channels. The two-way protocols define 50 kHz paging channels. The transmitters are designed to support one or more paging protocols. Few, if any, transmitters support them all.

Modern paging transmitters may either be linear or frequency modulated (FM). FM transmitters transmit on a single frequency at a time. Linear transmitters may transmit on multiple frequencies at the same time.

Transmitters receive messages from system controllers and transmit them at the assigned time. The transmitters must support the protocol used between the paging system controller and the transmitters. They must encode the messages using the transmission protocol, and transmit the data at the precise time indicated by the controller.

Transmitters also must support operations and maintenance functions including the ability to accept configuration changes and new software downloads. In some cases, the configuration data are sent over the same distribution links as the paging data. Sometimes the configuration changes can be made over dial-up links or directly at the transmitter site using a control panel on the transmitter.

Transmitters consist of several modules. The transmitter controller serves as the brain of the transmitter. It contains the processing logic needed to manage paging and control packets. It connects to the network to receive messages from the system controller. The transmitter also includes one or more power amplifiers to generate the signal that is fed to the transmitter antenna. It may have manual control interfaces that permit technicians to enter commands for conducting onsite service. Some advance paging protocols have very tight timing requirements that can only be met by using a GPS receiver. Data communication equipment, such as routers or satellite receivers, is also needed. These components will be located at the transmitter site.

Receivers

Receivers are used in two-way messaging networks. Like transmitters, they operate in specific frequency bands. They are sensitive to weak signals because they must detect messages from paging devices that may be operating at power levels well below 1 watt. Receivers generally operate at data speeds ranging from 1,600 bps to 9,600 bps; however, the effective signal data rates may be much lower due to overhead needed for identifying packets, error detection, and so on. As was the case for transmitters, the higher the data rates, the closer they must be situated.

Receivers must support the paging protocols used between the receiver and the system controller. These tend to be IP-based protocols. Receivers must also support operations and maintenance functions including accepting new configuration settings and new software downloads. These functions may be supported over dial-up links, or packet data networks using IP-based protocols.

Like transmitters, receivers consist of many components including a controller with the necessary processing logic and circuitry to detect and process signals, data communications equipment, GPS receiver, and antennas.

Gateways and Servers

Two-way messaging requires portals to access the Internet, e-mail systems, databases, and other repositories of electronic data. These portals act as gateways and are network servers. Their functions include managing protocol translation, routing information between external systems and the paging infrastructure, and executing database engines.

Servers frequently perform specialized functions. For example, a paging carrier may use a server to accept messages from an external e-mail server using Simple Mail Transfer Protocol (SMTP), and translates the e-mail request into a form and protocol understood by the paging infrastructure. Examples of such a configuration are shown in Figure 11–6.

Terminal and Distribution Networks

The networks used to connect paging terminals to each other, paging terminals to system controllers, and system controllers to transmitters and receivers can be very complex. The networks may be terrestrial wireline or wireless.

Networks may exist between any components in the paging system. Except for small test or on-premises systems, a distribution network always connects system controllers to transmitters and receivers. Terminal networks may exist to interconnect paging terminals in large, geographically dispersed paging systems.

INTERNET SERVERS AND GATEWAYS— PORTALS TO A PAGING SYSTEM

Network links often must carry more than one type of information and must support more than a single protocol. For example, many paging systems send paging traffic to transmitters using proprietary protocols based on User Datagram Protocol/Internet Protocol (UDP/IP), and use the same networks to manage the transmitters using Simple Network Management Protocol (SNMP), File Transfer Protocol (FTP), Telnet, and others.

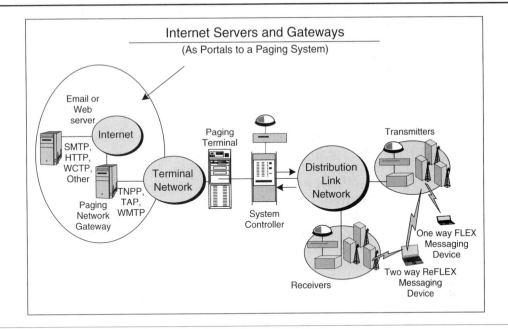

Figure 11–6
Servers and Gateways in a Paging System

Subscriber Devices

Wireless subscriber devices can include pagers, cellular or PCS phones, personal information managers (PIMs), and specialized terminals. They typically operate in a limited frequency band and support a single or limited number of OTA protocols. This means that a given device may have a very regional focus.

The protocols and networks they support may limit device capabilities. Most paging devices are designed to operate on a single frequency that is factory programmed into the pager. The frequency corresponds to a frequency that a specific carrier is licensed to use. If a carrier acquires new frequencies, either new pager subscribers must purchase new pagers or existing pagers must be reprogrammed.

Paging subscriber devices run the gamut from simple one-way paging to complicated two-way messaging devices. In fact, they may even fall somewhere in between one and two way. So-called one-and-a-half-way pagers are devices that receive messages and automatically acknowledge reception. These devices have both transmitters and receivers, but are not capable of sending user data. Some pagers permit the user to send preprogrammed messages only and others support voice messaging.

Subscriber devices have numerous features that distinguish one from the other. Some are numeric-only, one-way devices; others support alphanumeric text; some support binary data; some have just a few simple buttons, while others have complete QWERTY keyboards. Some are single-line text displays; others feature graphics-capable displays. Color displays are common now, but can be costly in devices. Some paging subscriber devices are fully programmable and have capabilities similar to miniature computers. In general, these devices are all two way.

To receive a message, pagers have a unique address encoded that is used to identify the specific device. This address may be called the capcode or radio identification code (RIC). In addition to the individual address, many messaging devices support one or more broadcast addresses. These are shared addresses that permit many wireless devices to receive the same message. This is useful for receiving an information service such as news, weather, or sports by many devices. Many wireless messaging devices can be programmed over the air to add or delete broadcast addresses, change home address assignments, enable or disable features, and so forth.

It is important to note that a single paging network may host a variety of pagers, from the simple to the complex. The configuration of the paging network is influenced to some degree by the types of subscriber devices that must be supported. For example, some older paging protocols such as POC-SAG are not constrained to send messages on frame boundaries. Newer higher speed protocols have frame boundary and timing constraints. When the paging network must support both, the paging channel must be divided between the two. Some paging infrastructure can be configured to favor one protocol and hence one class of pager over the other.

Protocols

This section discusses many of the protocols used in paging systems. It is not necessary to understand the details of the protocols. The intent of this section is to provide an overview of the protocols and a general sense of the complexity involved in sending information across paging networks.

Protocols are very simply the rules of the road that must be followed when exchanging information. They define the order of, content of, and constraints on information to be sent, plus they identify the optional and mandatory information that must be exchanged between two entities. They also define features that may be supported. They greatly influence or determine attributes such as latency, capacity, efficiency, robustness, immunity to errors, and the type and amount of data that can be exchanged. It is important to be aware of the role protocols play and to understand that protocols may limit the capabilities of a paging application.

Protocols tend to evolve over time. Individuals working on technical committees write protocols and update them as new capabilities are defined, or as clarifications are needed. Different paging carriers may support different paging protocols or different versions of the same protocols; thus, their networks have different capabilities and may support different features.

Protocols define the limits of what is allowed. Paging infrastructure suppliers are free to implement system components as long as they do not exceed bounds set by the protocols. However, it is very often the case that paging infrastructure suppliers do not implement or support all the features and capabilities allowed by the protocol.

It is not safe to assume that just because a paging system supports a particular version of a protocol that it supports all the features allowed by that version of the protocol. For example, some paging protocols such as TNPP define codes used to identify various types of pagers. A paging application may support only a subset of the pagers defined in the protocol specification.

There are numerous protocols used at different points in a paging system. It is important to understand that every interface point where one protocol is handed off to another is a point of potential congestion and message latency. At each of these points, the system usually must store the messages in a queue prior to processing them.

Before beginning our discussion of protocols, we should define a few terms and roles. First, the individual or system that accesses the paging system is referred to in the following text as the caller. The person or wireless device the caller wants to reach is the subscriber. Callers may or may not be subscribers to a paging service.

Figure 11–4 identifies various protocols used in paging systems. This section will not discuss all of these protocols, but will focus on those commonly encountered in a paging system. The protocols shown in the figure are not all-encompassing. Many others have been defined over the years and are supported in various paging systems.

From PSTN to Paging Infrastructure

Several paging protocols have been defined over the years to deliver messages or information from devices and external systems to a paging system. The primary ones include Telocator Network Paging Protocol (TNPP) and Telocator Alphanumeric Protocol (TAP).

In addition to these machine-to-machine protocols, a caller who responds to an automated voice response script can enter messages into a paging system from a telephone. The paging terminal answers an incoming call, and through a series of voice prompts, obtains the information needed to send a message to a subscriber. The caller enters the requested information from the telephone keypad using DTMF tones. The methods and protocols used to manually enter page requests from a telephone will not be discussed further in this section.

Telocator Network Paging Protocol

Telocator Network Paging Protocol (TNPP) is an ASCII character oriented protocol originally intended for transmission over RS-232 asynchronous data links utilizing speeds ranging from 300 bps to 9,600 bps. Infrastructure suppliers have actually implemented the protocol over many types of data links besides RS-232 serial links, at speeds in excess of the maximum 9,600 bps data rate allowed by the specification.

TNPP includes provisions for delivering information between two paging nodes and for forwarding information to other nodes. That is, TNPP supports intermediate nodes that act as routers or forwarding agents. Both possibilities are illustrated in Figure 11–7.

TNPP supports direct messaging between connected nodes, such as between paging systems 1 and 2, 2 and 3, 3 and 4. It also supports forwarding of messages between systems that are not directly connected. For example, TNPP permits paging system 1 to send packets to paging system 4 by routing them through paging systems 2 and 3, as shown by the broken line in Figure 11–7.

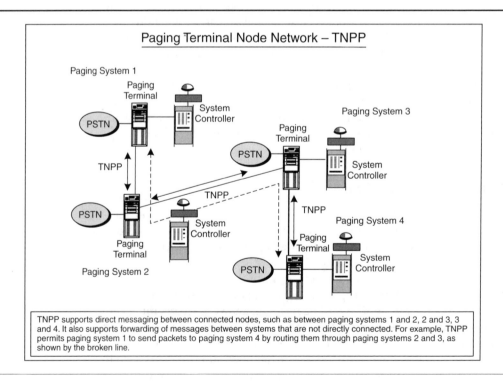

Figure 11–7
Message Routing in Paging System Using TNPP

TNPP was not designed for two-way messaging or for sending binary data. These are the main shortcomings of the protocol. Nevertheless, it is one of the most widely used terminal-to-terminal and terminal-to-control protocols currently in use in the paging industry.

Telocator Alphanumeric Protocol

Telocator alphanumeric protocol (TAP) is commonly used to send alpha text messages from devices such as PCs and page entry devices to pagers. It is an ASCII-based protocol that supports both an automated and a manual mode. Each block of information sent to the paging system must be no longer than 256 characters, comprising not more than 250 information bytes plus 3-control characters and a 3-character checksum. Multiple blocks are allowed by the protocol. The protocol does not impose limits on numbers of blocks; however, user or carrier systems may impose limits on total message size or total number of blocks that may be sent in a single connection or session.

The protocol supports sending messages to pagers. Each page message block includes a pager ID and a message. The pager ID is typically the PIN of the subscriber who is to receive the message. Any ASCII character with a value less than or equal to $0 \times 7F$ (DEL) may be used in the message. Some older systems do not support nonprintable ASCII characters less than 0×20 (SPACE). TAP does not support binary data types.

TAP is a simple one-way paging protocol. It may be used to send information to either one-way or two-way messaging devices, but it does not support the receipt of information from two-way messaging devices.

Simple Network Paging Protocol

Some carriers use the Simple Network Paging Protocol (SNPP) as a method of submitting pages from PCs or other systems on the Internet to their paging network. It is an ASCII-based protocol that uses IP as the network protocol. The protocol permits a system to send pages and to query their delivery status.

Two-way messaging extensions have been defined in version 3 of the protocol to permit the sender to set multiple-choice responses (MCRs) in the pager, to confirm message delivery and to query for responses.

Mail and Internet Protocols

Mail and Internet protocols are commonly used to support Internet and e-mail applications. Gateways that exchange information with paging systems may support these protocols. Most paging terminals do not support these protocols directly.

Simple Mail Transfer Protocol

Simple Mail Transfer Protocol (SMTP) is a protocol that is widely used to send mail across a network. It is documented in RFC-821. The protocol permits the exchange of standard ASCII text messages. The protocol provides for a sender to establish a connection to a receiver, which may be the final or an intermediate node in the destination path. The sender uses various SMTP commands to open a connection, request that mail be sent, specify message recipients, define the message content, and close a connection. The recipient replies to all messages with error and return codes indicating its ability to process the request.

The protocol includes provisions for forwarding mail to hosts other than the one specified by the sender. Provisions are made for expanding mailing lists, verifying mailing lists, and sending mail directly to a recipient's terminal screen.

HyperText Transfer Protocol

HyperText Transfer Protocol (HTTP) is a stateless, connectionless, object-oriented protocol. It is well suited for Internet browsing. While it is commonly used to send ASCII text or HTML pages from one Internet site to a web browser or other user agent, the protocol is not limited to sending these formats. It supports MIME-type encoding and user-specified character sets.

HTTP requires a reliable connection and is typically implemented to run over TCP/IP networks. However, any network that guarantees in-order message delivery such as TCP can support HTTP.

Few if any paging terminals implement HTTP directly in currently deployed systems. HTTP is implemented in some gateways that are used as portals to paging systems, however.

HTML is not well suited for wireless devices. Alternatives based on XML have been proposed for use in wireless devices. These include hand-held markup language (HDML), wireless markup language (WML), and voice markup language (VoxML). HTTP can be used to exchange information encoded in any of these markup languages. Chapter 4 addresses this issue in more detail.

Protocols within Paging Infrastructure

Terminal to Terminal

Telocator Network Paging Protocol. TNPP, as described earlier, can be used to submit messages to a paging system from an external system that has implemented the protocol, such as an operator-assisted paging (OAP) system. It can also be used to route paging messages between terminals and to send cap pages from paging terminals to paging system controllers.

Generally, TNPP ID pages are sent to paging terminals where the subscriber ID is converted to the pager's capcode. TNPP cap pages are sent to the system controller for encoding and distribution to transmitters. If an external system has its own subscriber database that contains the subscriber's capcode, then it can send TNPP cap pages directly to the system controller, bypassing the paging terminal. These possibilities are shown in Figure 11–8.

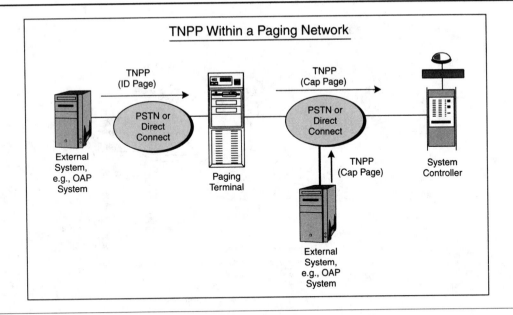

Figure 11–8
TNPP within a Paging Network

Wireless Messaging Transfer Protocol. Wireless Messaging Transfer Protocol (WMTP) is a fairly new paging system protocol developed by Glenayre in consultation with Motorola. The protocol is designed for both one-way and two-way messaging. It was initially used to support voice messaging (using the InFLEXion protocol) in some U.S. markets. Later it was used in Asia for one-way FLEX roaming systems. It is now being used to support two-way data (using ReFLEX) in several U.S. markets. The protocol supports most one-way and two-way OTA paging protocols.

WMTP was designed to support the following features.

- Acknowledged one-way message delivery
- Acknowledged two-way message delivery
- Group delivery
- Targeted messaging to specific cells
- User response messages from a pager
- Automatic roaming
- Output congestion control

WMTP is an IP-based protocol used to exchange numeric, alphanumeric, and binary messages between terminals and system controllers in a paging system. It currently specifies TCP/IP as the transport and network layer protocols, but the application layer protocol is independent of underlying layers.

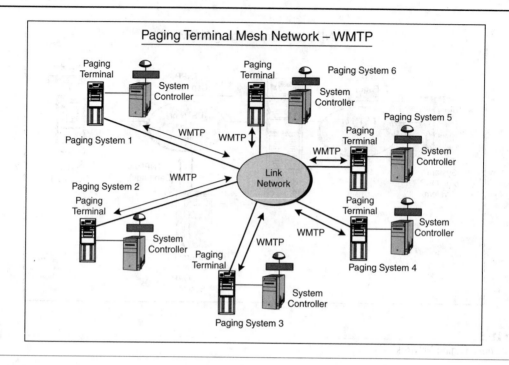

Figure 11–9
Paging Terminal Mesh Network for WMTP

Any data link protocol that is able to carry TCP/IP may be used for WMTP, including Point-to-Point Protocol (PPP), Serial Link Interface Protocol (SLIP), Frame Relay, or X.25. WMTP supports both peer-to-peer and broadcast messaging.

Systems that support WMTP are set up as mesh networks. Every paging terminal is able to directly access every other paging terminal in this network. That is, the paging terminals do not act as intermediaries between one paging terminal and another, as shown. This is in Figure 11–9.

Figure 11–9 shows six different paging systems interconnected via a link network. Each paging terminal contains a subset of the whole paging subscriber database. Callers may dial into any of the paging terminals to send messages to subscribers whose profile records may be in any of the terminals. Figure 11–9 shows each paging terminal connected to a single system controller. This is not a requirement of the protocol, but in fact, most existing paging systems that support WMTP are configured this way. Figure 11–9 omits the transmitter and receiver components, which connect to the system controllers. Each system controller manages a set of transmitters and receivers.

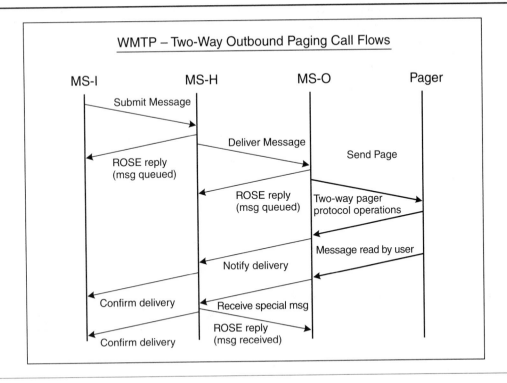

Figure 11–10
WMTP One-way Outbound Call Flow

Messaging through a paging system to subscriber devices using WMTP requires complex interactions between subsystems that comprise the paging system. Sample flow diagrams are shown in Figures 11–10 through 11–12 to illustrate this complexity. Similar exchanges of information are necessary for other paging protocols.

MS-I is typically the paging terminal that answers the call, MS-H is the paging terminal that holds the subscriber database, and MS-O is the system controller that encodes the message and sends it to transmitter(s). Often, MS-I and MS-H are the same node. Some paging systems may combine the MS-H and MS-O in the same node.

The point of the call flows is that considerable activity occurs within a paging system in support of a single inbound or outbound page. All of this increases message latency. Combining short messages into longer ones and limiting total message size should minimize message exchanges.

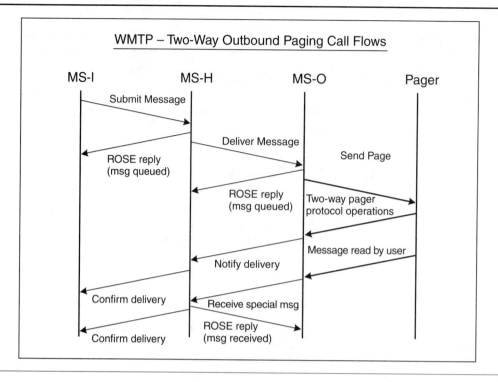

Figure 11–11
WMTP Two-way Outbound Call Flow

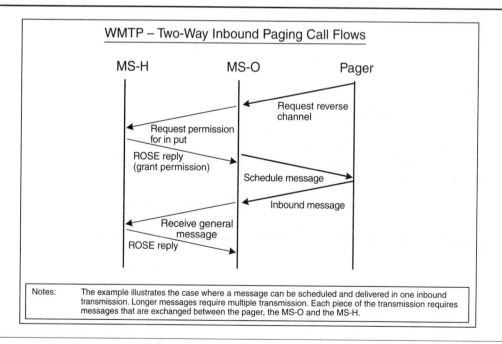

Figure 11–12
WMTP Two-way Inbound Call Flow

Terminal-to-System Control

Telocator Network Paging Protocol. TNPP, as described earlier, can be used to submit messages to a paging system from an external system that has implemented the protocol, such as an OAP system. It also can be used to route paging messages between terminals and to send CAP pages from paging terminals to paging system controllers.

Wireless Messaging Transfer Protocol. As mentioned, WMTP is a terminal-to-terminal protocol. It also supports messaging between paging terminals and system controllers.

The system controller performs a role in the protocol as the output switch (MS-O). Some infrastructure suppliers combine the terminal and system controller roles in a single integrated system. Others separate these functions in different system components.

Call flows shown in Figures 11–10 through 11–12 illustrate the involvement of the system controller (MS-O) in the messaging flow in WMTP networks.

The example illustrates the case when a message can be scheduled and delivered in one inbound transmission. Longer messages require multiple transmission. Each piece of the transmission requires messages that are exchanged between the pager, the MS-O, and the MS-H.

Protocols between Paging
Infrastructure and Subscriber Devices

The following sections discuss OTA protocols used between the paging system and the subscriber devices. Subscriber devices generally support a single OTA paging protocol. Both one-way and two-way protocols are discussed.

One-way OTA Paging Protocols

This section discusses one-way protocols used in paging systems. Programmable devices used on paging systems are all two-way devices. Information on one-way protocols is nevertheless useful, because the two-way paging protocols are built on a foundation defined for one-way protocols. Most of the section deals with the FLEX protocol, Motorola's high-speed one-way paging protocol that has been adopted as a standard in most regions of the world. It is the protocol on which the predominant two-way protocols are based. However, we begin the discussion with other widely used one-way paging protocols.

Post Office Code Standardization Advisory Group. Post Office Code Standardization Advisory Group (POCSAG) is a one-way protocol that supports data rates of 512 bps, 1,200 bps, or 2,400 bps. This protocol has been in use for over 10 years. It supports the mixing of different paging protocols or different POCSAG speeds on the same paging channel. The protocol is still used in older one-way numeric and alpha pagers. It does not support binary data types or roaming. It is supported on any available paging frequency with an outbound channel bandwidth of 25 kHz.

European Radio Message System. European Radio Message System (ERMES) is a one-way protocol adopted most extensively in Europe. It supports numeric, alpha, and binary data types. Outbound data rates are 6,250 bps. ERMES operates in the 169.425 MHz to 169.800 MHz frequency bands in 25 kHz channels. Roaming is supported between ERMES systems.

FLEX. FLEX is Motorola's high-speed one-way paging protocol. Its framing structure is used in two-way paging protocols. FLEX supports numeric, alpha, and binary data types. It can operate on any available paging frequency and uses a 25 kHz outbound channel. It supports data rates of 1,600 bps, 3,200 bps, and 6,400 bps. It has a defined roaming capability, and defines a group call capability that permits a single message to be efficiently transmitted to many recipients.

FLEX uses a frame structure where each frame is 1.875 seconds long. Exactly 128 frames make up a complete paging cycle. A cycle therefore lasts 4 minutes (1.875 sec/frame × 128 frames/cycle).

FLEX pagers can be programmed to listen for messages in specific frames. Their battery life depends on how frequently they wake up and listen for messages. They can listen as frequently as every frame (128 times in 4 minutes), or as infrequently as once a cycle (one time in 4 minutes).

Two-way OTA Paging Protocols

The ReFLEX and InFLEXion protocols are built on the FLEX protocol foundation. These are the protocols supported in the current two-way pagers. The following sections will describe each of these protocols.

ReFLEX 50. ReFLEX 50 is a two-way protocol intended for data applications. The first versions of the specification were released in February 1994. The stated purpose of the protocol is to provide an asymmetrical high-capacity two-way signaling system that uses a paging-based protocol on the forward channel. The asymmetrical nature of the protocol is reflected by the different data rates supported on the inbound compared with outbound channels. The phrase "high capacity" must be understood in the context of the paging environment when the protocol was developed. The data rates supported by the protocol were a definite improvement over what was supported by earlier protocols. When compared with data rates available today in wired networks and anticipated for future 3G wireless networks, the term is misleading.

The protocol defines personal, information service, and global addresses. Personal addresses are used to deliver messages to a particular subscriber. Information service addresses permit the same information to be efficiently delivered to a group of subscribers; one message is sent to one address and is received and displayed by all pagers that have the information service address enabled. A single global address is defined that permits all pagers to receive a message, such as a system-wide alert.

ReFLEX 25. The ReFLEX 25 is a two-way data protocol, first released in June 1995. It is similar to ReFLEX 50. ReFLEX 25 systems operate in North America using the 929 to 932 MHz and 940 to 941 MHz outbound bands, and 896 to 902 MHz inbound bands. Outbound channels may be either 25 or 50 kHz; inbound channels are 12.5 kHz.

Much of the discussion of ReFLEX 50 applies to ReFLEX 25, so it will not be repeated. Differences in the two protocols include the following:

- ReFLEX 25 defines personal and information services addresses, but not a global address.
- ReFLEX 25 defines inbound data rates of 800 bps, 1,600 bps, 6,400 bps, and 9,600 bps, whereas ReFLEX 50 uses 9,600 bps.
- The slower inbound data rates (800 bps, and 1,600 bps) allowed for ReFLEX 25 permit a smaller number of bits in a data packet (128 bits) than can be supported with higher data rates (6,400 bps and 9,600 bps support 154 bits per data packet).

The amount of data that can be packed into a single inbound message depends on several system settings.

InFLEXion. InFLEXion is a two-way protocol designed for voice paging. Voice paging has not proven to be very successful in the paging market; demand did not meet industry expectations. Therefore, very little will be said about the InFLEXion protocol.

InFLEXion systems operate in North America using the 930 to 931 MHz and 940 to 941 MHz outbound bands, and 896 to 902 MHz inbound bands. Outbound channels are 50 kHz; inbound channels are 12.5 kHz.

A 50 kHz outbound channel can be subdivided into seven InFLEXion subchannels, each 6.25 kHz wide. The inbound channel is used for message delivery and read acknowledgments, as well as various control messages.

Messaging Options

Paging systems are store-and-forward messaging systems. They support several forms of messaging, each of which can be useful in different ways. This section describes these forms of messaging.

Broadcast Messaging

Many information services use broadcast messaging to deliver content to many subscribers with a single message. The information is sent to a broadcast address. Any pager that has been preprogrammed with this broadcast address will receive the message. This efficient form of messaging can be used to form groups of subscribers, sometimes called radio groups. Every pager that has this address programmed in its broadcast address list will receive a single message sent to a radio group address. This form of group messaging does not support confirmation of delivery; that is, the device does not acknowledge a broadcast message.

An alternate form of group messaging is also possible. This form, sometimes called terminal group messaging, is not actually broadcast messaging, although it may appear that way to the user. Terminal group messaging simply involves setting up a list of subscribers in a paging terminal and assigning this list of subscribers a unique subscriber number. The paging terminal knows that any message sent to this group subscriber should be sent to each subscriber in the list as a personal address message. A single inbound message request received by a paging terminal therefore results in multiple messages being sent to the system controller and transmitters. Since the device sees this message as a personal address message, it will acknowledge its receipt.

Broadcast messaging is regularly used within the paging infrastructure to send a single message, either a personal or a group message, to many transmitters. In the case of satellite distribution networks, all transmitters in the system will hear everything that is sent over the satellite link. The protocols used in these networks include transmitter addresses in the datastream which tell the transmitters whether the messages are for them. Some paging systems that incorporate IP packet switched distribution networks use broadcast IP to deliver messages or operations and maintenance commands to many transmitter sites.

Peer-to-Peer Messaging

Peer-to-peer messaging is used for all personal messages. It requires the message to be addressed to an individual subscriber's personal address. The unique address or capcode ensures that a single paging subscriber receives the message.

Peer-to-peer messaging is supported by two-way messaging protocols to send messages to a specific transmitter that is closest to a subscriber. Since two-way messaging devices have transmitters, they can tell the paging system where they are. They do this by identifying which transmitter's signal is the strongest. This theoretically permits the paging system to target delivery to a single transmitter site. Although two-way messaging protocols may support this targeted delivery, the paging infrastructure equipment used in some two-way systems may not support the capability.

Multicast Messaging

Multicast messaging is an efficient form of group messaging that is supported in some IP-based distribution systems. This form of messaging permits a carrier to group transmitters so that a single message sent to this group address is distributed to all transmitters that belong to the group. This capability generally requires that routers in the distribution network support the message replication and distribution. This form of messaging improves network link utilization, but is for the most part transparent to the subscriber and application developer. Multicasting is not supported in the OTA paging protocols used for two-way messaging.

CHALLENGES OF THE WIRELESS WORLD

Writing applications for a wireless environment presents numerous challenges. Although many of these challenges may exist in the wired world, they may be more pronounced in the wireless world. Application developers must be prepared to deal with these issues if they want their applications to behave reliably in a wireless environment.

Transmission Errors

Messages sent over wireless links are exposed to many forms of interference that can alter the content received by the target device or system. In fact, it is possible that messages are completely blocked due to some physical obstruction. The application developer must be prepared to handle both of these situations.

OTA protocols contain sophisticated encoding rules to support error detection and correction. Two-way protocols typically provide support for retries when messages are sent but not successfully acknowledged by the receiving system or device.

Transmission problems can occur at any point in the distribution path. An application server may be offline, distribution links could be down, problems may exist in the paging system, interference may exist between the paging system and the messaging device, the subscriber's device may be turned off, or its battery may be dead. Failures can occur at any point in the wireless transaction—before a request is initiated, in the middle of a request, or after one is completed. The application developer must keep these possibilities in mind when designing error-handling strategies.

Message Ordering

Message order is not guaranteed in all paging systems. Many paging systems implement sophisticated scheduling algorithms to maximize distribution links and OTA RF channel utilization. This goal requires the system controller to pack messages in the outbound queue as efficiently as possible. To do this, the controller may order message transmission based on message length and time in queue. The result is that messages may be reordered by the system controller. If message order is important, the application developer must handle it.

Coverage Problems

RF coverage is not universal in paging systems. Operators build out their paging systems by markets. They add additional markets and increase coverage in existing markets according to business plans driven both by revenue opportunities and availability of capital. Coverage is extended to include primary transportation routes connecting major markets before secondary routes are covered.

Some carriers have expanded one-way paging systems to support two-way messaging by adding receivers and upgrading paging system software. This may mean that coverage is more complete for outbound (paging system to subscriber device) messaging than for inbound. The effect is that a subscriber may be able to receive messages, but may not be able to send replies. Of course, the opposite is also possible, depending on the location of transmitters and receivers.

Message Latency

Message latency is the time it takes to deliver a message. Many factors affect message latency. These include the queuing time in systems that handle the message, data rates across distribution networks, and processing time needed in various nodes along the path from point of origin to destination.

In paging systems, the queuing time in systems that handle the messages is usually the predominate source of message latency. Paging systems are store and forward systems. Paging protocols provide features that help maximize the use of available RF spectrum, an asset that is the most precious commodity to paging system operators. The protocols also tend to support features that help extend battery life in the subscriber devices. Unfortunately, it is not possible to achieve maximum RF utilization, maximum battery life, and minimum message latency at the same time. Trade-offs must be made. Sophisticated message scheduling algorithms are used in the paging system to achieve an acceptable balance between these performance and utilization goals.

Paging system operators control the system configuration parameters that affect message latency, battery life, and network and RF link utilization. The configuration parameters are set to optimize response and performance for the most typical application which is peer-to-peer messaging. Applications developed for paging systems must operate within performance and response constraints set by paging system operators.

Message latency is influenced by the specific paging protocols used. Some one-way, low-speed paging protocols may actually achieve lower latency because the messages are not required to be sent in particular frames.

Higher speed paging protocols such as those used for two-way messaging impose requirements on the paging system to send messages in the frames where the subscriber device expects them. If sent at the wrong time, the subscriber device will be asleep and will miss the message. The amount of the paging cycle that a subscriber device is awake listening for messages is configurable. Devices that are programmed to wake more often can send and receive messages more frequently. This reduces message latency and increases overall throughput, but shortens battery life.

A two-way paging system may be configured to send messages to subscriber devices as frequently as once a frame, or as infrequently as once a cycle. For the ReFLEX paging protocol, this represents a variation of between a few seconds to several minutes. Most paging systems try to keep this delay to less than 30 seconds.

It is possible that a message cannot be delivered to a subscriber at the time it is sent, due to transmission problems, coverage problems, or the fact that a subscriber's device is turned off or has a dead battery. Two-way paging systems usually store nondelivered messages for quite some time, waiting for the subscriber device to register with the system. Some systems simply retry transmission at various times until either the subscriber device acknowledges receipt or some maximum retry value is reached. This possibility means that a device, and hence an application running on it, may receive stale information.

Depending on the application, this could be as undesirable as not receiving the information at all. Consider the confusion caused by receiving a stock quote that is 2 days old, especially if the user is not told that it is 2 days old!

Network Capacity and Data Rate Limitations

Paging carriers will avoid hosting applications that result in excessive use of limited network resources. The most precious resource is usually available RF spectrum. Operators will attempt to block sources of data that place too much of a burden on their networks. Alternatively, paging carriers may price their service such that data-intensive applications will be cost prohibitive to subscribers. What is too much may vary from carrier to carrier, or may vary within a carrier's markets.

Paging protocols and RF channels support outbound data rates that range from less than 1,200 bps to 6,400 bps. Inbound data rates range from 1,600 bps to 9,600 bps. Effective data rates are lower than these numbers due to overhead needed for routing, packetizing, error detection, and correction.

A general rule of thumb is that effective data rates are around 50 percent of raw data rates in wireless systems. Another observation is that data rates in wired environments have always remained at least an order of magnitude greater than those in wireless environments.

The process of sending a message through a paging system is more complicated than just sending a single packet of information from source to destination. It involves a call flow where several messages are exchanged between various subsystems in the paging system to control the handoff and delivery of the information. The complexity of sending messages coupled with the overhead incurred in sending a message means that it is sometimes more efficient to combine many small pieces of information into a single larger message than to send many small messages. The optimal message size may vary by carrier system.

Network Dependencies

The behaviors and features supported by a paging system may vary from one paging system to another, both between carriers and within a single carrier's system. These differences are caused by several factors.

First, systems employ equipment provided by different paging infrastructure suppliers. Some paging carriers have developed certain paging system components on their own. Different equipment supports different sets of features, capabilities, and interfaces. Some of these have implications on how applications must be written to access information or route messages through the system. For example, an application may have to use one technique to access information from an application server when using one carrier's network and use a different technique when using a different carrier's network.

Most of the features supported in a paging system are provided through the software that runs in the various nodes that make up the paging system. Software version differences certainly exist across paging systems. This means that two paging systems built with the same hardware supplied by a single vendor may behave differently, because the software releases loaded into the subsystem components are not the same.

A paging system is highly configurable to permit operators to manage such factors as performance, utilization, and battery life. Even if all paging carriers used exactly the same hardware and software from a single infrastructure vendor, their systems may be configured differently, because they may need to support different classes of subscribers or give preference to different applications.

All these network dependencies mean the performance and suitability of an application can vary from system to system. The application developer should be aware of this fact. Note that while it is possible an application could be suitable for some paging systems but not for others, it is much more likely that a poorly designed application would not be suitable for any paging system.

Protocol Capability versus Network and Device Capability

As we discussed in the section on protocols, it is very likely that some components that make up a paging system do not support the full range of capabilities allowed by the protocols used in the system. Some application developers may have a comprehensive understanding of some or all the protocols used within a paging system. They may assume that because a protocol says it supports a certain feature, the paging system must support it. Most of the time this is a valid assumption. However, it is wise to check with a carrier to make sure this is the case, especially for features that have crucial application design implications.

Security Issues

Any information that is transmitted in any form is subject to interception and alteration. For some applications, this is a critical concern.

Paging system operators, protocol developers, paging infrastructure suppliers, and subscriber device suppliers tackle the issue of security in various ways. The solution needed depends to some degree on the physical configuration of the network, and the processes running on the network which support the application. The sensitivity of the information being transmitted will determine the level of security needed.

Complete end-to-end security in an application almost certainly will require the application developer to implement it both at the server and the client (subscriber device) side.

Parts of a system can be made secure by physically or logically limiting access. For example, a database server may be placed behind a corporate firewall. Access to this server may be limited to a particular gateway. In this case, the information between the gateway and the database server need not be encrypted. Encryption would be implemented at the gateway and used from that point to the subscriber device.

Some paging devices, such as Motorola's PageWriter, support encryption. The encryption method used in the PageWriter is 128-bit RC4. The application developer must provide similar encryption/decryption support for the server side of the system. One solution is to use Motorola's FLEX Messaging Server (FMS) product.

OUTLOOK FOR THE FUTURE

By their nature, future trends are mostly speculative. Hopefully they provide useful forecasts of what the landscape might look like in the not too distant future.

New Protocols

Trials are underway for higher speed second generation data protocols such as GPRS on GSM systems. This protocol theoretically can operate at data rates which are an order of magnitude greater than current paging networks; however, achieving these higher data rates involves reserving significant numbers of GSM voice channels for data, a possibility that is not likely to happen initially. After all, voice is still the number one killer app for GSM. Nevertheless, initial deployments of GPRS should boost effective data rates by a factor of 2 or 3 over current paging systems. The success of systems that support GPRS will be determined in part by availability of wireless devices and applications.

Work is underway to define and finalize new protocols for third generation wireless networks. cdma2000, WCDMA, and others hold out the promise of much higher data rates. These efforts should help spur acceptance of wireless Internet opportunities, but past experience with other new technology rollouts shows that it will take some time from the point the standards are finalized to when systems are deployed to fully support the standards. Then it could take several years for the technology to be fully accepted in the marketplace. Many hurdles must be overcome before commercial service is offered, not the least of which are billing and customer service issues.

Data-oriented protocols and technologies that extend the Internet to wireless devices are also important focuses. Protocols such as WAP, WML, HDML, and VoxML are already supported in various servers and wireless devices. This effort will continue and accelerate as more content providers offer WML or HDML versions of their HTML pages. At the same time, the hope is that customers in ever increasing numbers will be attracted to these new devices and services.

New Devices and Operating Systems

Device suppliers are busy designing products that are ubiquitous (can be used on all types of networks). Several recent product announcements, such as Motorola's announcement of a new DSP that is capable of supporting all the new 3G protocols, bring this promise closer to reality. However, wireless device product development cycles are fairly long. It will take time to develop and market these devices.

Device manufacturers are working with major players in the software industry to develop operating systems that are suitable for small wireless devices. Future devices are much more likely to support these standard operating systems than proprietary ones. This makes it easier for developers to write applications that will work across different vendors' products. Migrating to new operating systems in devices no doubt will take time.

Support for Standards

All the key players in the wireless Internet market recognize the need for standards. They are moving away from proprietary components to open standards that enable interoperability. Standards will be developed and adopted for network protocols, for Internet content markup languages, and for operating systems. It is even feasible that standard SDKs will emerge which would simplify application development across wireless devices and server platforms.

Manufacturers will continue to move toward using common hardware components and platforms as a way to reduce costs and development times. This will help ensure that applications will be portable across devices.

The future appears to be bright for developers who embrace the challenges of writing applications for the wireless Internet. Future directions all point to continued emphasis on wireless Internet opportunities. Success depends on bringing together all the necessary components: infrastructure, networks, devices, billing and customer care systems, protocols, applications, and the people who develop and manage everything.

■ SUMMARY

In this chapter, you learned how paging systems operate, who the major players are, and what applications paging systems can be used for. Information in this chapter is courtesy of Motorola.

REVIEW QUESTIONS

1. Paging carriers are _____.
   a. providers of content
   b. operators of infrastructure
   c. suppliers of infrastructure

2. An example of a zone is _____.
   a. Los Angeles, California
   b. Macon, Georgia
   c. SkyNet
   d. PagNet
   e. a and b only
   f. a and d only

3. Zones can _____.
   a. be overlapping
   b. cover a wide area
   c. can be serviced by one paging vendor
   d. all of the above

4. True/False Paging terminals are usually the entry point to the paging system.

5. True/False Paging messages entered from the PSTN are usually numeric.

6. True/False Paging terminals process requests for paging services regardless of the point of origin.

7. True/False Paging systems simulcast a message over a wide area.

8. Paging transmitters range in power from less than _____ to _____ watts.
   a. 25; 50
   b. 5; 50
   c. 100; 300

9. Receivers are used in _____ messaging networks.
   a. two-way
   b. one-way
   c. many

10. Portals act as _____ to the Internet, e-mail systems, and others.
    a. gateways
    b. databases
    c. language compilers

HANDS-ON EXERCISES

1. Using your browser and a search engine such as Google (www.goggle.com), locate paging system providers on the Internet.

2. Taking the information from question 1, produce an Excel spreadsheet table (or similar) listing these providers, their rates, and the areas they serve.

3. Using your browser and a search engine, locate other paging tutorials on the Internet and review them. Produce a short paper using Microsoft Word or Corel's WordPerfect (or similar), discussing this information.

4. Using a graphical package such as Microsoft Paint or Adobe Photopaint, produce a picture showing a simple paging network.

12 Trends

In the previous chapter we discuss paging systems. In this chapter we will conclude by discussing emerging trends in the wireless arena.

OBJECTIVES

After reading this chapter you will be able to:

- Understand what wireless trends are emerging
- Know the major players
- Imagine what the wireless market may look like in the near future

OBERTHUR CARD SYSTEMS DEVELOPS NEW CARD APPLICATION— IS THIS THE WAY FORWARD FOR 3G WIRELESS COMMERCE?

Bankcard issuers will be able to use a new smart card application for the convenient storage of data on the smart Visa cards. The "Convenience Storage" application was developed by Oberthur Card Systems and is based on the smart Visa Framework, which Oberthur developed with Visa U.S.A. earlier this year. The smart Visa Framework is a comprehensive set of commands, data structures, security protocols, and access methods that help to generically manage personal data. Also, pending Visa certification, Oberthur plans to introduce its new CosmopolIC Lite low-cost Open Platform card featuring the smart Visa Framework in ROM in 2003.

"The key advantage we provide is that the card gives cardholders total control of any data they might want to store on the card, from payment products to drivers' license information, through a global PIN management system," noted Oberthur executive vice president of sales and field marketing Thierry Burgess. "Personal data is securely stored directly on the smart card."

Cardholders will be able to use Oberthur's PC-based Convenience Storage application to store and retrieve any type of data from their card. Typical uses include the storage of personal information such as birthdays, Social Security numbers, or frequent flyer account numbers.

The smart Visa Framework applet suite was designed by Visa U.S.A. and developed by Oberthur. It integrates standard functions, commands, and data structures for personal data applications, whether for authentication or convenience purpose. The smart Visa Framework also offers a standard security framework for the management of credentials—such as pass codes and digital certificates, and standardized functions such as common PIN management. The Convenience Storage application utilizes the smart Visa Framework to introduce a tag-based data structure allowing for the storage of a wide range of personal information on a smart card.

"We are very pleased with the result of our collaboration with Oberthur," said Patrick Gauthier, senior vice president of smart card applications for e-Visa, a division of Visa U.S.A. "With these new products, Oberthur demonstrates their broad understanding and ability to deliver on the value proposition for smart cards."

Source: http://www.3g.co.uk/3GHomeSearch.htm

TECHNOLOGY AND CULTURE

Many wireless technologies begin from Europe. As mentioned, this is due to several factors, primarily technological and cultural. In the past decade, Europe's telecommunications operators were on a boom cycle (Hard choices 2001). These firms believed they would receive huge profits from the rise of the Internet and from even newer technologies such as WAP. However, these beliefs did not hold water, because most incumbent operators accrued large debts, primarily from the millions spent on research and development in wireless technologies. Over the period of 2000–2001, for example, European companies spent $46 billion for 3G mobile licenses in Germany and $36 billion for licenses in the United Kingdom.

WAP

Meanwhile, WAP so far has failed to take off. The gap between the best and the worst performers in the market widened across business lines, making it even more difficult for the laggards to attract capital to pursue their growth strategies. Without a truly significant breakthrough that would generate additional revenue, it is difficult to see how some integrated incumbents will regain healthy growth rates or even survive.

The wireless sector, both in Europe and the United States, may have to undergo a significant fundamental restructuring. The present market structure in Western Europe (e.g., five large integrated incumbents and ten smaller integrated companies) probably cannot be continued. Most companies will have to embark quickly on the unpalatable task of shedding their assets and stepping away from areas they thought were core businesses. Then, Europe will be left with two or three large integrated telecom companies holding majority stakes in data, wireless, and wireline services. In the United States, this is also expected to happen, though the outcome is less certain. It does seem clear that the few U.S. vendors with significant resources such as Sprint will thrive and swallow up smaller, less financially stable firms. As an example, it is unclear whether the niche market player, Palm, can survive on its own.

One of the ways wireless firms may come out of their slump is to rely upon the introduction of easy to use and inexpensive wireless products (Wong 2001). One of the frontiers not readily used in the United States is Internet-enabled (WAP) cellular phones. In Europe, where dial-up access is less certain, WAP phones are commonplace. In the United States, however, they are rarely used, because Americans are used to inexpensive dial-up access. So why put up with a microbrowser in a WAP phone when you can access the Net on a regular-sized computer monitor? One of the innovators in this marketplace, Openwave Systems CEO Don Listwin sees better days ahead in this arena. New cell phone technology, improved networks, and better software will make web browsing via cell phones a more popular communications feature, according to Listwin.

Openwave, a company formed through the merger of Phone.com and Software.com, makes wireless web software that runs many of the features found on cell phones, such as wireless Net access.

The United States has lagged behind Japan in wireless Net access, but Listwin expects this to change as more telecommunications carriers launch so-called 2.5G networks. Listwin believes new cell phone technology, such as a mouse for web surfing, will help popularize the notion of the mobile Internet. As a result, buying a book online, which previously took up to seventy clicks on a cell phone, may now require just three clicks to make the deal. Instead of drab green or gray screens and text-only outputs, the new generation of cell phones can support graphics and color, he added.

Openwave unveiled new software that will allow cell phone users to send instant messages to MSN Instant Messenger buddies. More partnerships may be on the way, the company reports. Openwave technology supports the messaging standard SMS.

People can also preset fifty or sixty of their favorite sayings on a phone. To save time typing repetitive phrases, such as "Good morning, how are you?" people can type the phrase into the phone once, save it, and then send it again by pressing the number "1" on the keypad, for example.

"We announced instant messaging as the key new technology that will bridge the telephony world and the Internet world," Listwin said. "I believe the convergence of two powerful environments will begin now. It will begin in earnest as the devices are here, the networks are here and the messaging technologies are here."

More movement is afoot as well in this arena. Opera Software says it will support the recently released WAP 2.0 standard in future versions of its browser. WAP 2.0, unveiled in 2001 by the WAP Forum, integrates technologies native to the Web (Festa 2001). Opera's pledge comes as browser makers angle to provide software that can surf the Web via mobile computing devices such as cell phones and handhelds. The U.K.-based mobile software unit of Psion selected Opera as the browser for its handsets. The agreement came shortly after Opera announced its arrangement to supply IBM with small browsers.

One of the bright spots in the wireless arena is the announcement of the new release of WAP (Charny 2001). Among other things, the latest version of WAP supports "WAP push," which allows a wireless device user to receive news alerts without being asked, as well as other services. In a move certain to boost the acceptance of WAP, three of the world's top four handset makers—Nokia, Ericsson, and Motorola—said they plan to support the latest version in their next generations of handsets.

3G

The other major technology, 3G technology on cell phones, is emerging as well. The emergence, and subsequent acceptance, of this wireless technology may be the single most determinant of how far wireless will be accepted by the public. This technology is undergoing rapid change. As an example, Nokia, the world's leader in handset market share, said in June 2001 that it would ship the 3G mobile phones to the United States by 2002, reiterating what it has said for months. Nokia spokeswoman Megan Matthews said European markets would get the phones sometime in the third quarter (Charny 2001). This has proven accurate; however, market penetration is still weak.

Nokia said it was trying to reassure a public that has grown skeptical of any claims about when the next generation of phone services might arrive in the United States and elsewhere. This year, carriers throughout the world, including NTT DoCoMo in Japan, have been delaying the introduction of the always-on, broadband Internet-capable cell phones for a variety of reasons, including a delay in the manufacture of handsets that will work on the new networks.

Nokia expects higher speed 3G cell phone networks to take off "exponentially" in 2003, and the company has set a goal of serving 35 percent of the customers who use those networks (Shankland 2001).

The company began trials of these 3G networks in the first half of 2002, and early customers will begin using them in the second half. In the United States, advancement is happening as well. U.S. telecommunications company Sprint is racing to build 3G network services. The technology is promised to be fast enough to send video clips, music, and other content.

Nokia, with more than 700 million cell phone customers, has set a goal of claiming a 40 percent share of the cell phone handset market, said Anssi Vanjoki, executive vice president of mobile phones at Nokia. Aggressive adoption of 3G and other technologies is one way the company hopes to achieve the goal (Shankland 2001).

Nokia unveiled a digital music player and four new cell phones to push the new technology. The Nokia Music Player will play digital music encoded with the popular MP3 format or the comparatively unknown advanced audio coding (AAC) format, which allows music companies to protect copyrights, Vanjoki says. It also includes an FM radio.

For the cell phones, the new 3330 includes support for the WAP standard for adding more sophisticated Internet services to cell phones. While the 3330 uses version 1.1 of WAP, the new 6310 for corporate customers uses version 1.2.1, according to Vanjoki. More significantly, it comes equipped with general packet radio service (GPRS) technology. Nokia expects to sell millions of GPRS phones this year, he added.

Nokia also recently unveiled the type of phone that will hit the U.S. market. The phone is the 8390 model, which among other things supports the newest version of WAP, a technology used to help cell phone users view Web pages. The phone has all the trappings of what is considered the next generation of phones capable of broadband Net access and other features such as downloading software or music; but it also is capable of having data beamed into it using an infrared port, similar to the way a Palm hand-held computer works, the Finnish wireless company announced.

The handsets are expected to work on new phone systems being built by VoiceStream Wireless, AT&T Wireless, and Cingular. These systems are based on the standard GSM. The Nokia phones in question are designed to work on GPRS, which will serve as the new network when the three carriers upgrade, offering the next generation of cell phones with always-on mobile Internet access at broadband speeds.

How can we explain why Americans are not using the new WAP-enabled phones? As mentioned, the availability of dial-up plays a major role. Simply put, some cell phones are so hard to use that most people are abandoning the fancy features such as e-mail, according to two studies that put the blame on WAP (Charny 2001). This fact is important because more than 90 percent of the handsets on the market contain WAP programming, a set of standards for cell phones. There are about 18 million WAP users worldwide, and close to 200 carriers have launched WAP or are in final testing, according to the WAP Forum, an industry group representing about 95 percent of the world's handset makers.

Usability is another major concern. The Meta Group (www.metagroup.com) found between 80 and 90 percent of corporate customers of WAP phones have "indicated a wholly unsatisfactory experience with the level of effort required to obtain information exceeding the threshold for perceived value."

A survey by J.D. Power and Associates (www.jdpa.com) discovered that one in four WAP phone users in the United Kingdom were using WAP phones for something other than making phone calls or sending short text messages. The year 2000 survey found one of every three WAP phone users were using their phones to do more than just make a phone call. This needs to become the norm for the U.S. market for WAP to truly succeed.

This is a contentious issue, although perhaps WAP is not entirely to blame for the dissatisfaction. Most networks maintain a constant connection between the two devices that are communicating with each other, but that hogs the network, causing slowdowns for others. One of the complaints about WAP phones is that they operate too slowly.

Many carriers are now upgrading their networks, in the hopes of capturing some of the $1 trillion in revenues forecast by 2010 (Charney 2001). Most will be using networks that will send the call or data in packets of information, which will not require a constant, bandwidth-hogging connection. This more efficient method may ease the delay problems.

All this is occurring in a changing environment where the number of sites on the Internet has slipped, along with a recent decline in registrations of addresses ending in domains such as .com (Mariano 2002).

The Web Server Survey from Internet consultancy Netcraft found that the number of websites dropped by 182,142 from November to December 2001.[1] That decline marks only the second time the company's survey, first released in 1995, has found fewer sites online in a monthly period. Although the drop may seem insignificant considering the more than 36 million sites found online, the survey highlights a shift in web address renewals. Netcraft attributed the drop to the decline in new domain name registrations; it also found that some current web addresses are being abandoned.

The company's latest findings come as the Internet Corporation for Assigned Names and Numbers (ICANN), which sets the standards for web addresses, attempts to expand the number of domain name suffixes. In 2000, ICANN added seven suffixes—.museum, .biz, .info, .aero, .name, .coop, and .pro—and chose a handful of companies to administer them (Jacobus 2000). But the new addresses have become caught in a web of setbacks.

WIRELESS IN LOCAL PROXIMITY ENVIRONMENTS

In another arena, new wireless devices are on the upswing. For example, Hewlett Packard introduced in January 2002 a new wireless Pavilion notebook and a kit that aims to ensure it plays well with an existing desktop PC (Spooner 2002). HP's latest Pavilion notebook, the zt1000, also offers either 802.11b or Bluetooth wireless networking technology for sharing data between devices without cables. This should prove to be a major boost for both wireless LANs in general and Bluetooth, as HP is such as major player. HP, it so happens, is the top retail notebook seller (Spooner 2001). HP also is not the only vendor to offer this capability. Both IBM and Compaq Computer are joining the race to integrate wireless technology into corporate notebooks (Fried 2001).

Sources close to Compaq say the company plans to introduce laptops with multiport, an add-on technology for wireless networking (Fried 2001). The first multiport modules, with technology from Intel and Ericsson, will add two hot wireless technologies—802.11b wireless networking and Bluetooth—to their Armada portables.

Meanwhile, IBM is moving ahead with plans for its first corporate laptop with built-in wireless networking. IBM says the ThinkPad T23 will add integrated wireless networking to the company's line of thin-and-light portables.

IDC analyst Alan Promisel said companies such as IBM and Compaq are adding built-in wireless in part because of demand, but also because their competitors are offering the same. It is important to the two companies to not appear as if they are falling behind, Promisel says.

---

[1] For more information, go to www.netcraft.com/survey.

Dell Computer and Toshiba have already added wireless networking to their notebooks, following the lead of Apple Computer, which has been pitching wireless networking under the brand name AirPort for some time.

Toshiba also is active with its latest notebook PCs. The new Portege 4000 and Tecra 9000 notebook models come with built-in 802.11b wireless networking technology and Bluetooth wireless support. The new notebooks also add a secure digital media slot and an IEEE 1394 connection (Spooner 2001).

Demand for wireless networking is picking up. In some cases, however, it is more a bet on the future than an immediate need. Often, companies are buying laptops with built-in wireless networking even before they have a wireless network established.

Bluetooth got off to a slow start, but it has picked up some steam with manufacturers such as Toshiba, IBM, and Dell Computer all offering it in various forms (Spooner 2001). Though some believe the two wireless technologies compete, others say they will ultimately coexist by performing different jobs.

This is not the first time that Bluetooth has been expected to burst onto the technology scene. Many industry watchers have expected it, but problems getting products to market and demand for 802.11b wireless networking technology has slowed its adoption (Wilcox 2000). Market researcher Dataquest predicts that demand for Bluetooth will result in the shipment of about 4 million chipsets. The company predicts shipments will rise to 36 million in 2002, and 186 million in 2003. An even more bullish forecast by Cahners In-Stat estimates that manufacturers will ship 13 million Bluetooth chipsets this year. Annual shipments are expected to rise to as many as 780 million units by 2005, the firm's research arm said in a recent report.

Still, those figures are lower than what Cahners had predicted, when it said Bluetooth shipments would reach nearly 15 million this year and 955 million in 2005 (Shim 2001). This time last year, Cahners was even more optimistic, predicting shipments of 1.4 billion in 2005 (Fried 2000).

As you discovered in Chapter 5, Bluetooth's technology allows data to be transferred between devices that are up to 30 feet away from one another and at speeds up to 1 Mbps. 802.11 supports a range of about 150 feet and data rates up to 11 Mbps. As a result, Bluetooth will probably be used primarily to send data between devices, while 802.11 can connect a PC to a network and allow it to share an internet connection and download large files.

It is clear from these developments, and others, that wireless is here to stay. The fastest growing wireless sectors, cellular and wireless devices, are expected to continue as usability issues become resolved and the public accepts wireless devices as commonplace.

■ SUMMARY

In this chapter we discuss trends in the wireless communications spectrum. We determine emerging trends, discuss the major players in the field, and speculate on the future of the wireless market.

APPENDIX A

SET UP A NETWORK TO SHARE A BROADBAND INTERNET CONNECTION USING A SOFTWARE ROUTER

Using a Mac or PC gateway with Two Ethernet cards and routing software

When configuring router software to share a single Internet address you will have the option of using one Ethernet card or two Ethernet cards on the gateway computer. Using two Ethernet cards has security advantages if you use personal file sharing since it isolates your local network from the cable modem network.

The following is a network using a gateway computer with two Ethernet cards and routing software to share a single Ethernet address.

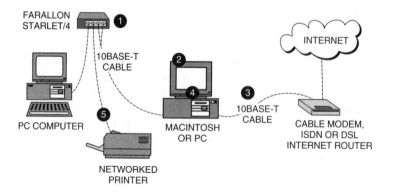

1. Run a 10Base-T Ethernet cable from each computer and printer to the Farallon Starlet hub.
2. Install a second Ethernet card on the Macintosh that will be acting as the gateway to the Internet. If your Mac does not have built-in Ethernet you will need to purchase two Ethernet cards for the gateway machine. If your Mac has built-in Ethernet you will just need one additional card.
3. Run a 10Base-T cable from the card on the gateway Mac to your high-speed Internet device, such as a cable modem, ISDN, or DSL router.
4. On the gateway computer, install and configure software such as Vicomsoft's *SurfDoubler* to allow you to share the single Internet address.

Source: www.proxim.com.

SETTING UP A NETWORK WITH BROADBAND MULTI-USER INTERNET ACCESS

The following steps are for a network using **multiple Internet addresses** and assumes you have an account with an ISP. For a network using a single Internet address see *Designing a network to share a single-user Internet access with a broadband gateway.*

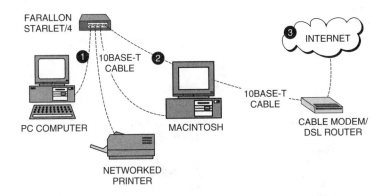

1. Run a 10Base-T Ethernet cable from each computer and printer on the Ethernet network to the Starlet hub. *See also Building an Ethernet Network with a Mac, PC, and Printer.*
2. Run a 10Base-T cable from your high-speed Internet device, which is either a cable or DSL modem router that provides multiple addresses, to the Starlet hub.
3. Now your Macs and PCs will all have Internet access and, with help from software such as Miramar System's *MacLAN* or Netopia's *Timbuktu Pro*, they can exchange files and share the printer.

Both MacOS and Windows 95/98/NT have built-in networking software that allows the Mac or PC to share files with another Mac or PC. This only works in a Mac-only or PC-only network. In this configuration, you will not need to purchase additional software for file transfers.

HOW TO SET UP A NETWORK TO SHARE A SINGLE-USER HIGH SPEED INTERNET ACCESS

Using a Mac or PC gateway with One Ethernet card and routing software

When configuring router software to share a single Internet address you will have the option of using one Ethernet card or two Ethernet cards on the gateway computer.

The following is a network using a gateway computer with one Ethernet card and routing software to share a single Ethernet address.

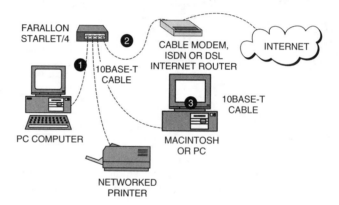

1. Run a 10Base-T Ethernet cable from each computer and printer to the Farallon's Starlet hub.
2. Run a 10Base-T cable from your high-speed Internet device, which is either a cable modem, ISDN or DSL router to the Uplink port on the Starlet.
3. On the gateway computer install and configure software such as Vicomsoft's *SurfDoubler* to allow you to share the single Internet address.

Source: www.proxim.com.

APPENDIX D

HOW TO SET UP A NETWORK TO SHARE A BROADBAND SERVICE

Using Netline Broadband Gateway and Ethernet Hub or Switch

The following is a network using a gateway to share a single IP address.

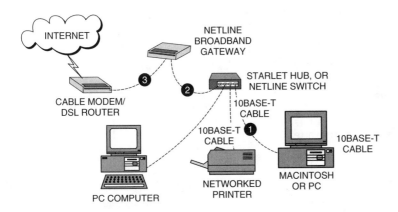

The following steps assume your computers are Ethernet-equipped.

1. Run a 10Base-T Ethernet cable from each computer and printer to a Farallon Starlet 10Base-T hub or a NetLINE 10/100 Switch.
2. Run a 10Base-T cable from the hub/switch to the gateway.
3. Run a 10Base-T cable from the gateway to your cable modem/DSL router.

Source: www.proxim.com.

MOTOROLA 9505 PORTABLE SATELLITE PHONE

What Is It?

Now smaller, lighter, and more resistant to water, dust, and shock than Motorola's previous offering, this newest addition to the Motorola Satellite Series (TM) portfolio is ideal for industrial or rugged conditions, yet appealing to the traveling professional.

Why Do You Need It?

Whether for business, adventure, or pleasure, you can stay in touch across seas and time zones, in remote locations and on the go. One telephone number, through the use of a mini personal subscriber identity module (SIM) card, can help keep you connected.

Features:

- Smaller and lighter
- Packed with features, yet simple to use
- Quick Access Interface
- Water, shock, and dust resistant for rugged environments
- Data Capable—Use your satellite phone to transmit data with an optional RS232 adapter
- IrDA Port (for future applications)
- 21 Language Choices for Prompts
- Crisis Calling
- Vibracall Alert
- Illuminated Holographic Display
- Color—Silver

Talk Time Features:

- Provides up to 24 hours of standby time with standard battery
- Provides up to 2 hours of talk time with standard battery

Display:

- 4 × 16 character Illuminated Graphic Display
- Illuminated Holographic Display

Calling Features:

- Auto Redial Notification
- Call Barring

- Call Forwarding—Unconditional, mobile subscriber busy, subscriber not reachable
- Clear Last Digit/Clear All Digits
- Fixed Dialing
- International Access Key Sequence (+ key)
- Mailbox for Numeric and Text (120 characters)
- Quick Access Interface
- Selectable Keypad tone (3 choices)
- Selectable Ringer tone (10 choices)
- Keypad Disable
- Unanswered Call Indicator
- Volume Adjustment (earpiece or ringer)

Memory:

- 100 Alpha and Numeric Memory Storage
- Last 10 Numbers Dialed
- Name Storage (recall by name or location)
- Memory Scroll by Location
- 32-Digit Number Capacity Phone Book
- 16-Digit Name Tag
- One-Touch Dialing
- Subscriber Identity Module Card (additional memory storage)

Indicators/Alert Features:

- Battery Meter
- Signal Strength Meter

Visual/Audible Features:

- Battery Meter (always shown in display)
- Illuminated Keypad
- Low Battery Warning
- Status Review of Features
- Visual Display of Volume Settings

Usage Control Features:

- Automatic Display Call Timer
- Automatic Lock
- Call Restrictions
- Display Call Timers (last call/total calls)
- Display/Change Unlock Code
- Electronic Lock
- Programmable Audible Call Timer
- Subscriber Identify Module PIN Availability (security code)

www.netzero.com
www.ieee.org
www.atmforum.org
www.press.umich.edu/jep/works/AnaniaFlat.html
www.osi.org
www.ansi.org
www.cnet.com
www.zdnet.com
www.gartnergroup.com
www.nokia.com
www.gsmworld.com
www.wapforum.org/new/20020612433new.htm
www.wapforum.org
www.pcmag.com
www.bluetooth.com
www.123wapinfo.com
www.microsoft.com/hwdev/tech/network/blvetooth/default/asp
www.zdnet.co.uk/news/2000
www.news.cnet.com/news/ 0-1006-200-5520038.html
www.gsmworld.com/news/press
www.cnetnews.com
www.fcc.gov
www.cellular.co.za/celltech.htm
www.computerworld.com
www.lifestreams.com
www.protocols.com
www.techguide.com
www.umtsworld.com
www.networkworld.com
www.infoword.com
www.nydmv.state.ny.us/dmvfaqs.htm#cell
www.nokia.zimismobile.com/nokiasms/nokiasms.html
www.whatis.com
www.panicpendant.com
www.mobitex.com
www.eecis.udel.edu/~dra/pl.html
www.fcc.gov/wth/publicsafety/plans.html
www.commnetcomm.com/commnet.html
www.pswn.gov
www.proxim.com/learn/library/casestudies/pdf/princesscruise.pdf
www.wlana.org
www.wlana.com
www.solveit.com
www.solveIT.com
www.proxim.com

www.hydra.carleton.ca/info/wlan.html
www.orinocowireless.com
www.nokia.com/corporate/wlan/
www.wirelesslan.com
www.nortelnetworks.com
www.cisco.com
www.3com.com
www.homeRF.org
www.solveIT.org
www.google.com
www.noaa.gov
www.stratosglobal.com
www.echostar.com
www.msn.com
www.globalstar.com
www.globalstar.com/telit_sat-550.html
www.geo-orbit.org
www.howstuffworks.com
www.howstuffworks.com/framed.htm?parent=satellite.htm&url=http:
    //spacelink.nasa.gov/Instructional.Materials/Multimedia/Satellite.
    Tracking/.index.html
www.howstuffworks.com/satellite7.htm
www.nasa.gov
www.iridium.com
www.qualcomm.com/globalstar/about/satellites.html
www.teledesic.com
www.orbcomm.com
www.loralskynet.com
www.directv.com
www.obake.peacesat.hawaii.edu/info/2-miss.html
www.un.org
www.aquilamsl.com
www.maptech.com
www.battelle.org/transportation/case-studies.stm
www.msnbc.com/news/748505.asp?0dm=C1757"beefing upsatellites for
    space wars"
www.lucent.com/minds/infotheory/what5.html
www.csep10.phys.utk.edu/astr161/lect/earth/atmosphere.html
www.pdana.com
www.vodafone.com
www.goggle.com
www.3g.co.uk/3GHomeSearch.htm
www.metagroup.com
www.jdpa.com
www.netcraft.com/survey

abbreviated handshake   a creation of a new connection state based on an existing secure session.

absorption   the ability of some objects to absorb RF.

access point   a device that transports data between a wireless network and a wired network (infrastructure).

acknowledgment frame   a frame sent to the sender of data to acknowledge receipt of a data frame.

Advanced Mobile Phone System (AMPS)   the earliest cellular network; used primarily in the United States. It is analog based.

Advanced Networks and Services   a nonprofit corporation formed by MERIT, MCI, and IBM to take control of NSFNET.

Advanced Research Projects Agency (ARPA)   an agency of the Department of Defense; progenitor of ARPANET, a predecessor to the Internet.

amplitude modulation (AM)   the process of varying the amplitude of a sound, often periodically. AM is essentially the modulation of a carrier's wave's amplitude by the program signal being transmitted.

analog   refers to electronic transmission accomplished by adding signals of varying frequency or amplitude to carrier waves of a given frequency of alternating electromagnetic current.

ANSNET   name for the upgraded NSFNET; operated by ANS.

antenna   a specialized device that converts energy from one form to another, that converts radio frequency fields into alternating current or vice versa.

attenuation   an event where signal loss occurs due to several factors including distance and resistance.

apogee   when a satellite follows a non-circular orbit around the earth, the satellite's path is an ellipse with the center of the earth at one focus.

Such a satellite has variable altitude and variable orbital speed. The point of highest altitude is called apogee. The term also applies to the maximum distance in kilometers or miles between the satellite and the center of the earth.

application layer   highest layer of the OSI reference model; contains all networking applications such as e-mail, file transfer, and remote login capability.

ARPANET   network developed by ARPA in the late 1960s to support DOD research activities.

asynchronous   events transpiring at different times.

Asynchronous Transfer Mode (ATM)   high-speed networking technology developed in the 1990s that uses a hybrid packet/virtual circuit approach to data transfer.

author   a person or program that writes or generates WML, WMLScript, or other content.

band   a specific range of frequencies in the RF spectrum, which is divided among ranges from very low frequency (VLF) to extremely high frequency (EHF).

bandwidth   the width of a band of electromagnetic frequencies.

base station   where radio signals originate. This can be in the case of a MSC for a cellular telephony cell or an emergency dispatching system at a fire or police station or many other such applications.

baseband   specifies the medium access and physical layers procedures to support the exchange of real-time voice and data information streams and ad hoc networking between Bluetooth units.

bearer network   used to carry the messages of a transport layer protocol between physical devices.

Bluetooth   a wireless communications link, operating in the unlicensed ISM band at 2.4 GHz using a frequency-hopping transceiver. The link protocol is based on time slots.

Bluetooth host a communications channel used to communicate with other Bluetooth hosts; a computing device, peripheral, cellular telephone, or access point to PSTN are examples.

Bluetooth unit voice/data circuit equipment for a short-range wireless communication link.

capability refers to the session layer protocol facilities and configuration parameters that a client or server supports.

capability negotiation the mechanism used for agreeing on session functionality and protocol options.

CDPD (Cellular Digital Packet Data) a specification for supporting wireless access to the Internet and other public packet-switched networks. This makes it possible for mobile users to get access to the Internet at up to 19.2 Kbps.

cell component of the cellular telephony networking scheme in which geographical areas are divided into areas, called cells. Each cell has a base station controlling all activities within that cell.

cell site similar to MSC and MTSO, this term describes the central switch and transmission center of a cell in a cellular telephony network.

cell sites on wheels (COWS) temporary cellular telephony stations set up for special events such as sports events or emergency situations. They are typically cellular telephony equipment carried in a truck and driven to specific locations for a short period of time.

cells on light truck sites (COLTS) identical to cows.

cellular digital packet data (CDPD) an AMPS overlay packet radio service.

channel the path over a wire or fiber, or in wireless systems, the slice of radiospectrum used to transmit the message through free space.

client a device (or application) that initiates a request for a connection with a server.

client-server communication communications between a client and a server.

coaxial cabling specialized cabling used for both voice and data transmission; excellent bandwidth handling and EMI shielding.

code division multiple access (CDMA) a form of multiplexing that allows numerous signals to occupy a single transmission channel, optimizing the use of available bandwidth.

connection state the operating environment of the record protocol; includes all parameters needed for the cryptographic operations.

connectionless session service an unreliable session service; only the request primitive is available to service users, and only the indication primitive is available to the service provider.

connectivity devices any device on a network that provides connectivity. These include hubs, switches, or routers.

content subject matter (data) stored or generated at an origin server; typically displayed or interpreted by a user agent in response to a user request.

content encoding as a verb, indicates the act of converting content from one format to another.

coverage area the area where two Bluetooth units can exchange messages with acceptable quality and performance.

crosstalk the result of signals of one wire leaping over and mixing with signals in an adjoining wire.

data link layer layer 4 of the OSI reference model; handles end-to-end reliability issues.

datagram transport a transport service that does not guarantee that the sent transport SDUs are not lost, duplicated, or delivered out of order.

decryption the process of deciphering the original encrypted data.

de facto standard standard approved by those without national or international recognition.

de jure standard standard approved by a nationally or internationally recognized organization.

device a network entity that is capable of sending and receiving packets of information and has a unique address.

device address the unique network address assigned to a device and following the format defined by an international standard.

device OS the standard operating system in a device; not part of what is specified by WAP.

digital describes electronic technology that generates, stores, and processes data in terms of two states: positive (1) and nonpositive (0).

Digital-AMPS (D-AMPS) a digital version of the earliest AMPS network.

direct sequencing continuously distributes the data signal across a broad portion of the frequency band.

directional antennas where the signal is transmitted only in one direction at a time such as in a microwave antenna.

downlink the link from a satellite down to one or more ground stations or receivers.

electromagnetic spectrum the complete range of the wavelengths of electromagnetic radiation, beginning with the longest radio waves (including those in the audio range) and extending through visible light all the way to the extremely short gamma rays that are a product of radioactive atoms.

elliptical orbit the way some satellites revolve around the earth.

encryption the process of creating unrecognizable data that are transmitted to the recipient.

entity the information transferred as the payload of a request or response.

extensible markup language (XML) a World Wide Web Consortium (W3C) proposed standard for Internet markup languages, of which WML is one such language. XML is a restricted subset of SGML.

fault tolerance measure of the degree to which network data are likely to be lost.

File Transfer Protocol (FTP) part of the TCP/IP protocol suite; optimized for bulk file transfers.

FLEX a one-way paging protocol developed to optimize channel efficiency, battery life, and cost per bit for transmitting messages over a wide geographical area.

frequency an important aspect of wireless communications, where the frequency of a signal is mathematically related to the wavelength.

frequency band a frequency set aside by the FCC and used for a specific purpose such as the 800 MHZ public services band.

frequency division multiple access (FDMA) the division of the frequency band allocated for wireless cellular telephone communication into thirty channels, each of which can carry a voice conversation or, with digital service, digital data.

frequency hopping moves a signal from frequency to frequency in a fraction of a second to make the signal invulnerable to radio interference or jamming.

frequency modulation (FM) the process of varying the frequency of a signal, often periodically. FM, like AM, is essentially the modulation of a carrier's wave's amplitude by the program signal being transmitted.

frequency reuse reusing the same frequency in nonadjoining cells.

full handshake a creation of a new secure session between two peers.

gamma ray the highest energy, shortest wavelength electromagnetic radiations.

general packet radio service (GPRS) defined in GSM 2.60 and 3.60, provides a packet data service overlay to GSM networks.

geodesy a section of applied mathematics that deals with the determination of the size and shape of the earth and the exact positions of points on its surface.

geodetic datums define the size and shape of the earth and the origin and orientation of the coordinate systems used to map the earth.

geostationary earth orbit (GEO) a geostationary satellite orbits the earth directly over the equator, at an altitude of 22,000 miles. It remains over the same spot on the earth's surface at all times, and stays fixed in the sky from any point on the surface from which it can be seen.

geosynchronous orbit (GEO) highest orbit satellites, at 22,500 miles above the earth's surface.

gigahertz (GHz) a unit of frequency equal to 1 billion hertz.

Global Positioning System (GPS) a satellite-based system providing precise three-dimensional locations to users.

Glonass a Soviet space-based navigation system comparable to the American GPS.

ground segment gateways that include the gateway control centers and gateway earth stations and the network control center located in the United States.

handshake the procedure of agreeing on the protocol options to be used between a client and a server.

handshake protocol the protocol that carries out the handshake.

header contains metainformation, which is general information about a session that remains constant over the lifetime of a session.

hertz (Hz) a unit of frequency of one cycle per second.

heterodyne a note of two different tones.

high definition television (HDTV) the new generation of broadcast television service; provides enhanced picture and sound quality.

host terminal interface the interface between Bluetooth host and Bluetooth unit.

iDEN Integrated Digital Enhanced Network.

iDEN circuit-switched data provides a point-to-point connection between the device and the network.

iDEN packet data provides a packet data radio service to the iDEN system. This packet data service utilizes mobile IP as the mechanism to enable mobile devices to roam within iDEN.

IEEE 802.X a set of specifications for LANs from the IEEE.

independent WLAN a network that provides peer-to-peer connectivity without relying on a complete network infrastructure.

inductor a passive electronic component that stores energy in the form of a magnetic field. It is typically a wire loop or coil.

information theory concept invented by Claude Shannon proposing a linear schematic model of a communications system.

information content providers firms that produce the requisite information that is displayed on paging devices.

infrared radiation lies between the visible and microwave portions of the electromagnetic spectrum. Infrared waves have wavelengths longer than visible and shorter than microwaves, and have frequencies which are lower than visible and higher than microwaves.

infrastructure WLAN a wireless network centered around an access point.

infrastructure suppliers manufacturers of the infrastructure equipment used by paging carriers.

interoperability the ability of public safety personnel to communicate by radio with staff from other agencies, on demand and in real time.

initiator the WTP provider initiating a transaction.

inquiry a Bluetooth unit transmits inquiry messages to discover the other Bluetooth units that are active within the coverage area.

intensity the quality of a signal upon reception.

ionosphere the outermost layer in the earth's atmosphere.

Integrated Services Digital Network (ISDN) digital communications technology developed in the 1980s and optimized for nonvoice traffic such as multimedia.

International Standards Organization (ISO) international standards body that creates international telecommunications standards.

Internet an interconnected series of networks, commonly accessed by users and businesses for information sharing.

Internet Engineering Task Force (IETF) organizations given control by the DoD for development of TCP/IP enhancements and revisions.

IS-136 General UDP Transport Service (GUTS) a general-purpose application data delivery service.

IS-136 Packet Data provides a packet data radio service in IS-136.

IS-136 R-Data a two-way narrowband transport mechanism that is supported on the digital control channel (DCCH) and digital traffic channel (DTC).

isochronous events transpiring at the same time.

isochronous user channel used for time-bounded information such as compressed audio (ACL link).

JavaScript a de facto standard language that can be used to add dynamic behavior to HTML documents.

killer applications (killer apps) applications written by software developers that make so much of an impact on users that they purchase the device on which the killer app resides.

latency transmission delay.

layer functionality grouped into a bundle, or layer, used in a reference model.

layer entity in the OSI architecture, the active elements within a layer that participate in providing layer service.

local area network network of computers usually confined geographically. Examples are Ethernet and Token Ring.

logical channel   the different types of channels on a physical layer.

low earth orbit (LEO)   this satellite system employs a large fleet of "birds," each in a circular orbit at a constant altitude of a few hundred miles.

macrocell   the larger cellular telephony cell which can be divided into smaller cells in order to utilize frequency re-use.

Marconi, Guglielmo (1874–1937)   Italian physicist and inventor; one of the earliest inventors of wireless systems.

master   the Bluetooth unit that initiates the connection.

maximum packet lifetime (MPL)   fixed by the used carrier (the network system).

medium access control (MAC)   typically referred to as Ethernet addresses; can be found on network interface cards.

medium earth orbit (MEO)   a satellite with an orbit within the range from a few hundred miles to a few thousand miles above the earth's surface.

megabits per second (Mbps)   a metric used to measure network performance; millions of bits per second.

megahertz (MHz)   a unit of frequency equal to 1 million hertz.

mesosphere   the third highest layer in the atmosphere, occupying the region 50 to 80 km above the earth's surface.

method   the type of client request as defined by HTTP/1.1.

metropolitan area network   network of computers operating in a campuslike setting or within 5 km distance; hybrid between LANs and WANs.

microcell   a bounded physical space in which a number of wireless devices can communicate.

military network (MILNET)   the strictly military components of the original ARPANET, created in 1983.

mobile switching center (MSC)   describes the central switch and transmission center of a cell in a cellular telephony network. Typically, this is a telephony switch, antenna and tower that controls transmission within the domain of its cell.

mobile telephone switching office (MTSO)   same as MSC in that it describes the central switch and

transmission center of a cell in a cellular telephony network.

mobile device   refers to a device, such as a phone, pager, or PDA, connected to the wireless network via a wireless link.

mobile traffic switching office (MTSO)   the base station managing calls within a cell in a cellular telephony network.

multipath   the signal variation caused when radio signals take multiple paths from transmitter to receiver.

network layer   provides routing of messages over the subnet; referenced as layer 3 in the OSI reference model.

network type   refers to any network classified by a common set of characteristics and standards.

Nordic Mobile Telephone (NMT)   cellular telephony system deployed in the Scandinavian countries of Northern Europe.

omnidirectional   broadcasts in all directions simultaneously.

omnidirectional antennas   antennas where the signal is transmitted in all directions at once such as in broadcast radio or cellular telephony.

optimized handshake   a creation of a new secure session between two peers. The server looks up the client certificate from its own source without requesting it over the air from the client.

orbital velocity   the velocity needed to achieve balance between gravity's pull on the satellite and the inertia of the satellite's motion.

origin server   the server on which a given resource resides or is to be created; often referred to as a web server or an HTTP server.

OSI reference model   seven-layer model incorporating the totality of functionality required to handle communications of a computer; all functionality of like purposes are grouped into a layer.

packet   format of aggregated bits that can be transmitted in one, three, or five time slots.

paging   a Bluetooth unit transmits paging messages to set up a communications link to another Bluetooth unit that is active within the coverage area.

paging carriers   corporations that operate the paging infrastructure.

peer process   a process on another computer that performs an identical function to the process on the first computer.

peer to peer   direct communication between two terminals typically thought of as clients without involving an intermediate server; also known as client-to-client communication.

perigee   the point of lowest altitude is called perigee. The term also applies to the minimum distance in kilometers or miles between the satellite and the center of the earth.

personal communications system (PCS)   the latest, digital-based, cellular telephony network; used primarily in the United States.

personal digital assistant (PDA)   recent innovation developed as a hand-held device providing scheduling, calendar functions, calculator ability, and even Internet access, all in a device that typically can fit in a pocket.

Personal Digital Cellular (PDC)   cellular system deployed in Japan.

physical channel   synchronized RF hopping sequence in a piconet.

physical layer   handles raw bit transfers between entities on a network; referenced as layer 1 in the OSI reference model.

physical link   connection between devices.

piconet   in the Bluetooth system, the channel is shared among several Bluetooth units. The units constitute the piconet.

point-to-point communications   typically refers to line-of-sight applications such as microwave.

point-to-point short message service (SMS)   a narrow bandwidth data transport mechanism typically available in cellular and PCS networks.

port   used as a subaddressing mechanism inside a device.

presentation layer   handles data conversion and formatting between peer applications; referenced in the OSI reference model as layer 6.

Programmable Read-Only Memory (PROM)   a chip in a device used to instruct the device how to operate.

propagation   involves deviations from a state of rest or equilibrium.

propagation delay   the amount of time it takes for a signal to go from one point to another.

protocol control information (PCI)   information exchanged between WTP entities to coordinate their joint operation.

protocol data unit (PDU)   a unit of data specified in the WTP and consisting of WTP control information and possibly user data.

protocols   rules or conventions on how entities may operate and/or exchange information.

proxy   an intermediary program that acts both as a server and a client for the purpose of making requests on behalf of other clients.

Public Safety Wireless Network (PSWN)   fosters communications interoperability among local, state, and federal public safety agencies.

Public Switching Telephone Network (PSTN)   the primarily in-ground telephone system used by millions.

push and pull data transfer   common vernacular in the Internet world used to describe push transactions and method transactions, respectively. A server "pushes" data to a client by invoking the WSP/B push service, whereas a client "pulls" data from a server by invoking the WSP/B method service.

push transaction   a two-way request-acknowledge communication initiated by the server to push data to the client.

radio frequency (RF)   when supplied to an antenna, RF gives rise to an electromagnetic field that propagates through space; sometimes called an RF field or radio wave.

ReFLEX   two-way paging protocol developed to enable the efficient delivery of messages and content over the air in both the outbound and inbound directions.

repeater   a device that receives a signal, amplifies it, and then retransmits it.

resource   a network data object or service that can be identified by a URL.

responder   the WTP provider responding to a transaction.

RFCOMM server   an application that awaits a connection from an RFCOMM client on another device.

RFCOMM server channel   this abstraction is used to allow both server and client applications to reside on both sides of an RFCOMM session.

roaming movement of a wireless node between two microcells; usually occurs in infrastructure networks built around multiple access points.

satellite a specialized wireless receiver/transmitter that is launched by a rocket and placed in orbit around the earth.

scatternet two or more piconets colocated in the same area (with or without interpiconet communication).

secure connection the WTLS connection that has a connection state. Each is identified by the transport addresses of the communicating peers.

secure session a session that is negotiated on a handshake.

server a device (or application) that passively waits for connection requests from one or more clients; it may accept or reject a connection request from a client.

server centric content from different sources is terminated in the server.

service data unit (SDU) unit of information from an upper level protocol that defines a service request to a lower layer protocol.

service discovery the ability to discover the capability of connecting devices or hosts.

service primitive an abstract, implementation-independent interaction between a WTP user and the WTP provider.

session a long-lived communications context established between two programs for the purpose of transactions and typed data transfer.

session layer layer 5 of the OSI reference model; provides sessions or dialogue control between peer applications.

session resume situation where a new secure connection is established based on a previously negotiated secure session.

session service access point (S-SAP) a conceptual point at which session service is provided to the upper layer.

session service provider a layer entity that actively participates in providing the session service via an S-SAP.

session service user a layer entity that requests services from a session service provider via an S-SAP.

sextant an instrument for measuring angular distances used especially in navigation to observe altitudes of celestial bodies, as in ascertaining latitude and longitude.

shared secret authentication an authentication method based on a shared secret.

signal an electric current or electromagnetic field used to carry data from one location to another.

Simple Mail Transport Protocol (SMTP) the TCP/IP protocol utilized to send and receive electronic mail messages.

simplex mode allows transmission of data in only one direction at a time.

sine wave a wave that is unique in that it represents energy entirely concentrated at a single frequency.

slave the Bluetooth unit that receives the connection.

SMS (Short Message Service) a service for sending text messages to cellular phones that use Global System for Mobile communication. GMS with SMS is found primarily in Europe.

space segment the constellation of satellites. This is one of the three main components of the ORBCOMM system.

spread spectrum a form of wireless communications in which the frequency of the transmitted signal is deliberately varied.

Sputnik the first satellite (of Russian origin).

stratosphere the atmosphere above 10 km.

subscriber communicators hand-held devices for personal messaging as well as fixed and mobile units for remote monitoring and tracking.

subscriber device suppliers manufacturers of paging devices.

switched mobile radio (SMR) wireless system used to dispatch emergency vehicles and the like.

terminal a device providing the user with user agent capabilities, including the ability to request and receive information; also called a mobile terminal or mobile station.

third generation (3G) the newest generation, or introduction, of cellular telephony technologies. 3G includes broadband data and transmission and multimedia capabilities over a cellular network.

throughput a measure of performance in telecommunications; usually the amount of data transmitted in a given amount of time.

time division multiple access (TDMA)   a technology used in digital cellular telephone communication that divides each cellular channel into three time slots in order to increase the amount of data that can be carried.

time slot   the physical channel is divided into 625-second-long time slots.

topology   physical or logical blueprint, or layout, of a computer network.

Total Access Communications System (TACS)   spinoff from AMPS used primarily in the UK.

transaction   the unit of interaction between the initiator and the responder; it involves (1) a method transaction, (2) a push transaction, and (3) a transport transaction.

transmission   a collection of one or more packets from a source to a destination.

Transmission Control Protocol/Internet Protocol (TCP/IP)   layers 3 and 4 of the reference model created by the IETF; used widely in the Internet arena.

transponder   a radio that receives a conversation at one frequency and then amplifies it and retransmits it back to earth on another frequency.

transport layer   layer 4 of the OSI reference model; divides data and passes the fragments to the network layer.

troposphere   the lowest part of the atmosphere.

trunked radio   the automatic sharing of multiple radio channels.

twisted-pair wiring   telephone wires twisted together to reduce crosstalk and EMI. Most widely used of communications media. Inexpensive to purchase and install; high attenuation factor.

two-way radio systems   dedicate a single radio channel to a particular group of users who then share the channel; for example, a trucking firm.

unbounded networks   another term for wireless networks.

underlying bearer   a data transport mechanism used to carry the WDP protocols between two devices.

unstructured supplementary service data (USSD)   a narrow bandwidth transport mechanism; a GSM supplementary service.

uplink   the link from a ground station up to a satellite.

user   a person who interacts with a user agent to view, hear, or otherwise use a resource.

user agent   any software or device that interprets content (e.g., WML).

user data   the data transferred between two WTP entities on behalf of the upper layer entities for whom the WTP entities are providing services.

wavelength   the distance between identical points in the adjacent cycles of a waveform signal propagated in space or along a wire.

wide area network (WAN)   network of computers operating over a wide geographical area; typically operated city to city, coast to coast, or nation to nation.

Wireless Application Protocol (WAP)   a specification for a set of communication protocols to standardize the way wireless devices, such as cellular telephones, can be used for Internet access.

wireless local area network (WLAN)   network that typically operates at effective throughput rates of between 1 and 2 Mbps.

wireless markup language (WML)   a programming language that allows portions of web pages to be presented on cellular telephones and personal digital assistants (PDAs) via wireless access. WML is part of the Wireless Application Protocol (WAP) specification.

wireless network   communications not using a bounded medium such as copper wiring. Examples are cellular telephones and satellite communications.

wireless node   a user computer with a wireless network interface card (adapter).

WMLScript   a scripting language used to program the mobile device.

WTA context   the complete set of variables, with content and the state of the WTA user agent.

WTA event   a notification, in the form of content, that conveys a change of state of the sender.

WTP provider   an abstract machine that models the behavior of the totality of the entities providing the WTP service as viewed by the user.

WTP user   an abstract representation of the totality of those entities in a single system that make use of the WTP service.

zone   area in which paging activities are performed.

BIBLIOGRAPHY

1999 Federal Radio-Navigation Plan. 2000, February. Washington, D.C.: U.S. Department of Transportation and Department of Defense. www.navcen.uscg.mil.

Anonymous. 2002. 3GPP—Market representation partners. www.gsacom.com.

_____. 2002. GSA—Global mobile suppliers association: Member profiles. www.gsacom/members/index.html.

Apstar1. 2001. Apstar1, www.tbs-satellite.com/tse/online/sat apstar 1.html.

Arizona Vending. 2001. www.mini-bank.com.

[Author, I.] 1995. Global positioning system standard positioning service specification, 2nd ed. www.navcen.uscg.mil.

American Mobile Telecommunications Association. 1997. *Specialized wireless communications: An industry overview*. Washington, D.C.

Barnes, Cecily. 2000. Trekkie game players to use cell phones as "communicators". Retrieved December 5, www.cnetnews.com.

Beefing up satellites. Retrieved at www.msnbc.com/news/748505.asp?0dm=C17J7.

BellSouth to hang up payphone business as wireless grows. 2001. Retrieved February 2, Bloomberg News/CNET News.com.

Berk, J. 1998. Intel Corp.: A brief history of PCS technology development in the United States. Retrieved April 15, www.pcsdata.com/history.htm.

Bluetooth—A revolution in personal connectivity, www.indiainfoline.com/cyva/feat/blue.html.

Bluetooth boost gives UK a firm lead, www.zdnet.co.uk/news/2001/ns-20013.html.

Bluetooth camp hesitates, www.zdnet.co.uk/news/2001/4/ns-206585.html.

Bluetooth exiled from LAN of hope and glory, www.zdnet.co.uk/news/2001/13/ns-22073.html.

Bluetooth official site, www.bluetooth.com.

Bomford, G. 1980. *Geodesy*. Oxford, England: Claredon Press.

Borland, J. 2000. Wireless phone tracking plans raise privacy hackles. Retrieved November 10, CNET News.com.

Borland, J. 2001. Mobile markets fall flat for many consumers. Retrieved January 18, CNET News.com.

Borland, J. 2001. NetZero, Juno to unite in merger. CNET News.com.

Buckingham, S. 2000. *SMS tutorial*. Newbury Berkshire, UK: Mobile Lifestreams Limited. www.iec.org/online/tutorials/wire_sms/.

Budde, P. 2001. Contents, services and applications, www.budde.com.au/TOC/TOC1465.html.

Calhoun, G. 1988. *Digital cellular radio*. Norwood, Mass.: Artech House.

Carlson, C. 1997, November 17. SMR breaks subscriber records. *Wireless Week Magazine*, 21.

Cellular communications. Retrieved February 13, 2000, www.theweb.badm.sc.edu/rover/mgsc890/mobile.htm.

Charny, B. 2001. Cell phone service costs drop. Retrieved April 9, CNET News.com.

_____. 2001. Bust to boom for telecoms by 2010. Retrieved May 1, CNET News.com.

_____. 2001. Studies give wireless standard WAP a slap. Retrieved May 24, CNET News.com.

_____. 2001. Nokia says next-generation phones on time. Retrieved June 28, CNET News.com.

_____. 2001. WAP 2.0 makes debut. Retrieved August 2, CNET News.com.

Charny, B., and C. Grice. 2001. Satellite phones: Lost in space. Retrieved March 8, CNET News.com, www.news.cnet.com/news/0-1004-200-5071066.html.

Ciampa, M. 2001. *Designing and implementing wireless LANs*. Boston: Course.

Comdex: Bluetooth taking off, www.zdnet.co.uk/news/2000/45/ns-19139.html.

Consumer Bluetooth on the way this year, www.zdnet.co.uk/news/2001/10/ns-21576.html.

Data on 3G—An introduction to the third generation. 2000. www.3gnewsroom.com/html/whitepapers/data_on_3g.shtml.

Datapro Information Services Group. 1997. *U.S. specialized mobile radio: Overview.* New York: McGraw-Hill.

Defense Mapping Agency. 1977. American practical navigator, publication no. 9. Washington, D.C. Defense Mapping Agency Hydrographic Center.

Department of Defense World Geodetic System. 1997. *Its definition and relationships with local geodetic systems,* 3rd ed. Bethesda, Md.: National Imagery and Mapping Agency.

Different PCS technologies. Retrieved February 13, 2000, www.pcsdata.com/PCSTechs.htm.

Dixon, R. 1984. *Spread spectrum systems.* New York: John Wiley & Sons.

Dreher, R., L. Harte, S. Kellog, and T. Schaffnit. 1999. *The comprehensive guide to wireless technologies: Cellular, PCS, paging, SMR, and satellite.* Boston, MA: APDG Publishing.

Dulaney, K. 2000. *Wireless Web access: Where in the hype are we?* Stamford, Conn.: The Gartner Group. www.gartnergroup.com.

Dziatkiewicz, M. 1997, April 28. The battle to provide solutions. *Wireless Week Magazine.*

Early Bluetooth lacks bite, www.zdnet.co.uk/news/2001/1/ns-20070.html.

Electronic news. 2000, April 17. *Electronic News Magazine* 46(16).

E-mail catches up to snail mail. 2001. www.usatoday.com/life/cyber/tech.

Emielinski, T., and J. C. Navas. 2000. Communications of the ACM. Retrieved April 28, www.proquest.com.

Ericsson form company to focus on Bluetooth, www.zdnet.co.uk/news/2000/46/ns-19257.html.

Europe: GPS in the new old world. 2001. www.trimble.com/intnl/europe.

Eurotechnology.com. 2001. www.eurotechnology.com/imode/index.html.

Farley, T. 2000. Basics of digital wireless communication. Retrieved February 13, www.privateline.com/PCS/history.htm.

Faulkner Information Service. 1997. *Selecting alternatives to cellular telephones.* Pennsauken, NJ.

_____. *Specialized mobile radio tutorial.* Washington D.C. www.pswn.gov.

Federal Communications Commission. 2001. Enhanced 911, www.fcc.gov/e911/.

_____. Specialized mobile radio service fact sheet, www.fcc.gov/wtb/specrds.html.

Feher, K. 1995. *Wireless digital communications.* Upper Saddle River, N.J.: Prentice Hall.

Festa, P. 2001. Opera to wrap into browser. Retrieved September 5, CNET News.com.

Feuerstein, M., and T. Rappaport, eds. 1993. *Wireless personal communications.* Norwell, MA: Kluwer Academic Publishers.

Fiber Optics Research. 2002. www.strategis.ic.gc.ca/sc mang.ecomevnt/engdoc/sld011.htm.

Forschungszentrum Julich. 2001. www.kfa-juelich.de/nmt.

Fowlie, M. 2001. The hard reality behind 3G services. www.cnet.com.

Free Internet access fires Brazil's online market. 2000. www.zdnet.com.

Fried, I. 2000. Will all hail Bluetooth? Retrieved December 4, CNET News.com.

_____. 2001. Compaq, IBM prepping new wireless laptops. Retrieved May 10, CNET News.com.

Future for PCS data computing. Retrieved February 13, 2000, www.pcsdata.com/future.htm.

The Gartner Group. 2001. www.gartner.com.

Globalstar Reports. 2001. Results for first quarter, www.globalstar.com/EditWebNews/195.html.

Goldin, D. 1998. Statement made before Subcommittee on Science, Technology, and Space Committee on Commerce, Science, and Transportation, U.S. Senate, September 23.

GPS Overview. 2000. www.utexas.edu.

GPS Primer. 2000. The aerospace corporation. Retrieved April 28, www.aero.org/publications/GPSPRIMER.html.

Graven, A. 2001, April 2. 3G: The next wave. *PC Magazine.*

Green, J. H. 1997. *The Irwin handbook of telecommunications.* Chicago: Irwin Publishing.

Grice, C. 2000. Funding stabilizes satellite firms. Retrieved July 11, CNET News.com, www.news.cnet.com/news.html.

_____. Iridium owners optimistic about new satellite focus. Retrieved December 12, CNET News.com.

_____. Pentagon signs phone deal with Iridium. Retrieved December 7, CNET News.com.

Grice, C., and B. Charny. 2001. Wireless jungle still waiting for its king. Retrieved February 2, CNET News.com.

GSM official website, www.gsmworld.com.

Gurley, J. W. 2000. Making sense of the wireless Internet. Retrieved August 14, CNET News.com.

Hard choices for Europe's telcos. 2001, December 23. *McKinsey Quarterly.*

Hecht, J. 2000. A fiber optic chronology. www.sff.net/people/Jeff.Hecht/chron.html.

Horak, R. 2000. *Communications systems and networks.* Foster City, Calif.: M & T Books.

HP unwraps its Bluetooth technology, www.zdnet.co.uk/news/2001/11/ns-21824.html.

IBM's persuasive software ready to roll, www.zdnet.co.uk/news/2000/32/ns-17376.html.

India Mobile. 2002. Introduction to SMS. Yahoo! Incorporated, www.in.mobile.yahoo.com/smsintro.html.

Intel Corporation. 2001. Proprietary research and white paper.

Jacobus, P. 2000. Net name body OKs seven new domains. Retrieved November 16, CNET News.com.

Kanellos, M. 2001. I-mode cell phones could rival PCs exec says. Retrieved February 5, CNET News.com.

Kary, Tiffany. Globalstar's pain may hurt partners, www.cnet.com/news/0-1004-202-4494765-0.html.

Kee, W., and W. Lee. 1995. *Mobile cellular telecommunications: Analog and digital systems.* New York: McGraw Hill.

Khoundary. A. 2000. Microwave health effects, www.newton.dep.anl.gov/askasci/gen99/gen99445.htm.

Konrad, R. 2001. Automakers steer cars toward connectivity. Retrieved January 7, CNET News.com.

_____. 2001. Ford to study dangers of electronic gadgets in cars. Retrieved January 10, CNET News.com.

Linux PDA featuring Bluetooth on sale by summer, www.zdnet.co.uk/news/2001/11/ns-21774.html.

Liu, B. 2001. IEEE to determine 802.11g WLAN specification. Internet News, www.internetnews.com/bus-news/article/0,,3_766861,00.html.

Mariano, G. 2002. It's a smaller World Wide Web after all. Retrieved January 2, CNET News.com.

Market conquest eludes Bluetooth technology, www.zdnet.co.uk/news/2000/49/ns-19737.html/.

Masini, G. 1976. *Marconi.* New York: Marsilio Publishing.

Mehrotra, A. 1994. *Cellular radio: Analog and digital systems.* Norwood, Mass.: Artech House.

Microsoft drops Bluetooth from Windows XP, www.zdnet.co.uk/news/2001/1113/ns-22043.html.

Motorola gets first approval for Bluetooth products, www.zdnet.co.uk/news/2000/37/ns-17998.html.

Motorola gets its teeth into Bluetooth, www.zdnet.co.uk/news/2000/38/ns-18088.html.

Motorola, IBM brush up on Bluetooth, www.zdnet.co.uk/news/2000/44/ns-18914.html.

Motorola official site, www.motorola.com.

Naisbitt, J. 1994. *Global paradox.* New York: Avon Books.

Nextel official website, www.nextel.com.

Nokia. 2001. www.nokia.com.

OSHA. 2002.

Our satellites in the sky. 2001. www.dishtv.com/contents/aboutus/index.shtml.

Palmer, M. 2000. *Hands on networking essentials.* Cambridge, Mass.: Course.

Parks Associates. Retrived at www.parksassociates.com.

PCS applications. Retrieved February 13, 2000, www.bus.1su.edu/isds/faculty2/walsh/isds7520983/pcs/PCS%20Applications.html.

PCS versus cellular. Retrieved February 13, 2000, www.bus.1su.edu/isds/faculty2/walsh/isds7520983/pcs/analogvspcs.html.

Pelton, J. 1995. *Wireless and satellite telecommunications.* Upper Saddle River, N.J.: Prentice Hall.

Personal networks group. Boynton Beach, Fla.: Motorola, Inc.

PRG guide. 2000. www.prgguide.com.

Prohm, B. 2000. *A case study in the changing anatomy of a wireless handset.* Stamford, Conn.: The Gartner Group.

Public Safety Wireless Network. 2001. Achieving interoperability through cooperation and coordination, www.pswn.gov/strategic_plan_1_5_99.htm.

Ragaza, L. A. 2000. Say what? *PC Magazine.* Retrieved November 7, www.pcmag.com/article.

Reuters News Service. 2001. Satellite firms revived despite obstacles. Retrieved June 21, CNET News.com.

Ross, 2000. Lukewarm reception seen for wireless auction. Retrieved December 11, www.news.cnet.com/news/0-1004-200-4064919.html?tag= prntfr.

Ross, P. 2000. Short take: FCC extends time to comply with 911 regulations. Retrieved September 8, CNET News.com.

_____. 2001. FCC heeds industry pressure on wireless auctions. Retrieved February 1, CNET News.com.

_____. 2001. FCC names potential spectrum for next-generation wireless. Retrieved January 4, CNET News.com.

_____. 2001. Wireless auction licenses could face legal jeopardy. Retrieved January 26, CNET News.com.

Satellite navigations systems. 1997. *Encyclopedia of Science & Technology.* New York: McGraw-Hill.

Schwartz, C. R. 1989. *North American datum of 1983.* Rockville, Md.: National Geodetic Survey.

Schwartz da Silva, J., B. Arroyo-Ferncindes, B. Barani, J. Pereira, and D. Ikonomou. 2001. Evolution towards UMTS, European Commission. www.tbm.tudelft.nl.

Shankland, S. 2001. Nokia expects 3G takeoff in 2003. Retrieved March 21, CNET News.com.

Shim, R. 2001. Bluetooth nibbles at wireless market. Retrieved April 25, CNET News.com.

Smith, S. 2001. Short message service. Whatis.com? Retrieved November 7, www.searchnetworking. techtarget.com/sDefinition.html.

Spooner, J. G. 2001. Bluetooth's future looking rosier. Retrieved December 14, CNET News.com.

_____. 2001. HP jumps ahead in notebook race. Retrieved December 14, CNET News.com.

_____. 2002. HP notebook part of cheap-wireless plan. Retrieved January 3, CNET News.com.

_____. 2002. Toshiba notebooks add wireless options. Retrieved September 18, CNET News.com.

Sprint, Cingular to launch 3G service. 2001. Retrieved March 20, Bloomberg News/CNET News.com.

Stallings, W. 1987. *Handbook of computer-communications standards.* New York: MacMillan.

State of SMR digital mobile radio. Washington, D.C.: The Strategis Group.

Stetz, P. 1999. *The cell phone handbook.* Newport, RI Aegis Publishing.

The Strategis Group, 2001. www.strategis.com.

Strout, R. D. 1992, August. 9-1-1 communications control center at Salt Lake City, Utah 911. *Public Safety Magazine,* a supplement to *Mobile Radio Technology Journal.*

TACS Analogue Systems. 2001. www.mobiles.co.uk/analogue.htm.

Tanenbaum, A. 1996. *Computer networks.* Upper Saddle River, N.J.: Prentice Hall.

Tech crisis—D.C. drowning in e-mail. 2001. www.zdnet. com/zdnn/stories/news/0,4586,2697768.html.

Tech report: Study: Web exceeds 1 billion pages. 2000. www.usatoday.com/life/cyber/tech.

Tech TV: Call for answers. 2001. www.techtv.com/callforhelp/answerstips/story/0,23008,2453644,00. html.

Toffler, A. 1991. *Third wave.* New York: Bantam.

Toshiba to mass produce Bluetooth LSI chips, www.zdnet.co.uk/news/2001/9/ns-21354.html.

Trimble. 2000. www.trimble.com.

Tristram, C., and J. Foust. 1999. Technology review. Retrieved April 28, 2000, www.proquest.com.

United Paramount Network. 1998. Babylon 5 TV series.

United States Army. 1967. *TM 5-241-1 Grids and grid references.* Washington, D.C.: Department of the Army.

United States Coast Guard Navigation Center official site, www.navcen.uscg.mil.

Varral, G., and R. Vbelcher. 1998. *Data over radio: Data and digital processing techniques in mobile and cellular radio.* New York: John Wiley & Sons.

Virostek, S. 1997. To dispatch or not to dispatch—The age of wireless competition.

Walrand, J., and P. Varaiya. 2000. *High-performance communication networks.* San Francisco: Morgan-Kaufmann.

WAP Forum. 2001. Retrieved November 4, www.wapforum.com.

WAP Forum. 2002. Retrieved June 4, www.wapforum.com.

Web Pro Forums. 2000. Time division multiple access (CDMA) tutorial. Retrieved February 25, www.webproforum.com/tdma/topic06.html.

Webopedia. 2002. Short message service. INT Media group, www.webopedia.com/TERM/S/Short_Message_Service.html.

Weisman, C. 2000. *The essential guide to RF and wireless.* Upper Saddle River, N.J.: Prentice Hall.

What is PCS? Retrieved February 13, 2000, www.bus.1se.edu/issds/faculty2/walsh/isds7520983/pcs/pcshistory.html.

Wilcox, J. 2000. As Bluetooth nibbles, competition lurks. Retrieved September 15, CNET News.com.

Will all hail Bluetooth? www.zdnet.co.uk/news/2000/48/ns-19462.html.

Will cell phone sales reach industry forecasts? 2001, January 18. Bloomberg News.

Winch, R. 1993. *Telecommunication transmission systems: Microwave, fiber optic, mobile cellular radio, data, and digital multiplexing.* New York: McGraw Hill.

Wireless data frequency ranges in the United States; broadband PCS frequency allocation. Retrieved February 13, 2000, www.pcsdata.com/frequency.htm.

Wireless rentals: Japan cellular rentals. 2001. www.intouchusa.com.

WirelessNow Information Service, www.commonow. com/protect/per mns/Two-Way_Radio_Competition.

Wong, W. 2001. Ironing out cell phones' web wrinkles. Retrieved November 13, CNET News.com.

Young, H. 1999. *Wireless basics.* Overland Park, Kan.: Intertec Books.